Margaret Roberts

Indigenous Healing Plants

Margaret Roberts

Indigenous Healing Plants

SOUTHERN
BOOK PUBLISHERS

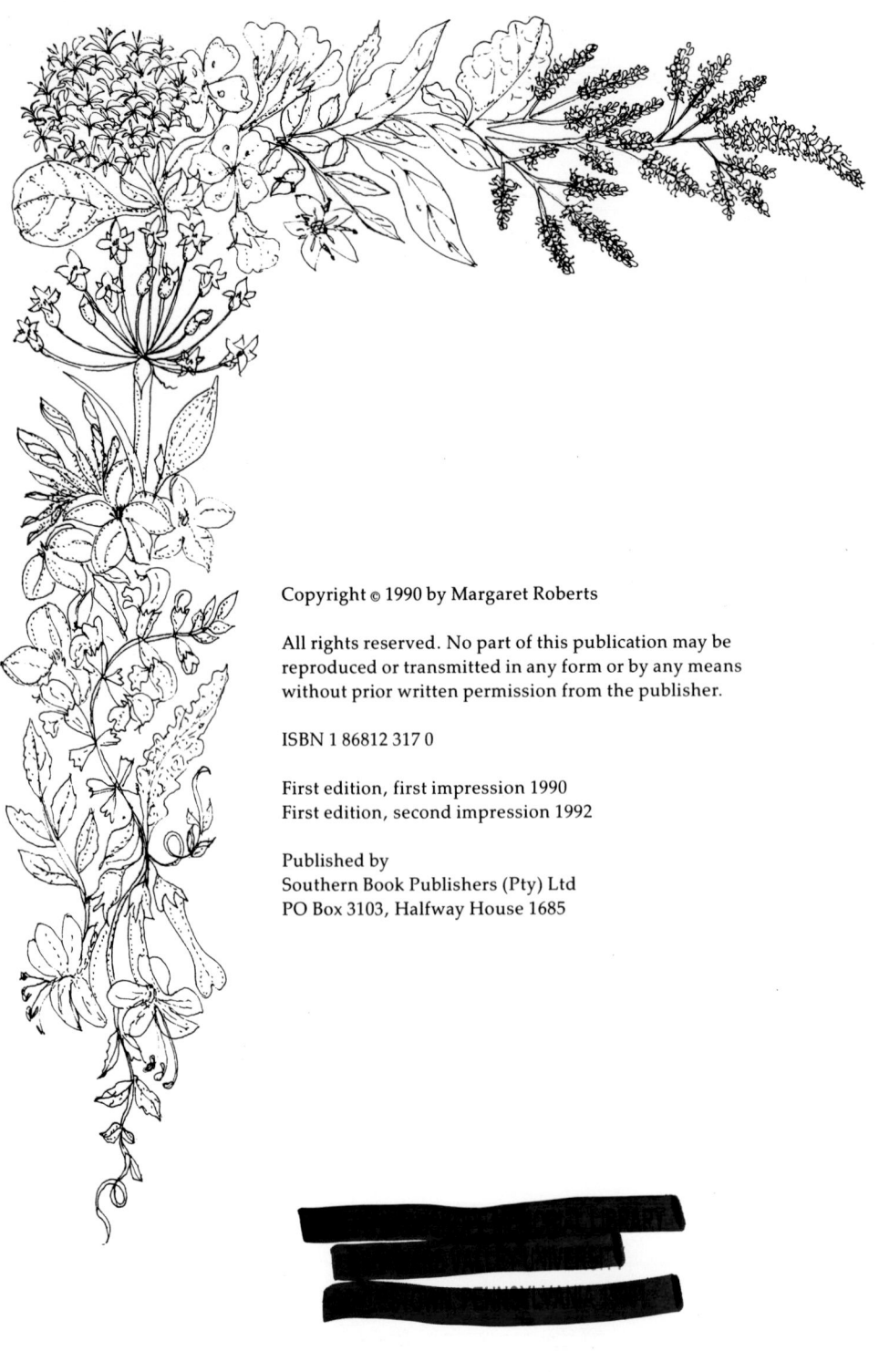

Copyright © 1990 by Margaret Roberts

All rights reserved. No part of this publication may be reproduced or transmitted in any form or by any means without prior written permission from the publisher.

ISBN 1 86812 317 0

First edition, first impression 1990
First edition, second impression 1992

Published by
Southern Book Publishers (Pty) Ltd
PO Box 3103, Halfway House 1685

Illustrations by the author
Cover design by Insight Graphics
Colour reproduction by Unifoto, Cape Town
Design and phototypesetting by Book Productions, Pretoria
Printed and bound by Creda Press, Cape Town.

Dedicated to South Africa and all her people

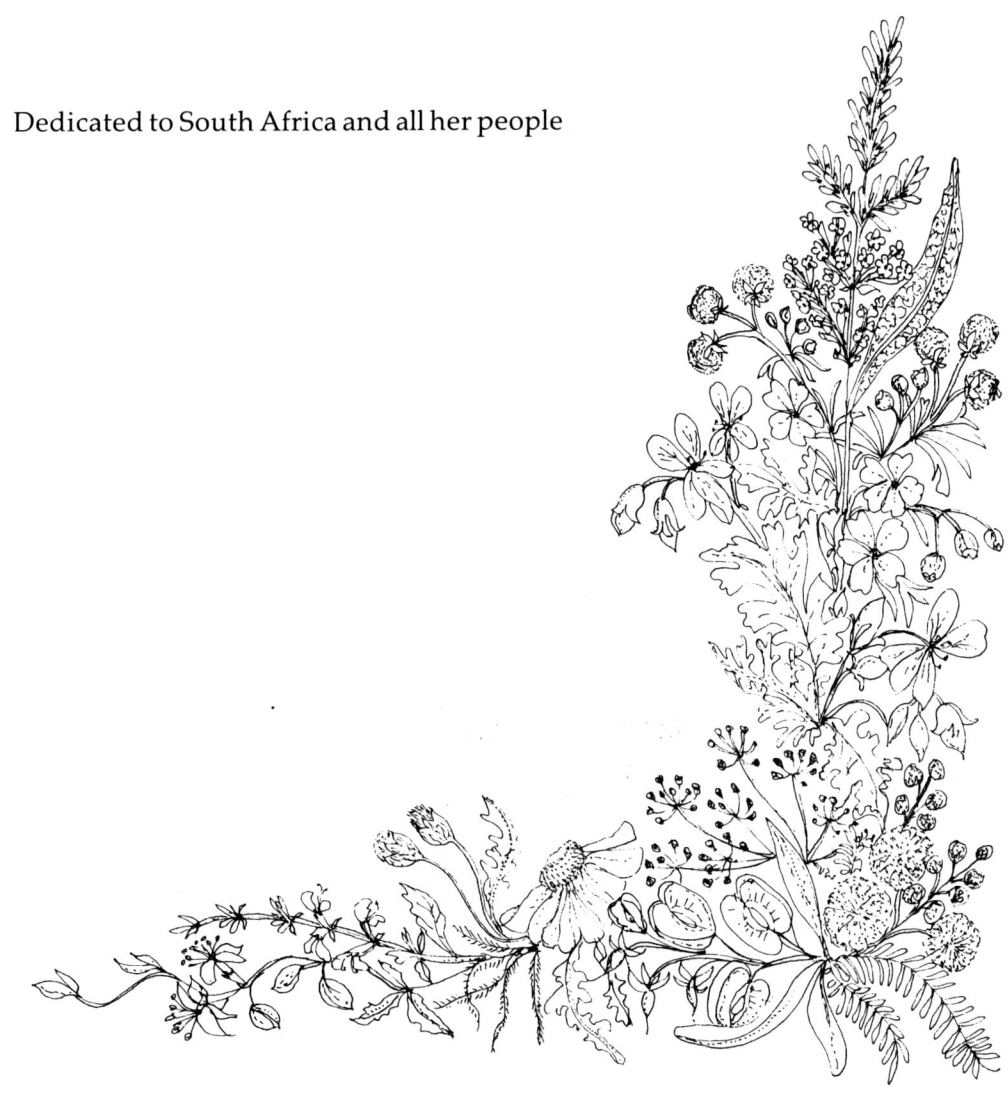

Acknowledgements

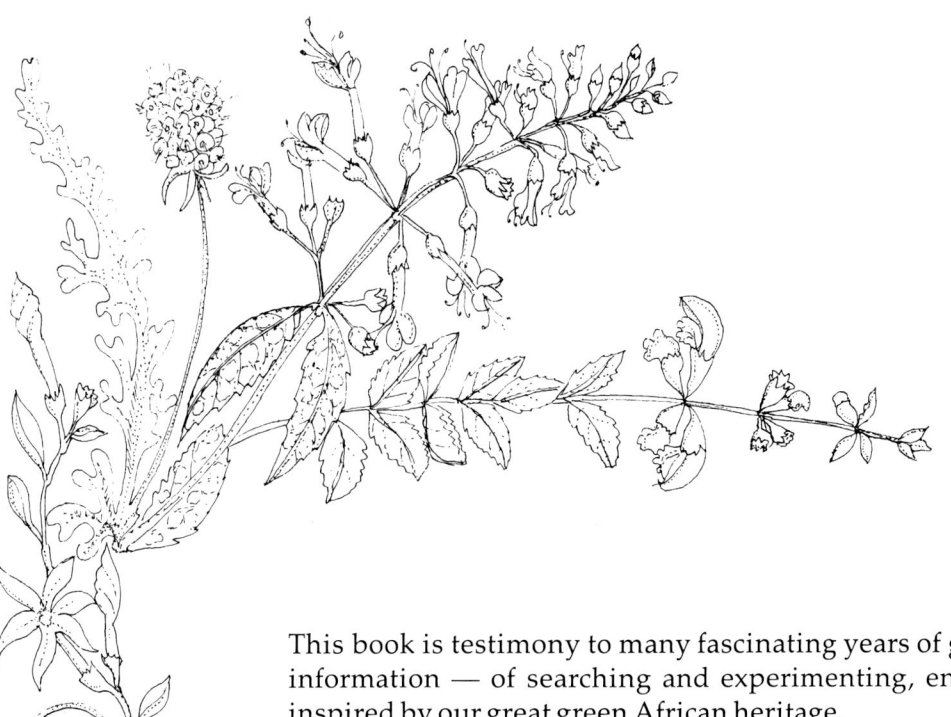

This book is testimony to many fascinating years of gathering knowledge and information — of searching and experimenting, enquiring and listening — inspired by our great green African heritage.

I do not know how to begin to express my gratitude for all the wonderful people and plants I have encountered on the way, for I have met more people and made more friends through plants than through anything else in my life.

First I have my parents and grandmother to thank for introducing me as a young child to the world of plants, an interest that was nurtured as a child in Gordon's Bay where my grandmother's housemaid would bring me a different wild flower every day from her home in the mountains. Later on, as a young bride on a farm at Olifantsbrug in the Transvaal, the farm workers taught me much, as did the Zulu gardeners at our seaside cottage at Umdloti in Natal. Country people everywhere — young and old — happily and willingly showed me their favourite remedies and shared with me their grandmothers' recipes. I am grateful to all the unnamed people who shared their knowledge and enthusiasm with me, and particularly to the staff at Kirstenbosch; Pat Morgan, who found plants and brought them to the Transvaal for me; Mary Gould, who helped me with the buchu section and who introduced me to the China flowers; Christien Malan, who encouraged me and whose knowledge of our wild plants astounds me; Elaine Kampher, who found plants, introduced me to the nursery staff at Kirstenbosch and offered her home, her time and her interest so unstintingly; Fiona Powrie for her help with the pelargoniums; Dr John Rourke for his interest and for making available to me that special little book, one of the first of Africa's recorded botanical references, Pappe's *List of*

South African Indigenous Plants Used as Remedies by the Colonists of the Cape of Good Hope; the garden staff at Kirstenbosch; and the herb sellers on the Parade in Cape Town — how we have talked and laughed, and how much we have enjoyed the plants together.

Thank you to Betty Louw of Babylon's Toren near Paarl who introduced me, the Transvaaler, to the Cape's renosterbos, agt-dae-geneesbossie, Hotnotskool and the luscious waterblommetjie. To Annette Rabie, who patiently deciphered and typed reams of enthusiastic outpourings and added her own store of knowledge and love of our wild plants, my gratitude knows no bounds. How wonderful it has been to work with someone so in rapport with this glorious subject.

To Barbara Tyrrell, without whose encouragement and teaching I would never have put paintbrush to paper, and to Brenda Clarke who guided me with technical advice on botanical drawing and painting, my profound and joyful thanks for letting me into your world and for giving me a new dimension in which to express my love of plants.

To my staff at the Herbal Centre who kept it all running smoothly while I worked on this book and to Dr Mariekie Hinsbeeck, my mentor, supporter and special friend, on whose farm I live and where so many of the plants I have written of grow — my deepest gratitude. Last but never least, to Southern Book Publishers, bless you for your faith in me; to Basil van Rooyen and Sally Antrobus who saw the need for this book, my editor and copy editor Nydia de Jager and Catherine Murray, and to Peter Sauthoff of Book Productions, my gratitude for a job well done.

God bless.

M.R.

Die Vesting
De Wildt
Transvaal

Author's note

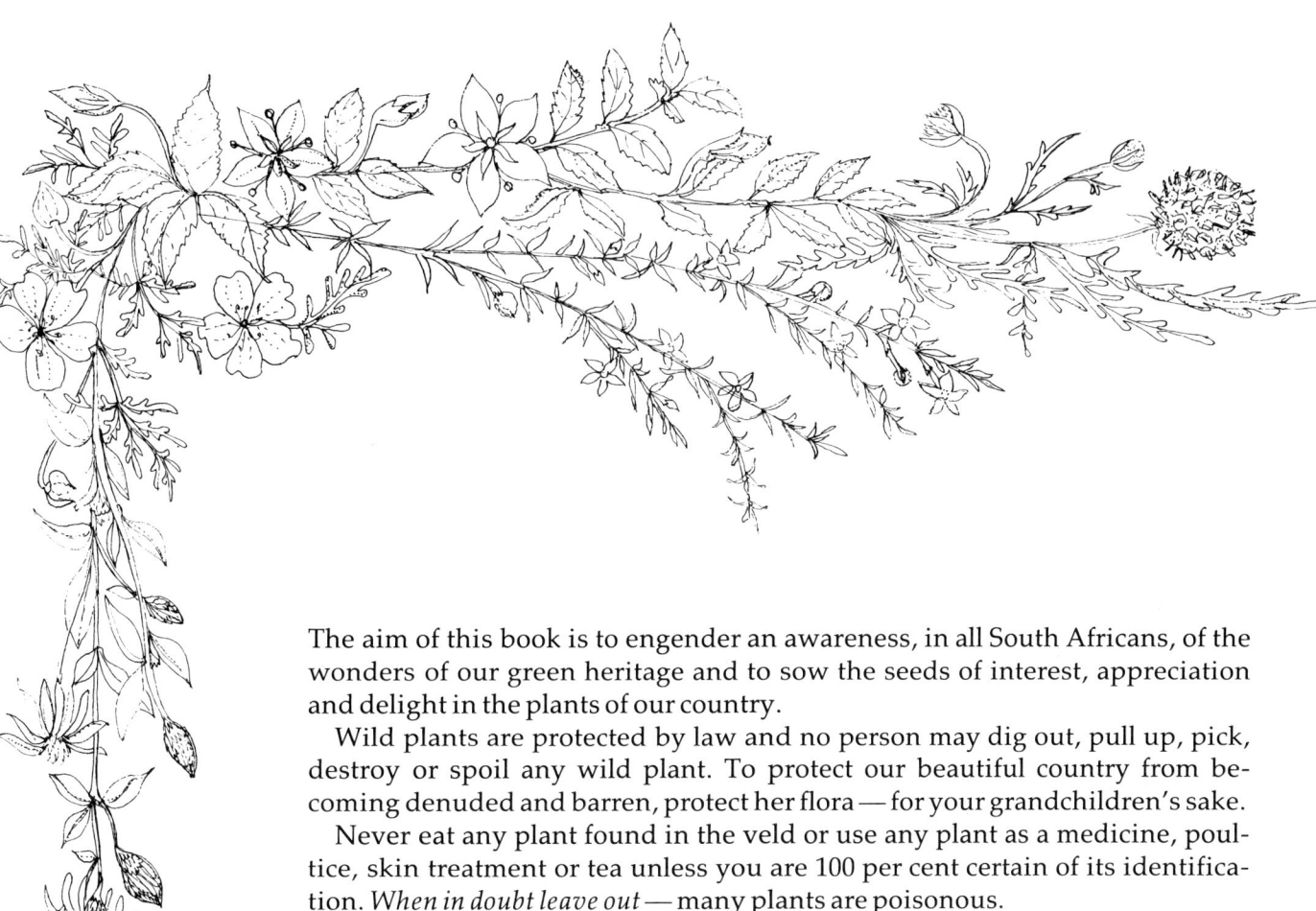

The aim of this book is to engender an awareness, in all South Africans, of the wonders of our green heritage and to sow the seeds of interest, appreciation and delight in the plants of our country.

Wild plants are protected by law and no person may dig out, pull up, pick, destroy or spoil any wild plant. To protect our beautiful country from becoming denuded and barren, protect her flora — for your grandchildren's sake.

Never eat any plant found in the veld or use any plant as a medicine, poultice, skin treatment or tea unless you are 100 per cent certain of its identification. *When in doubt leave out* — many plants are poisonous.

The author and publishers take no responsibility for any poisoning, illness or discomfort that may be incurred by the incorrect identification of a plant or the incorrect medicinal treatment of any person with a plant. You are strongly advised to consult your doctor before treating yourself or your family with home remedies. We make no claims as to the healing properties of plants and the curing of specific ailments, but rather indicate the traditional ways in which the plants have been used for many centuries. *You are strongly discouraged to experiment without the guidance of your doctor.* Remember — everything in moderation, especially so for plant medicines.

Contents

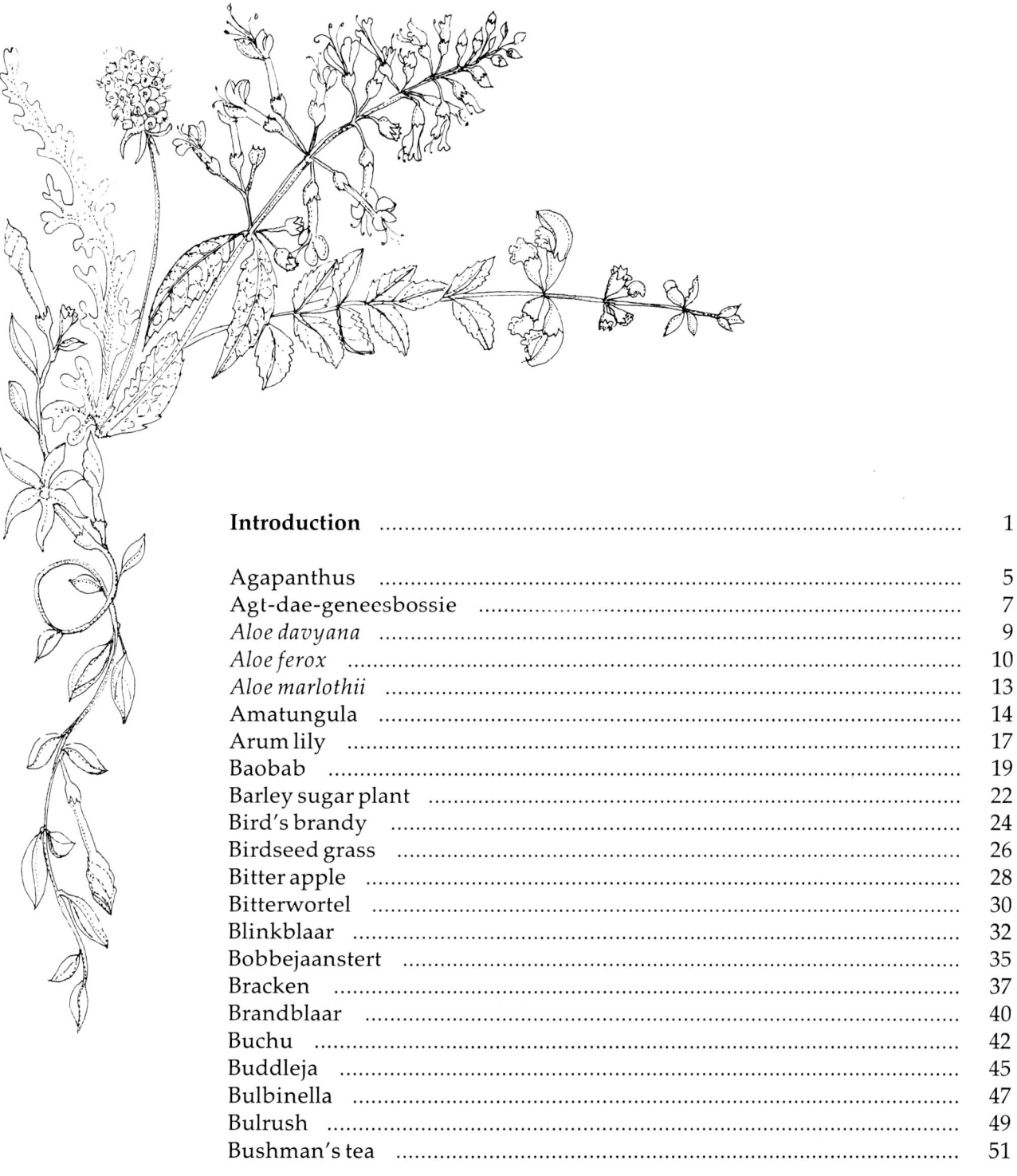

Introduction	1
Agapanthus	5
Agt-dae-geneesbossie	7
Aloe davyana	9
Aloe ferox	10
Aloe marlothii	13
Amatungula	14
Arum lily	17
Baobab	19
Barley sugar plant	22
Bird's brandy	24
Birdseed grass	26
Bitter apple	28
Bitterwortel	30
Blinkblaar	32
Bobbejaanstert	35
Bracken	37
Brandblaar	40
Buchu	42
Buddleja	45
Bulbinella	47
Bulrush	49
Bushman's tea	51

Bush tick berry	53
Cancer bush	55
Cape honeysuckle	57
Cape willow	59
Carpet geranium	61
Chamomile	63
Christmas berry	66
Clivia	68
Coral tree	70
Crinum lily	72
Cross-berry	74
Curry bush	76
Doll's protea	79
Dysentery herb	81
Elephant's foot	83
Flame lily	85
Ginger bush	87
Hard fern	89
Harebell	91
Heather	93
Honeybush tea	96
Horsetail	98
Hotnotskool	100
Hottentots' tea	102
Huilboerboon	104
Jasmine	107
Kattekruie	109
Kiepersol	111
Knobwood	113
Lavender tree	116
Lemon bush	118
Maidenhair fern	120
Marula	123
Melianthus	126
Melkbos	128
Mother-in-law's tongue	130
Nastergal	132
Nutgrass	135
Paintbrush lily	137
Papyrus	139
Parsley tree	141
Pennywort	143
Pig's ear cotyledon	145
Pincushion	147
Plectranthus	149
Plumbago	151
Pompon tree	153
Raasblaar	155
Raisinbush	157

Renosterbos	159
Resurrection plant	161
Sand olive	163
Sausage tree	165
Scented geraniums	167
Sickle bush	184
Silverleafed vernonia	186
Sorrel	188
Sour fig	190
Spekboom	193
Star flower	195
Stork's bill geranium	197
Sweet thorn	199
Toothache root	201
Traveller's joy	203
Tumbleweed	205
Waterblommetjie	207
Water lily	210
White cat's whiskers	212
Wild asparagus	214
Wild basil	216
Wild camphor tree	219
Wild cineraria	221
Wild dagga	223
Wilde als	226
Wilde wingerd	229
Wild foxglove	231
Wild garlic	233
Wild gladiolus	237
Wild medlar	239
Wild mint	241
Wild olive	247
Wild pear	249
Wild pineapple	251
Wild rosemary	252
Wild sage	255
Wild verbena	263
Yellow wandering Jew	265
Epilogue	267
Where to obtain indigenous plants and seeds	269
Bibliography	271
Index	272

Introduction

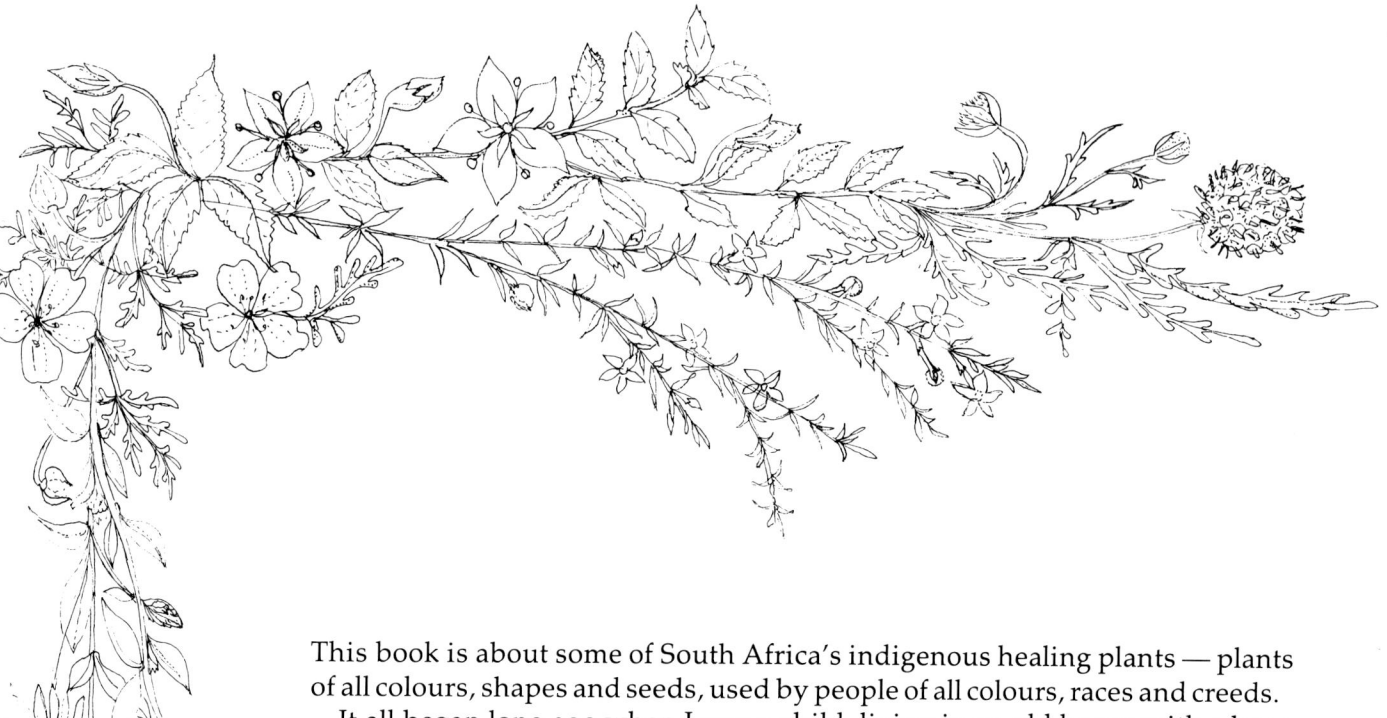

This book is about some of South Africa's indigenous healing plants — plants of all colours, shapes and seeds, used by people of all colours, races and creeds.

It all began long ago when I was a child, living in an old house with a large garden in Pretoria. My Scottish grandmother lived with us and it was she who taught me the names of the plants, and their many uses. My parents, too, were both avid gardeners. My father taught me the joy of good compost making, and good digging in the rich earth, and my mother taught me to see the beauty of each flower, each leaf, each seed through her artist's eyes, and inspired me at the age of 6 or 7 to draw, to observe and to appreciate. Plants were an integral part of our lives, and this laid the foundation of interest which has crept into every aspect of my life as the years have gone by.

For certain times of the year during the parliamentary sessions in Cape Town, my sisters and I lived with my grandmother in her cottage at Gordon's Bay. It was during those times that I planted and dug in the sea gardens with my grandmother and came to learn about some of the indigenous Cape plants and their uses. Sometimes the housemaids who worked for my grandmother and great-grandmother, seeing my joy in the plants, would bring me a flower or a leaf or a piece of grass from their home in the mountains, or picked along the path on their way to work. I would often have more than 20 flowers standing in small jars and bottles along the dressing table. I soon learnt their names, their uses, and to recognise their glorious smells.

We always took our dogs for walks through the veld, along the then unmanicured pavements, and so I got to know the wild plants; their colours, their shapes, their leaves, buds and seeds. I knew them by their common names, I

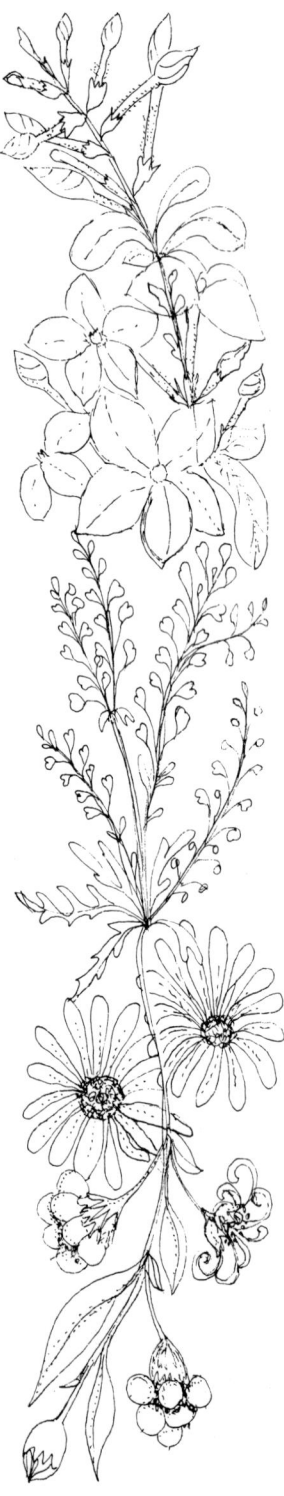

pressed them, I drew them and I loved them, and my grandmother, my parents and my sisters shared my interest.

When I grew up I trained as a physiotherapist — a long way from my botanical interest, yes, but my parents and grandmother wisely guided me into a profession that would form a stable background for my life's passion and work: natural health, medicinal plants and the precious plants of my own country.

At the age of 23, I married and went to live on a farm. My closeness to nature wove ever more deeply into my life. I brought up my three children with all the good natural things that farm life offers, and they too developed an interest in the things around them — the weather, the veld, the gardens and the crops. The herd boys and the farm workers taught us how to use the wild plants, and as I created the herb gardens, so the local African people, country people, Afrikaans visitors and overseas visitors all came to collect plants for medicine or simply to enjoy the herb garden. We sat under the thorn trees and talked, sharing knowledge as we shared plants. I knew then, as I increasingly know now, that plants link people, they link cultures, they link countries. As we sat talking, often not understanding one another in words, the precious plants formed a language that surged and flowed around us, and we went our separate ways with the plants uniting us in a common interest and friendship.

Then, shatteringly, 25 years later, my life on the farm ended, and I was forced to move away. In those next desperate and anxious years a new life had to be begun and the plants saved me. I again planted herb gardens, I dug and I sowed seeds and I took up the torn threads of my life and began anew. The plants soothed my pain, gave me a new interest, a new vitality and a new peace; and I created the most beautiful gardens I had ever made. I started a herb centre on a farm on the northern slopes of that great range of mountains, the Magaliesberg, with a climate different to that which I had known before, with a new set of indigenous plants to learn about. The Zulu, Venda and Ndebele who worked on the farm taught me new things, other sangomas and witchdoctors came to visit my new gardens, and people from overseas came specially to see the African indigenous medicinal section of my herb garden. The plants gave us all so much pleasure, linking us, comforting us and inspiring us.

So, come walk with me through the wild, inspiring and fascinating beauty of this beautiful and great garden, South Africa. Let me share with you some of the uses of the plants I have come to know. I have chosen only a few of our great wealth of medicinal plants, as these are the ones that I am familiar with. I have not made use of long and complicated, often changing botanical names, but still urge you to become familiar with them. I have not grouped the plants in genera, species or subspecies — I leave that to the botanists, and there are many beautiful books that will fullfil these needs. Instead I have written of familiar friends with 'nicknames' or endearing names that will spark recognition in the layman; and I have included their African names, which are to the best of my ability correct — spelling may change from tribe to tribe, but most will be recognisable.

I am not a botanist, nor am I a botanical artist, but I write and draw and paint from the sheer joy of knowing and loving the plants. Within these pages I hope to share some of that joy with you.

This book is the product of many years of interest in indigenous herbs, jotting down my findings and using the medicines made from the plants. However, I stress earnestly — *never treat yourself or anyone else with herbal medicines without consulting your doctor.* Untold harm can be done if dosages and the plants are not correctly used and identified. My father's words when I stepped into the adult world, 'Everything in moderation', hold good in everything I do, and particularly in my work with herbs.

When I completed my first book, *Margaret Roberts' Book of Herbs,* I ended my introduction with these words: 'Throughout this book I have aimed at accuracy, but to suppose that there are no mistakes would be arrogant; however, I trust there will be so few as not to detract from its usefulness.' I include these words here again as they so aptly sum up my feelings for this book as well.

At the end of our walk through the veld as you turn the last of these pages, my hope is that you too will be inspired to know more and more of our glorious green heritage, and perhaps discover more medicinal plants, and so continue to link the people of our country in our common interest — our precious wild plants.

Margaret Roberts

The Herbal Centre
De Wildt, Transvaal
January 1990

Agapanthus

Agapanthus species

AFRIKAANS	Bloulelie, krismislelie, keiserkroon, bruidslelie, bloukandelaar
SOTHO	Leta-la-phofu
XHOSA	Isilakati
ZULU	Ubani, icakathi

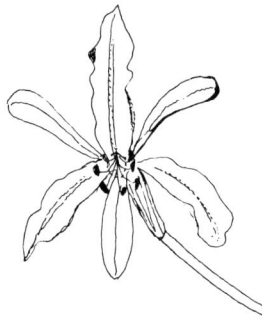

THE agapanthus is a much-loved garden plant. There are many species, ranging from a dainty miniature to the huge dark blue giant, from mauvy blue and light blue to white. This has made it a plant of considerable commercial value, as it is also long lasting when cut.

When I grew flowers for the market I had several long beds of white agapanthus. At Christmas time when there was a tremendous demand for white flowers for both weddings and for church flowers I found I could sell every single head at a good price, and my perennial rows of plants never needed any attention other than a good weekly watering, year after year.

I always encourage growers to consider agapanthus, for the plants grow easily in virtually any soil type, they do well in both sun and in shade, they flower at a time when their cool blues and whites are much appreciated — midsummer — and for the rest of the year their lush, green, strap-like leaves look handsome in the garden in any position. If there is heavy frost in winter the leaves will die down for a month or two, but the plant soon recovers and sends out bright new leaves in spring.

The agapanthus is much prized as a hothouse plant overseas. I have seen much admired and much prized tubs of agapanthus growing at Hampton Court in London, where the first plants flowered in 1692. They didn't believe me when I told them of our road island plantings, and great tracts of land under agapanthus out there in the sun, flowering year after year in the eastern parts of South Africa!

The agapanthus has always been considered a magical and medicinal plant by the indigenous people of South Africa. Xhosa women take a decoction of

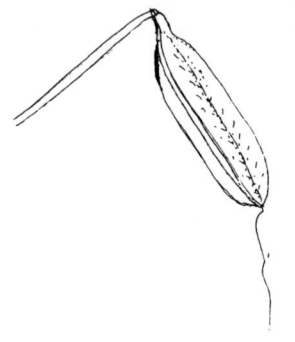

the root after the sixth month of pregnancy to ensure an easy birth, and the newborn child is washed in the same brew before being put to the breast for the first time, for health and strength. The women also make a necklace by threading pieces of the thick, fleshy root which they wear believing this to bring healthy, strong babies. Several tribes grow agapanthus near their homes in tins or old pots as it is considered to be the plant of fertility and pregnancy.

The Zulu also use agapanthus as a treatment for heart disease, paralysis, and the same infusion of the root in hot water is used as an emetic for coughs and colds, chest pains and tightness. I always find this amazing, as agapanthus is one of the plants suspected of causing poisoning in humans.

I was once shown by a Zulu tracker in the mountains of Natal how to put the leaves of agapanthus into your shoes to soothe your feet on a long hike! He also encouraged us to wrap our weary feet in the leaves to rest them when we got back to the camp at the end of the day. He sat there in front of the fire deftly weaving the soft leaves into a slipper shape over his feet, and for half an hour he sat peacefully letting the magical leaves work their spell. I tried it too, and was surprised to feel the silky smoothness soothe away the day's aches and pains, particularly around my heels.

The smooth, strap-like leaves make an excellent bandage to hold a dressing or poultice in place, and the leaves wound round the wrists will help to bring a fever down.

I grow agapanthus in all sorts of places in the garden, and derive constant enjoyment from the fresh, brilliant blues of their flowers. They have a long and abundant flowering period and I save some of their heads for dried arrangements. The dried flowers remain beautifully blue in a wild flower pot-pourri. (See page 268.)

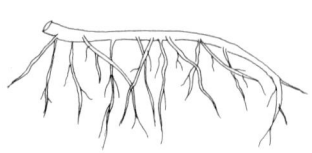

Agapanthus is a most worthwhile plant to grow in the garden and a fascinating collection can be made of the various varieties and colours. When planting, be sure to leave enough space — a metre all round — and dig in a good amount of compost. You will be rewarded by magnificent blooms for years to come.

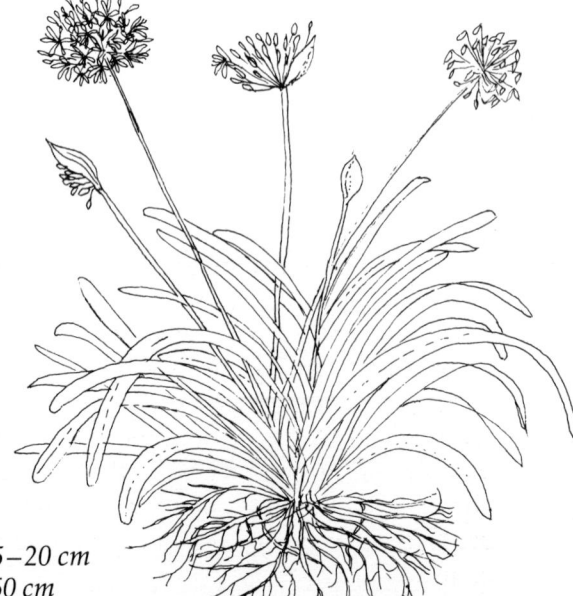

Height: Leaves 15–20 cm
Flowers 60 cm

Agt-dae-geneesbossie
Lobostemon fructicosus

ENGLISH   Eight-day healing bush
AFRIKAANS   Douwurmbos, lobos, luibos

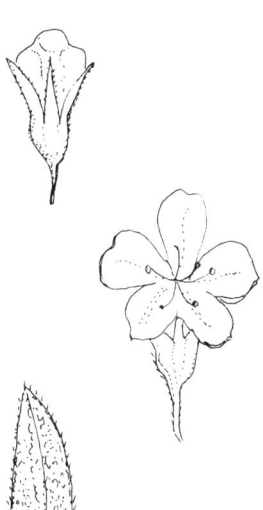

THIS remarkable blue and pink flowering shrub grows mainly in the south-western Cape. It belongs to the Boraginaceae family, and its grey-green, rough, hairy leaves have the feel of borage, its close cousin.

Much respected and used by the Khoikhoi, the settlers and the Malays in the last century, it had many uses, and was taken to Europe where it was cultivated in conservatories as it was believed this wonder plant could heal all sorts of ailments within eight days! The name douwurmbos was given as a tea made of the leaves drunk first thing in the morning was said to be a sure cure for ringworm in humans and in animals (¼ cup of fresh leaves to 1 cup of boiling water, stand for 3 to 5 minutes, strain, then drink).

An excellent ointment was made in the early days at the Cape (and is still popular today in some rural districts) by frying the leaves and flowers of the bush in butter with leaves of *Melianthus major, M. cosmosus* and the bulbs of *Cyanella lutea*, and using this for sores on the legs, particularly in women, and for the external lesions of syphilis. A lotion was made by boiling up the plants in water and using this as a wash.

The aromatic leaves made into a tea and used for skin diseases — rashes, eczema and sores — has been used for many years by people in the Cape, and bandages soaked in the brew are excellent as dressings for wounds, bites and scratches. The tea is considered to be cleansing during times of infection and blood poisoning, and for the first day of irregular menstruation in young girls. I have enjoyed the tea as a tonic in spring, when the exquisite flowers are out in profusion, and the roadsides are lined with the pinks and blues of this beautiful little shrub.

The outer edges of the flower's five petals are a soft pale blue, and the throat is pink. The flowers are borne in clusters at the tips and the overall effect with the soft grey green leaves is enchanting.

The plant grows from fresh seed, and in the hot Transvaal summer seems to do best in partial shade. The midsummer growth can be clipped back and added to pot-pourris.

Some nurseries offer the plants for sale so keep an eye out for it.

Height: 1 metre

Aloe davyana
Liliaceae family

AFRIKAANS Transvaal aalwyn
TSWANA Kxopane

*Height: Leaves 15–20 cm
Flowers 60 cm*

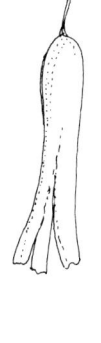
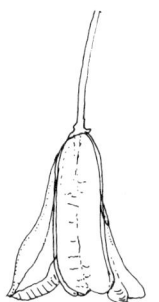

THE soft pink flower heads of the small *Aloe davyana* give colour to the winter buffs and browns of the Transvaal veld. It is a much used medicinal plant. Its jelly-like leaf pulp is a soothing dressing to burns, blisters, bites, stings, sores, wounds and venereal sores. The pulp is also applied to snakebites, and is even considered effective for the deadly mamba bite, the wound being rubbed continuously and very gently with it. *A. davyana* is also taken as a purgative tea. I edge my wild garden with it as it is so useful for sunburn, bites and stings and is so easy and undemanding in the garden.

My garden staff use the jelly-like pulp for blisters in their soccer shoes after a weekend match as it grows conveniently near at hand in the veld surrounding the soccer field. I have used it for sore, red mosquito bites, and find it quickly soothing — the jelly dries and forms a protective skin.

Great bunches of *A. davyana* flowers in copper urns make beautiful winter flower arrangements. The pretty pink colour is decorative and pleasing, and overseas visitors are enchanted with this unusual display.

Aloe ferox
Liliaceae family

ENGLISH	Bitter aloe, red aloe
AFRIKAANS	Bitteraalwijn, lapaalwyn, tapaalwyn
SOTHO	Hlaba, lekxala-la-quthing
XHOSA	Umhlaba, ikalene, khala
ZULU	Umhlaba

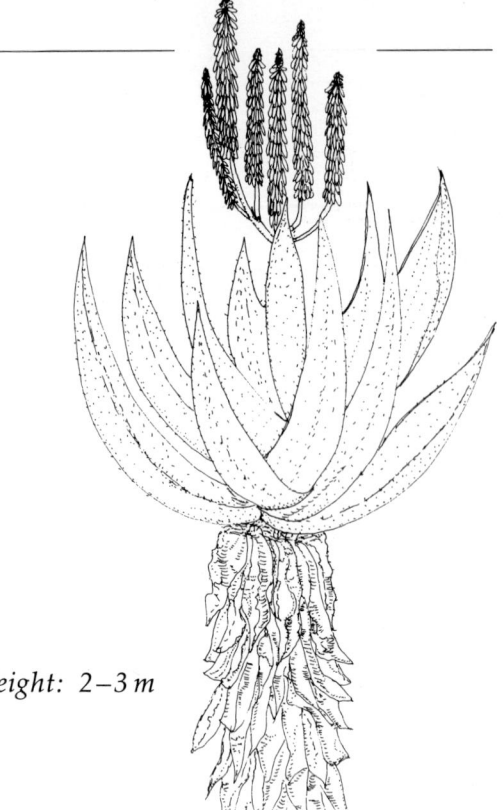

Height: 2–3 m

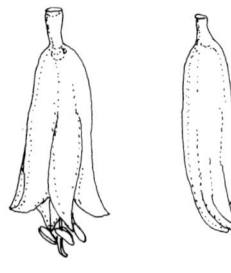

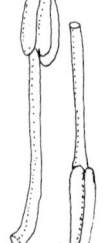

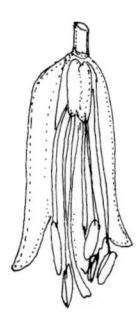

THE magnificent *Aloe ferox* stands majestic and regal on hillsides and in the veld in the southern Cape. The bright, fiery red flowers in winter and spring are laden with nectar and are much visited by birds and by children, who suck their sweetness. The nectar has been found to be narcotic and there have been several reported cases of impaired muscle control, weakness of the joints and even temporary paralysis by over-zealous sucking of the nectar.

For many years the *A. ferox* has been used as the best source of the famous medicinal Cape Aloes and exported as an article of commerce. The juice was often collected primitively by the coloured and black people by digging a hole in the ground, covering and lining the hole with a smooth skin, then placing cut leaves in layers all around the hole so that the juice could drip into the 'basin' thus formed. The dark resinous sap was then left to dry, and the lumps of raw aloe were — and still are — an important pharmaceutical ingredient containing remarkable and marketable substances such as aloin, resin, emodin, chrysamic acid and barbaloin.

Interestingly, *A. ferox* was introduced into European gardens in about 1700, cultivated from specimens that were brought to the famous Chelsea Physic Garden in London. The tapping process was developed in the early Cape colonial days and became an important industry in the southern Cape areas. The first consignment of Cape Aloes was exported in 1761.

The colonists, probably taught by the Khoikhoi, used the sap as a wound dressing, and the Dutch East India Company's gardens in Cape Town had several *A. ferox* plants in 1695.

The juice of the *A. ferox* is primarily regarded as an excellent if somewhat

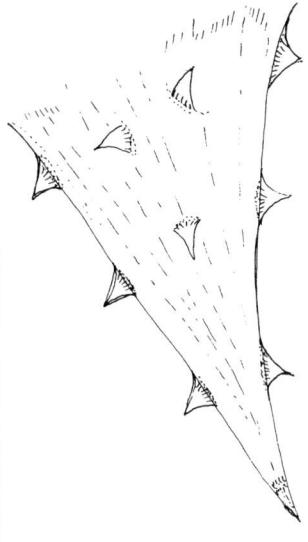

strong purgative for both man and beast, and the Xhosa and Zulu use it sometimes mixed with mealiemeal or with clay and eat it as a purge. Care and caution need to be applied here as it is also considered to be abortifacient.

The leaf juice is used to treat burns, to rid dogs and cattle of ticks, to treat scab on sheep, and the Zulu, Tswana and Xhosa apply it directly to the eyes of cattle in cases of opthalmia. Several African tribes also use the juice as a treatment for venereal sores, applying it directly onto the area at frequent intervals.

I have used the chopped leaf with *A. davyana* and khakibos and wild rosemary as an insecticide and a wash for the dogs during summer's tick and flea infestations and found it to be excellent. (See recipe below.)

Surprisingly, the leaves of *A. ferox* can be made into a delicious jam, and Betty Louw of the Paarl district gave me the very old recipe below which has proved to be very tasty — rather like watermelon konfyt, without the bitterness one would expect from the aloe. *A. ferox* is such a spectacular plant, it is worth trying to get one for your garden. It will grow over 3 m in height eventually, and needs about 2 m in width. Some nurseries do offer it for sale, so don't be tempted to remove one illegally from the veld.

Aloe ferox insect repellent

2 leaves Aloe ferox, *chopped*
6 leaves A. davyana, *chopped*
½ bucket fresh khakibos, roughly chopped
¼ bucket fresh, wild rosemary, roughly chopped (substitute ordinary rosemary if you have no wild rosemary)
enough boiling water to cover

Mix thoroughly and stand overnight. Next morning strain and add the same quantity of water again as it is strong. Use this as a rinse after bathing the dog, or as a splash on to plants, and cattle and dogs. Work it into the skin.

Aloe ferox konfyt

4 *large* Aloe ferox leaves
12 *finger-length new fig tree shoots*
lime
3½ kg sugar
9 cups water
2 tablespoons lemon juice

Make this konfyt in spring as you need the young shoots of the fig tree for flavour.

Cut the *A. ferox* leaves into manageable pieces under running water. Still under the running water, peel away the thorns and the skin. Leave overnight in a basin of cold water. In the morning rinse well and prick all over with a fork. Soak for 24 hours in lime water (1 tablespoon of lime to 2 litres of water). Then wash well under running water and drain. Wash the new fig tree shoots well and boil with the sugar in the water. Add lemon juice.

When the mixture comes to the boil add the *Aloe* pieces, and boil until thick — about 1½ hours. Leave overnight. Then remove the fig shoots. Reheat the syrup and pack the pieces into bottles while still hot. Seal well.

Aloe marlothii
Liliaceae family

ENGLISH	Mountain aloe
AFRIKAANS	Bergaalwyn
SHANGANA	Mhanga
SWATI	Inhlaba
TSWANA	Mogopa
VENDA	Binda mulsho, khopha
ZULU	Umhlaba

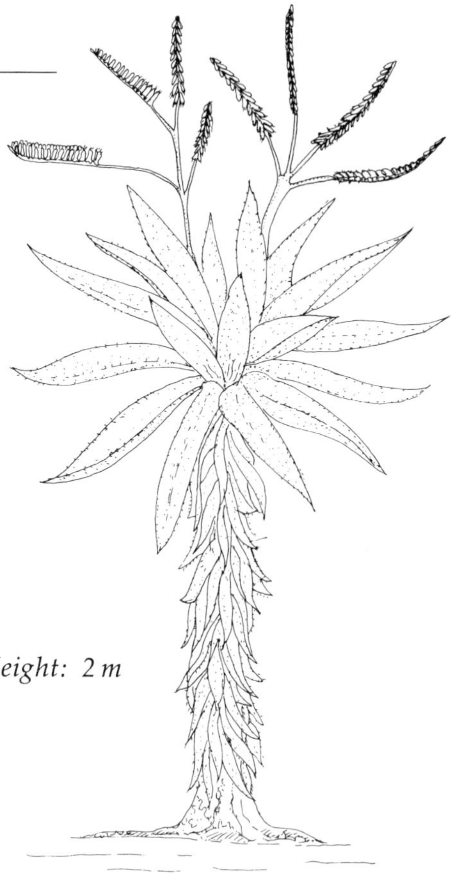

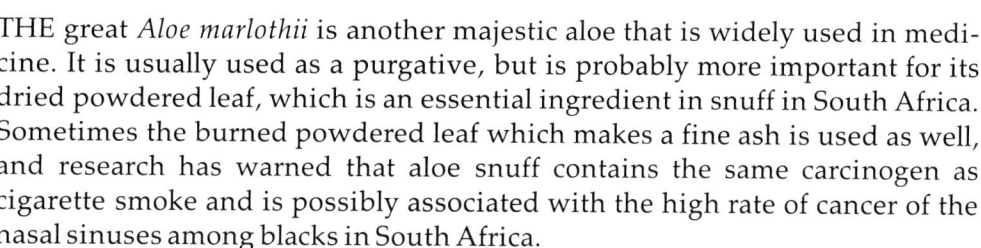

Height: 2 m

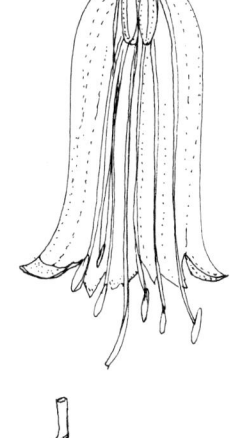

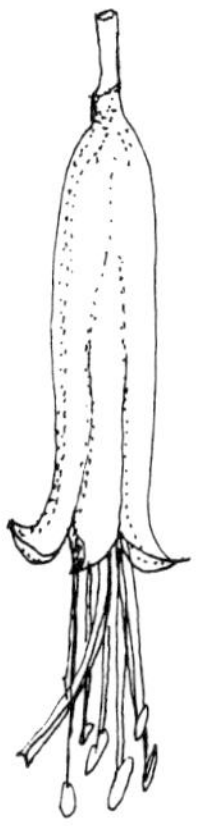

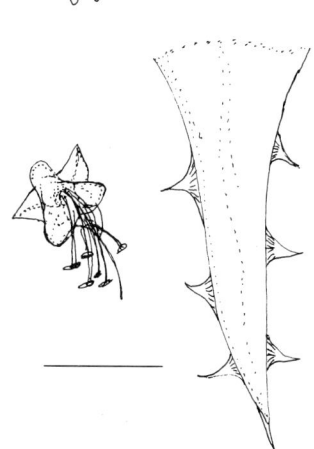

THE great *Aloe marlothii* is another majestic aloe that is widely used in medicine. It is usually used as a purgative, but is probably more important for its dried powdered leaf, which is an essential ingredient in snuff in South Africa. Sometimes the burned powdered leaf which makes a fine ash is used as well, and research has warned that aloe snuff contains the same carcinogen as cigarette smoke and is possibly associated with the high rate of cancer of the nasal sinuses among blacks in South Africa.

Pieces of leaf boiled up in sugar water are used to treat worm infestation — and the brew is considered to be excellent for tape worm. Zulu mothers rub the bitter juice over their breasts to hasten weaning, and the boiled leaf is the African remedy for horse sickness — usually 1 measure of chopped leaves is boiled in 4 times the measure of water for 10 minutes. This is allowed to cool and is then strained and fed to the animal with a bottle.

A brew of 1 tablespoon of chopped leaf in 2 cups of boiling water (half a cup is taken at a time) is good for stomach ailments.

This aloe is very decorative in the garden and its candelabra of golden yellow flowers in winter is much loved by the birds as it is sticky with nectar. It needs space as it grows up 2 m in height.

Aloe davyana, *A. ferox* and *A. marlothii* are only three of the great wealth of South African medicinal aloes. There is still much to be discovered about the medicinal uses of our aloes, and fascinating collections can be made to study their uses. South Africa offers one of the largest selections of aloes in the world and almost all of them can be used medicinally.

Amatungula

Carissa edulis

ENGLISH	Num-num, Arabian numnum, Natal plum
AFRIKAANS	Noem-noem
LOVEDU	Morogola
NDEBELE	Umlugulu, amatungulu
SHONA	Murambara, muraramombe, muruguru, mutsamviringwa, nzambara, esamviringa, umsamviringwa
TSWANA	Serokolo
VENDA	Murungulu
ZULU	Amatangoela

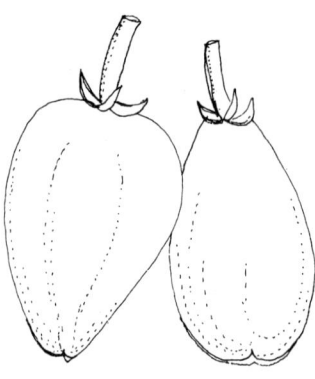

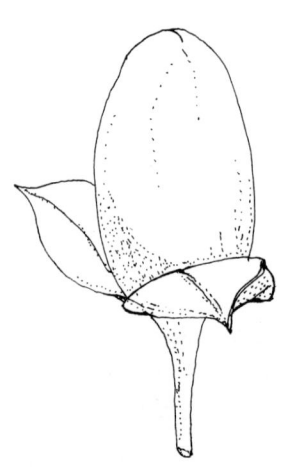

THE amatungula or num-num is an attractive, tough, dark green leaved spiny shrub with a number of species belonging to the genus *Carissa*. The beautiful deep pink and red fruits are edible and contain high quantities of vitamin C.

Often grown as a hedge to keep both people and animals out, it is commonest in Natal, and much loved by the Zulu. It grows well at the coast and I have always been thrilled to find, while travelling along the coastal roads in Natal, a bevy of small Zulu children selling a glorious, colourful basket of num-nums. The children pack the juicy fruits into banana leaf cones that they have neatly folded, and the colour and the beauty of the shocking pink fruits nestling in their green packet is stunning — I always have to stop and buy them!

The fruit makes a fragrant and delicious jam, and I add the sliced fresh fruits to fruit salads, jellies, bredies and meat loaves. It is deliciously astringent and adds an exciting taste to meat dishes.

Some varieties reach a metre and a half in height, and I grow a smaller, low-growing variety in my herb garden. The low-growing variety makes a good dark green covering for difficult terrain, binding the soil together. All bear fragrant star-like flowers that are beautiful in pot-pourris if you can bear to sacrifice the fruits in midsummer, and I find that the birds, the monkeys, the farm workers, the visitors and I are all continuously engaged in battle as to who will sample the fruits when they ripen!

Medicinally, the amatungula has a wide variety of uses. The root shaved and added to hot water (pour 1 cup of boiling water over 2 teaspoons of shaved root) is excellent for coughs, gastric ulcers, as a tonic and some tribes use it with

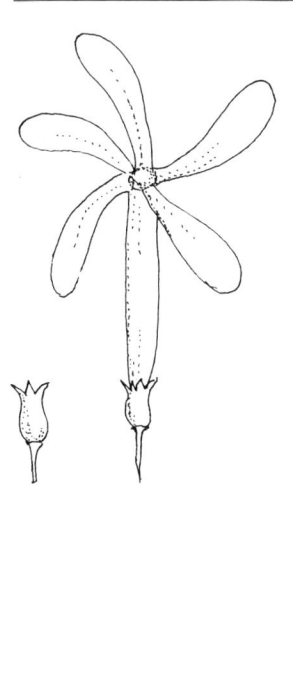

leaves and thorns to treat venereal diseases. I eat the fruit for a sore throat and find it wonderfully soothing.

A stick fixed to the roof of a hut is believed to repel snakes, and some tribes plant a few shrubs near their houses for protection.

The West Africans use a piece of root to impart flavour to stews and vegetables, and keep a root and a leaf sprig in their water containers to keep the water fresh. This is then used in cooking to impart an agreeable flavour.

The root can also be macerated, steeped in rum or gin, and so used as a bitter.

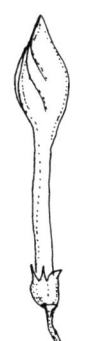
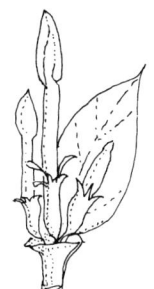
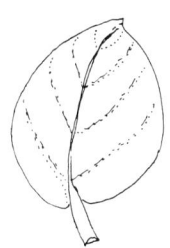
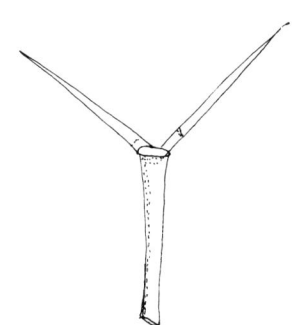

Height: 1–4 m

Amatungula jam

10 cups amatungulas
10 cups white sugar
2 cups water
1 thumblength piece ginger root

Wash and slice the fruit, leaving the skin on, into fairly thick pieces. Place in a heavy bottomed pot (not aluminium) with a lid and bring to the boil with the water. Stir in the sugar, a cup at a time. Add the ginger root. Use a wooden spoon to ensure that the fruit does not stick to the bottom of the pot. Turn the heat to low and cover the pot. Simmer slowly, stirring from time to time until the jam thickens. Owing to the fruit's high acidity it should take about 40 minutes. Pour into hot, sterilised jars. Cover with a circle of paper dipped into brandy and seal.

Amatungula syrup

3 cups sugar
3 cups water
5 cups fruit, roughly sliced
1 cinnamon stick
1 teaspoon ground cloves (optional)

This is a most glorious delicacy that is superb poured over ice-cream, rice pudding or added to soda water, ginger ale or Indian tonic water.

Boil the sugar in the water for 5 minutes. Add the fruit, cinnamon and cloves. Simmer slowly for 30 to 40 minutes, stirring every now and then. Strain the syrup through a fine mesh sieve. Pour into hot, sterilised bottles and seal.

Arum lily
Zantedeschia aethiopica

ENGLISH	Pig lily, jack-in-the pulpit, white arum, lily-of-the-Nile, trumpet lily
AFRIKAANS	Varklelie, varkore, varkblom, aronskelk
PONDO	Nyiba, nyibiba, nyuba
SOTHO	Mothebe
XHOSA	Inyiba, inyibiba, inyuba
ZULU	Intebe

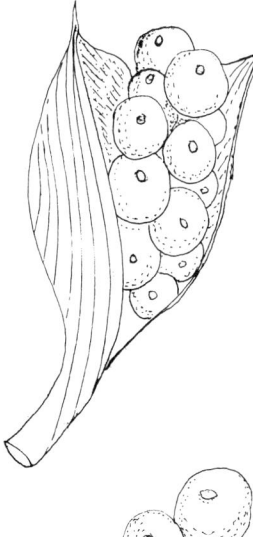

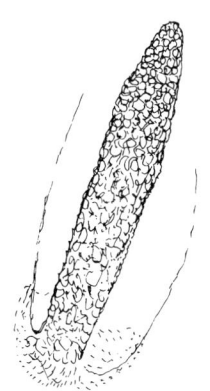

THE beautiful arum lily grows naturally in the Cape, Natal, Orange Free State and into the eastern Transvaal, presenting a glorious sight in spring when it blooms in abundance. In the Cape the arums are at their best in spring after the winter rains, and I am always amazed to see great swathes along barren open ground and up on hillsides around Stellenbosch, Somerset West and Darling, making a breathtaking sight.

The arum is one of South Africa's most famous indigenous plants and is much loved overseas, where nurseries offer it for sale at enormous prices. All varieties and colours are in great demand on the overseas markets and the cut flower sells for a good price in florists owing to its long-lasting qualities.

The arum is a much-loved medicinal plant. The leaves washed and warmed make an ideal dressing for wounds, sores, and minor burns, insect bites, stings and boils. The smoothness of the leaf makes a comforting dressing, and it can be held in place with a crepe bandage.

Gout and rheumatism sufferers have found a leaf dressing warmed in hot water and used as a soothing poultice eases the pain. I have also used it to help backache, and the coloured people in the Cape use it as a headache poultice, binding a large leaf around the head.

Ideally the arum lily likes damp, shady places, and with its great dark green leaves it is a handsome garden subject. I have grown glorious specimens with their starchy rhizomes right in the water at the edge of a stream, and find I can pick their long-lasting flowers virtually all through the year.

The seed is very fertile and I have had 85 plants from just one seed head — perfect germination! One can also split the plants from the rhizomes, and as

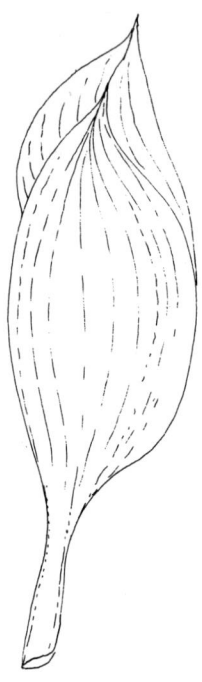

their leaves look good all year round they are so worthwhile growing in the garden, with the bonus of having a ready supply of dressing poultices and glorious cut flowers for the vase.

The name pig lily or varkblom is given because the arum is relished by pigs, who particularly enjoy digging up the fleshy rhizomes. They are not the only creatures who favour this delicacy. Living on a farm, I am often enfuriated by the porcupines who come from the mountain specially to dig up my arum lilies. They cannot leave them alone and even munch the stems of the leaves and flowers, which always amazes me as the juice is acrid, intensely burning to the skin, and causes swelling of the mucous membranes of the nose and mouth in humans.

In former times the Khoikhoi or Hottentots and Southern Sotho cooked a tasty bredie or stew of the young rhizomes, known as Hottentots' bread, and in times of drought this was considered a delicacy. Apparently the boiling or roasting of the rhizomes removes the toxicity, but I have never felt it worth trying as the reaction to eating the plant could be severe.

The early Cape colonists cultivated arums to feed their pigs, and as they grow so easily it was a profitable food source.

Height: 1 m

Birdseed grass

Bitter apple

Agt-dae-geneesbossie

Bird's brandy

Buddleja

Brandblaar

Bitterwortel

Bobbejaanstert

Baobab
Adansonia digitata

ENGLISH	Cream of tartar tree, lemonade tree, monkey bread tree
AFRIKAANS	Broodboom, kremetartboom, bobbejaanbroodboom
MANYIKA	Mubuyu
NDEBELE	Umkomo
SHONA	Mauyu, muuyu, mukomo, mumungu, tsongoro (seeds)
SWAHILI	Mbuyu
TSONGA	Shimuwu
TSWANA	Mowana
VENDA	Muvuhuyu
ZULU	Isimuhu, umshimulu

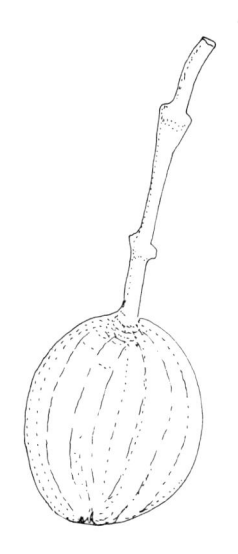

THE baobab is one of Africa's most magnificent and interesting trees. The huge trunk, sometimes 28 to 35 m in circumference, and the comparatively squat height — usually not more than 10 to 15 m — give it a strangely prehistoric look. In winter its bare branches look quite grotesque, and it is no wonder that so much myth and magic are associated with the tree by the indigenous people.

The baobab grows only in frost-free, tropical zones: the northern and eastern areas of the Transvaal and up through Botswana, Zimbabwe, Namibia and Zambia. It flourishes in hot, dry areas, and is a remarkable source of water in desert regions. In order to tap that great reservoir the indigenous people would drill a hole in the trunk, then tap a bung into the hole from which water could be drawn at will. The tree would probably contain 1 000 gallons of water, so with careful use long periods of drought could be withstood.

The bark on the great bole is pinkish grey and smooth and contains a tragacanth-like gum, and is thick with mucilage. Although bitter in taste it is nevertheless used as a food by the indigenous people. It has been used commercially as a medicine — it is excellent for bringing down fevers, and some tribes believe it to be a cure for malaria. Its strong, flexible, fibrous qualities make it an excellent rope, and it can be pounded into a strong string and woven into articles of clothing, mats, baskets and sacking. Long pieces of fibre up to a metre in length torn from the inner bark make a valuable fibre which in East Africa is woven into cloth; and a method for removing the bark from the standing tree has been evolved so as not to damage the tree.

The long pink ribbons from the inner bark yield a pulp which makes them useful for paper making. The paper is flexible, white and of a fairly good quality. The wood too is soft, fibrous, spongy and quite juicy, and is chewed by buck to quench their thirst, but is not suitable for any wood working. Its soft fibres also make it valuable in the paper industry (it gives strength to brown paper) and for ceiling boards, pressed wood boards or fibre boards.

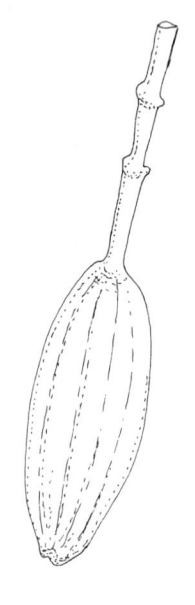

The wood is salty when burnt to an ash and this is valuable for tribes who have little contact with civilisation and thus no access to salt. It is also nourishing when mixed with food.

The leaf is edible, and is a valuable medicine. It can be eaten fresh or made into a spinach, the latter being most common. The water in which it is boiled is taken a little at a time to bring down fever, and for chest ailments, coughs and asthma. The leaf contains tannin, mucilage, sodium chloride and potassium acid, and is therefore valuable for treating fevers, and to check copious perspiration. It is also used by some tribes as a poultice around the chest and on sores and scratches.

Exquisite large white flowers are produced in summer. Five large, crinkled, waxy white petals surround a mass of golden stamens, and hang singly from the tips of the twigs. They have a strong, unpleasant scent, however, and bruise easily, and it is thought that they are pollinated by bats.

The large, hard-shelled fruit follows the flowers, and is about 12 cm long, ovoid, and covered in yellowy-grey fine hairs. The seeds are embedded in a white, powdery pulp which contains tartaric acid (hence the name cream-of-tartar tree). They are refreshing to suck, and when boiled or soaked in water make a pleasant, lemony flavoured drink that is taken to bring down fevers and relieve colds. Some tribes believe that to drink the water in which the seeds have been soaked will protect one from crocodiles!

The acid pith is a remarkable thirst quencher, and is also used with partially fermented mealie meal in the making of a much-relished acid porridge which is believed to make one strong and brave.

Young baobab seedlings have simple leaves and only after some years do they start to produce the characteristic three and then five foliate compound leaves. This may be the reason for the common African belief that there is no such thing as a young boabab tree — the young seedling is not recognisable as a baobab tree! Recent carbon testing has found that very large specimens may be well over 3 000 years old.

The vitality and resilience of the baobab is amazing. Some have been hollowed out to form dwellings, often being burnt or chopped, and some farmers have used them as storage barns, sheds and refuges; and in spite of the mutilation, the tree flourishes.

The baobab does not fall over — it collapses into itself, leaving great mounds of fibrous pulp. During a bad frost some years ago which killed many baobabs, there were several authentic reports that some of the dead trees burst into flames by spontaneous combustion. This fact, and the well known myth that God in error planted the trees upside down, has been used in African folklore and the tales are many and awesome. Several tribes believe that one should never pick a baobab flower as it is part lion (the smell is said to be similar) and its animal counterpart will search you out and devour you. I love that tale, for without humans to pick the flowers the fruits will then be prolific, and as they are devoured by baboons, monkeys, man and other animals, they need to be.

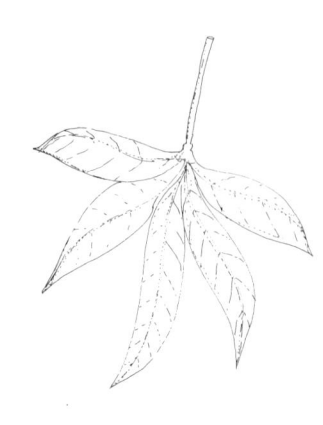

Many medical tests are being done on this great giant, and with all its qualities and remarkable uses, surely it could be of value to mankind now with our superior technology? As the seeds germinate so easily, we should be planting far more young trees than we are.

Baobabs do not like a lot of water, they need space, a warm, frost-free climate and well-drained, sandy soil. If you have the necessary conditions, do consider planting this magical tree of Africa — your interest in it will never wane, and your pleasure in it will grow as the tree grows — great, strong and magnificent.

Height: 15 m

Barley sugar plant
Pollichia campestris

ENGLISH	Wild sugar bush
AFRIKAANS	Teesuikerbossie, wilde druiwe, kafferdruiwe, suikertee
PEDI	Letsoai, monokotsoai-oa-makhoaba, sefakoana, tsimo-ea-manyokolo
SOTHO	Letswai, sefakwana
TSWANA	Sefakwanama
XHOSA	Behlungulu, utywala, amangabangaba
ZULU	Umhlungulu, ukudla kwamabayi

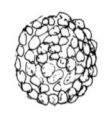

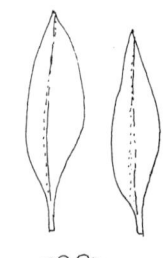

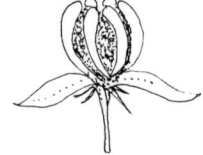

THE barley sugar plant is a grey green, low growing shrub with tiny, sharply pointed leaves growing in whorls or rings around the stem. The small, round, green 'flowers' like little berries are tucked into the axils of the leaves and around each flower is a bract which becomes juicy and fleshy and turns translucent white as it matures. This sweet-tasting 'berry' tastes much like a white mulberry, and is the barley sugar or 'teesuiker' that is so sought after by birds, animals and children. The whole plant is greedily devoured by stock, perhaps because of its high water content.

Medicinally this is an important plant and is used in several ways. It is used as a soothing inhalent for chest ailments such as tight chest, asthma, congestion of the nose and throat and bronchitis. Pour 2 cups of boiling water over 1 cup of stems and leaves; hold a towel tent over the head, place the bowl within the tent and inhale the steam. It is also said to be useful for treating rheumatism in the neck and shoulders. Twigs can be tied into a cloth or piece of pantihose and used in the bath to soothe rheumatism and as a wash to refresh and revive.

I have found that after an exhausting day on my feet a good rub of barley sugar plant tied into my facecloth and used in the bath around my legs and heels takes away all the aches and pains. This treatment is much used by old people in rural areas, who also place a compress of warmed, crushed leaves and twigs over painful areas such as bruises, swollen joints, a sprained ankle or an aching back. Pour hot water over the leaves and twigs, drain and cover with a hot cloth. Place the compress over the area, cover with a crepe bandage and leave on for 2 to 3 hours.

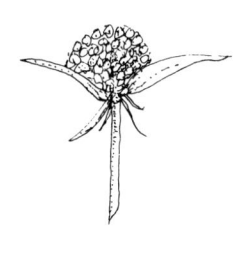

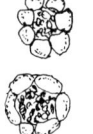

A tea made by pouring 1 cup of boiling water over ¼ cup of fresh chopped twigs and leaves and allowed to stand for 5 minutes, then strained and drunk sweetened with a few of the sweet white 'berries', is much favoured as a treatment for rheumatism, coughs, aches and pains, general fatigue, and to help give strength and stamina to the aged or those who have undergone physical exertion. This tea can also be used as a rub or lotion massaged lightly into aching legs, feet or shoulders or a bandage may be dipped into it, wrung out and applied to a painful area for gentle and comforting relief. I tried it around a sprained ankle and found it wonderfully soothing.

I have noticed Tswana workers after a long, hot day planting in the fields picking bunches of the barley sugar plant, chewing a few white berries, then pulping the leaves and stems in their hands and rubbing these over their heels and calves, and arms and shoulders, and pressing a few bits into their shoes as they set off for home. When I questioned them they said in doing so there would be no chance of stiffness setting in when they sat around the fire that night. Some also took home a bunch of leaves to boil in water as a wash to soothe aches and pains.

This remarkable, hardy little plant occurs in the eastern Cape, Natal, the Orange Free State, the Transvaal, Zimbabwe and further north. It propagates easily from cuttings and bits of rooted branch pressed into wet sand. Once planted in the garden the bush can grow to a metre in height, and as it is evergreen and perennial, I find it a charming garden subject. Seeds are sometimes available through indigenous nurseries and botanical gardens.

Height: 0,5–1 m

Bird's brandy
Lantana rugosa

ENGLISH	Chameleon's berry
AFRIKAANS	Wilde salie
LOVEDU	Khokukurwani, mokhukurwane
NDEBELE	Utshwala benyoni
PEDI	Mosunkwane, mokgotutane
SOTHO	Jwala-ba-dinonyana, mabele-mabutsoapele, mabele-mabutswa-pele, makhoabaa-matona, monokotsoaioa, monokotswai-oa-makhgwaba-matona
SWATI	Khitchukhurwani, inkobe
TSWANA	Selaole, dikobetsa badisana
VENDA	Tshidzomba-vhalisa
XHOSA	Utywala-bentaka, utyani-bentaka
ZULU	Imphema, ubukhwezane, ubukwelezane, ubungungundwane, uguguvama, umqebezane

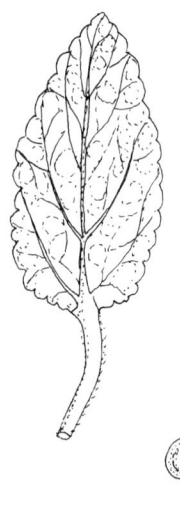

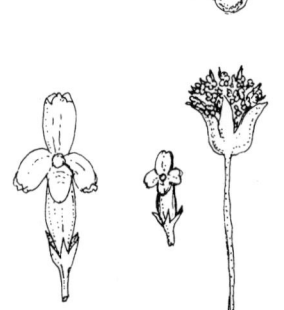

BIRD'S brandy is a small, low-growing, spreading shrub, which is an important medicinal herb. It grows all over South Africa and it is especially noticeable in summer when the small, clustered, purple berries cover the whole bush. If caged birds are fed too many of the over-ripe fruits they become intoxicated or stupefied, and this is where the common name 'bird's brandy' comes from.

The small, hairy, dark green leaves have a verbena-like smell, and in spring and early summer the fragrant mauve flowers make a wonderful pot-pourri ingredient.

Several African tribes boil a small quantity of leaf and stem in water and use it as a wash for sores, rashes and festering scratches and insect bites. A weak lotion is used by the Zulu and Tswana for sore eyes, and warmed leaves are packed behind and around the ear for earache.

The leaf is astringent, and is apparently used in the treatment of opthalmia in cattle — by squeezing the juice from the leaves directly onto the eye. Some tribes chew the leaf and use the saliva as a medicine for eye infections in cattle and sheep.

A tea made by pouring 2 cups of boiling water over half a cup of leaves and stems is taken, small quantities at a time, as a treatment for bronchitis, chest ailments and stomach ailments and as an eye bath for treating pink eye.

The plant also has magical uses and is much revered by several African tribes. As it is one of the earliest plants to fruit, it is burnt in the fields by the Basuto so that the smoke will ensure an early ripening crop.

The fruit is sweet and edible and is used mixed with the leaves and cooking oil as a body perfume and protector. It is also eaten in times of famine and is much loved by children.

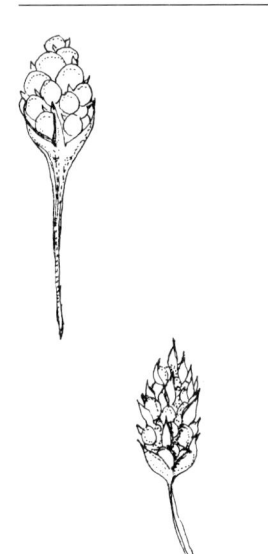

In some districts it is still used by whites as a treatment for skin infections and rashes. I make a bath preparation by boiling 2 cups of leaves and stems in 3 litres of water for 15 minutes. I then strain this and add it to the bath, which I find refreshing, skin softening and very soothing to rashes and scratches. This lotion can also be dabbed onto scratches — it is astringent and may smart initially, but it quickly soothes and heals.

Many years ago a farmer in the Klerksdorp district told me how he was taught by a Xhosa farm worker to rub the ripe fruits of bird's brandy over scrapes and cuts as an instant dressing. Both he and his Xhosa farmhand had used the leaves warmed in hot water as an excellent poultice for a sprain or rheumatic joint since they were young boys, having been shown how to use the plant by their respective grandmothers.

It is always fascinating to go back in time to trace who first discovered and experimented with the healing plants. Plants have the wonderful ability to link all nations together with their healing and health-giving properties and this pretty little wild plant does just that.

Height: 1 m

Birdseed grass
Lepidium africanum

ENGLISH	Pepperweed, pepperwort, pepper grass
AFRIKAANS	Sterkgras, peperbos
SOUTHERN SOTHO	Sebitsa, sebitsana
SWATI	Umatholisa

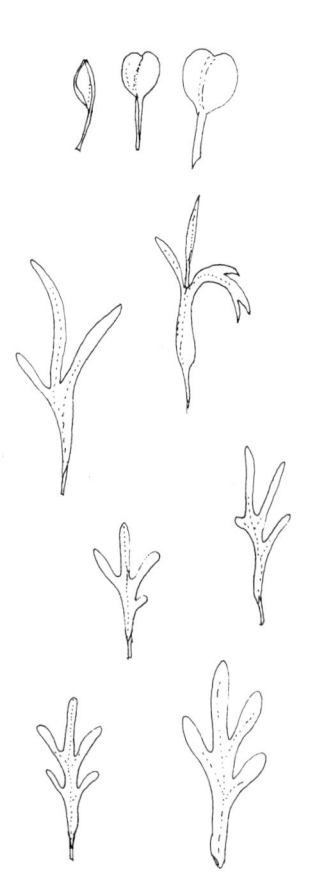

THIS rather dainty and attractive weed is found throughout South Africa. It is small, stiff, upright, with jagged-toothed leaves and spikes of first tiny white flowers followed by green seeds, giving the inflorescence the fine and pretty appearance suggestive of the spikelets of some grasses — hence the name peppergrass or sterkgras.

The leaves are edible as a spinach and are often combined with other weeds into a tasty 'moroggo', which is much loved by the Southern Sotho and the Zulu. The Tswana workers on my farm say they eat the leaves only if they have a bad cough or sore throat and not as a vegetable, but that they feed the plant to their chickens, cattle and goats to keep them disease-free if they have to be caged for any length of time.

The Zulu use the roots pounded into hot water (½ cup of fresh roots to 2 cups of boiling water) as a cough, sore throat and bronchitis remedy. The 'tea' is left to draw for about 5 to 7 minutes, then strained and drunk as a medicine or used as a gargle. Sometimes the root is dried and then ground and used in a tea, or the fresh root is shaved into porridge — about 1 tablespoon of the shavings makes one dose. It is difficult to ascertain the correct dosage as different families seem to have their own methods — so this is only approximate. I have often found some of the mixtures taught to me extremely strong and this was the case with peppergrass!

What I have found is that the seeds of this useful little plant added to sauces, stews, chutneys and soups give a most delicious taste that is spicy and peppery, so I pounce on the plants with delight every time I see them coming up in the garden or in the veld.

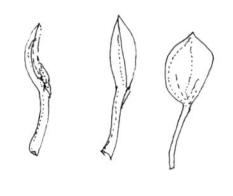

The dried plant makes a lovely fine effect in dried arrangements and can be cultivated easily as it reseeds itself several times during the season. I have grown beautiful specimens in a furrow where they get constant water. I have also used the green and dried seeds as a condiment in flavouring food for those suffering from stomach ulcers and tension, and everytime it was much enjoyed. It is wonderfully useful on a picnic if you should have forgotten to bring along the salt and pepper: use a sprinkling of peppergrass seeds on your hard-boiled eggs or braaied meat. The plants so often grow conveniently nearby!

Little bunches of peppergrass tied together and pegged into canaries' or budgies' cages is a great treat for them and this is probably how the name birdseed came about.

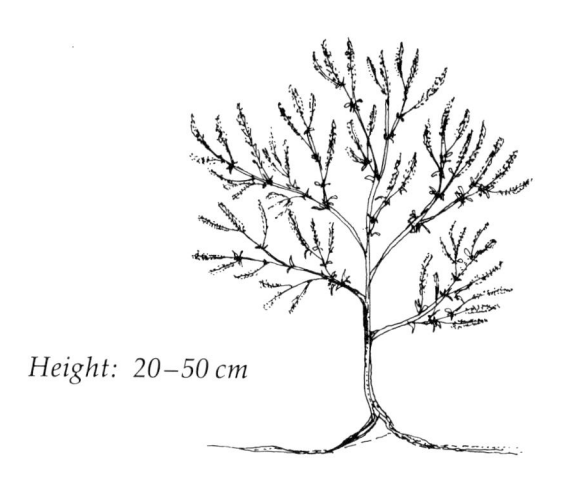

Height: 20–50 cm

Birdseed grass peach chutney

500 g dried yellow peaches
500 g raisins
250 g dried apricots
1,5 litre vinegar
4 onions, finely chopped
500 g brown sugar
2 teaspoons salt
2 teaspoons ground ginger
2 teaspoons mustard powder
2 teaspoons cayenne pepper
2 teaspoons allspice
2 tablespoons fresh birdseed grass seeds, stripped off their stems

Soak the peaches, raisins and apricots overnight in enough water to cover them. Next morning drain and chop roughly.

Bring the vinegar and sugar to the boil. Add the remaining ingredients, simmer, stirring occasionally for about 2 hours until the chutney is fairly thick. Ladle into hot bottles. Seal with a little wax, and a well-fitting lid. Label and date.

Serve as a relish with meat, fish, eggs, cheese, or as a spread on bread.

Bitter apple
Solanum sodomeum

ENGLISH	Apple of Sodom, Kei-apple
AFRIKAANS	Gifappel, bitterappel
SOTHO	Monyaku, morola
TSWANA	Thlare sa mpja, mokapana
XHOSA	UmThuma
ZULU	Isendelenja, umtuma

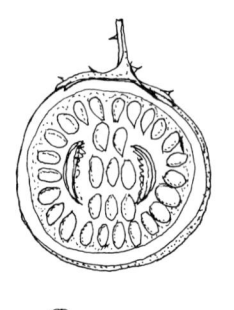

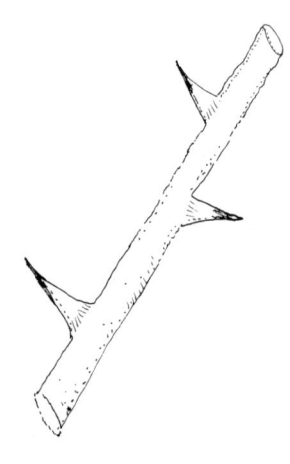

THIS spiny, branching member of the Solanaceae or potato family can sometimes be seen along the roadsides from the Cape to Zimbabwe. It seeds itself prolifically, particularly in disturbed soil, is extemely hardy and does not seem to be affected by drought.

The stems are purple or mauve when young, hairy and both the stems and leaves have thorns. The plant grows up to a metre in height. The fruit is a large berry, first green, then streaked with yellow when it ripens. Although it is poisonous, it has remarkable medicinal uses when applied externally. It is probably best known for treating ringworm in both animals and humans — a sliced fruit is merely rubbed on to the area two or three times a day and healing is rapid. Sandworm can also be treated in this way and I have seen a sandworm infection clear up overnight under the foot of a young Zulu girl after four applications of the ripe fruit. The plant is therefore much respected by blacks.

The Xhosa use the scraped root pounded and made into a tea for the relief of flatulence and colic and as a purgative, and the Zulu drink a tea made of the root to cause vomiting, and to aid fertility and impotence. Most useful of all, the fruit and leaf made into a paste can be applied to scab on sheep, anthrax pustules on cattle, mange on dogs, saddle and harness sores on horses, ringworm in cattle, dogs and horses and skin infections in humans! Some African tribes hold the crushed fruit in the mouth to relieve toothache (never swallowing it), and some chew the root, or the fruit, and spit the juice on to a wound to aid healing.

A tea of the root is used for treating coughs, tight chest, to clear bladder infections, and as a diuretic. The dosage is 1 tablespoon of crushed root to 1 cup of boiling water. Allow to stand for 5 minutes then strain and drink 3 times a day.

The plant also has magical uses which is also why it is so respected and favoured by African tribes. A piece of root is often tied round the waist, wrist or neck as a protection against poisoning or harm.

Young plants can be transplanted successfully if you do not disturb the root, and the mature plant in the garden with its yellow marble-like fruits is an interesting plant, but not very attractive. Although it is perennial it becomes straggly, so I either cut it back or set out new plants each year. I like to grow a plant or two for treating sores on the animals, and find it is much used by my gardeners for all sorts of ailments from burns to infected scratches and blisters. When the sangomas visit my herb garden in summer they always ask for bitter apple plants.

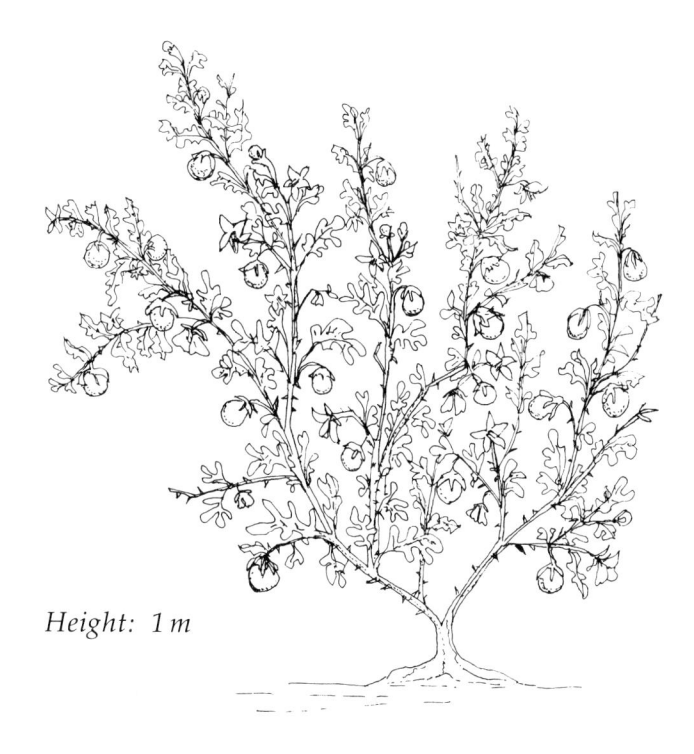

Height: 1 m

Bitterwortel

Xysmalobium undulatum

ENGLISH	Milk bush, milk wort, wild cotton
AFRIKAANS	Bitterhout, melkbos
HERERO	Ombaruru
PONDO	Ishongwe
SOTHO	Leshokhoa, lethokxwa, pohotshehla
VENDA	Mubva, vhubva
XHOSA	Ishongwe, iyeza elimhlope
ZULU	Nwachaba, ishongwe (plant), inshongwane (fruit)

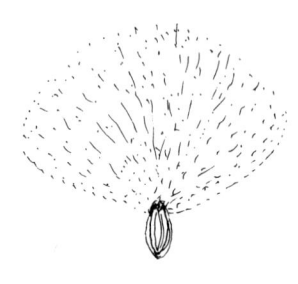

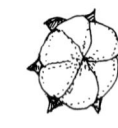

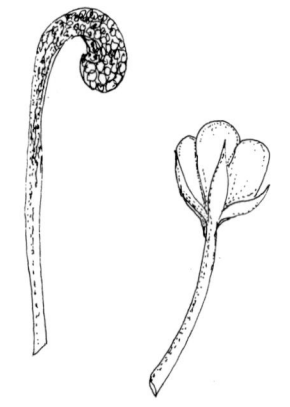

THIS strange and remarkable plant belongs to the Asclepiadaceae family. It is found throughout South Africa and grows up towards Central Africa in the open veld. It seems to grow well in all sorts of soils, and it is propagated by seed. It is probably one of the most prized indigenous medicinal plants and is used for a huge variety of ailments.

The rough, hairy leaves are edible although somewhat bitter, and can be made into a spinach, which is much loved by the Sotho and the Xhosa. The plant contains a milky latex, and this is used to treat warts, skin eruptions, scratches, wounds and corns by applying it directly to the area. I have found that it burns somewhat, so do use it with care.

The flowers are light green and hairy and form in groups around the stems, giving way to big, hairy, balloonlike seed pods. When ripe these are filled with fine silky seeds that are wind distributed, and are excellent for stuffing pillows, and of course, for lining birds nests! The Zulu and the Xhosa boil the flowers and seeds with the leaves in water as a colic remedy.

The root is the most used part of the plant and has a remarkable variety of uses. Dried, ground and powdered, it is used by the Zulu as a snuff to relieve headaches. They also sprinkle the ground root and stem on to skins and hides to prevent dogs from gnawing them! The powdered root is also taken in an infusion for dysentery, diarrhoea and stomach ailments by the Zulu, Xhosa, Mpondo and Tswana. A Venda builder on my farm makes a tea of the powdered root for colic, colds and coughs, and uses the cooled tea as a lotion for wounds and infected sores and as a poultice by soaking a cloth in it and binding it over the area.

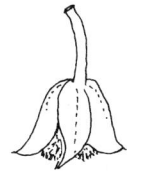

An infusion of the fresh root may be used as a tea to bring down fevers.

Collect the silky seeds from the bush once the hairy pod has become dry and brittle. Sow them in trays filled with wet river sand and keep moist until they are big enough to prick out. Transplant into compost-filled pots and replant in prepared beds when they reach 5 to 8 cm in height.

Height: 1 m

Blinkblaar
Ziziphus mucronata

ENGLISH	Buffalo thorn, shiny leaf, wait a bit
AFRIKAANS	Blinkblaar wag-'n-bietjie, bokgalo, buffelsdoring, wag-'n-bietjie-doring, wag-'n-bietjiebos
HERERO	Omukaru
KWANGALI	Mukekete
LOZI	Mukala, mukwata
NAMA	Omukaru
NDEBELE	Umpafa, umpakwe, umpasamala
SHANGAAN	Nphasamhlala
SHONA	Chinanga, muchecheni, mupakwe, mupasamala
SOTHO	Mokhalo
TSWANA	Bokxalo, mokgalo, mokxalo
VENDA	Mutshetshete
XHOSA	Umphafa
ZULU	Umlahlankosi, umphafa

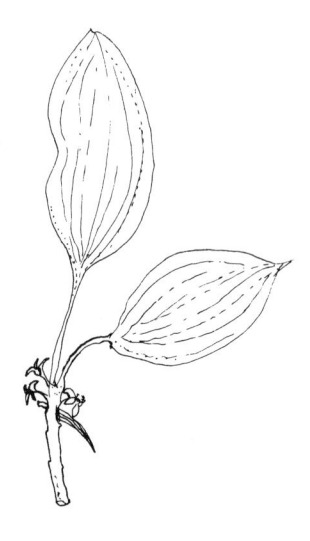

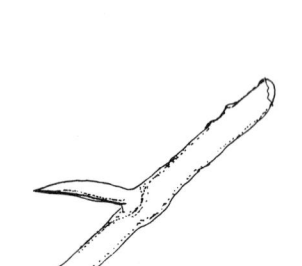

THE blinkblaar is one of Africa's great medicinal trees. It occurs in a wide area across South Africa, with the exception of parts of the Cape, although it is found in the upper eastern Cape towards Natal and the Orange Free State.

It is usually a shrubby, medium-sized tree but in old specimens it can reach 10 to 12 m in height and is a beautiful shade tree with a wide crown and sweeping branches. In the Transvaal it is often planted as a specimen tree, and looks good all the year round. In winter its bare thorny branches are attractive and the glossy brown marble-sized berries attract birds and monkeys.

The bark is grey, fissured and rough textured. The branches are grey and smooth and often droop under the weight of the leaves and berries. The ovate leaves are smooth, shiny and bright green, giving the tree its name blinkblaar, meaning shiny leaf.

Small, yellowy, inconspicuous flowers appear in spring and summer and their sweet scent and copious nectar draw the bees. They are followed by the russet brown fruits which remain on the tree during winter and often into spring and summer — so that the berries can be found all year round. The berry has a hard central stone, surrounded by a meal-like pulp, which is a wonderful thirst quencher. Berries sucked on a long walk through the veld will keep you from flagging, and stock and wild animals browse happily on fallen fruits as well as the leaves.

The blinkblaar is considered a magical tree by African tribes and is believed to be an important tree to grow near the home as it will ward off evil spirits and lightning. It is also considered to be immune to lightning, so those who shelter under it in a storm will be safe. If the branches are cut after the first summer rains it is believed that a drought will surely ensue, so the wood may only be cut at certain times of the year.

The wood is used for many purposes. It is carved into bowls, spoons and yokes, the flexible branches are peeled of their bark and thorns for oxwhips, and the thick branches are used for fencing posts, roof struts, grain mortars and gates. The wood is fairly durable and elastic and therefore much sought after. It matures to a beautiful light brown with a sheen.

Medicinally the tree is much respected and used by blacks and rural whites. The roots, baked then crushed and powdered, are widely used as a remedy for pain. The powder is made into a poultice, held in place with bandages, and this is believed to draw out the pain. To ensure that the pain does not return the whole poultice is eaten after a time by some tribes. Others bury the poultice and make a fresh one each day which they apply to the area until the pain eases. I have seen this to be very effective for backaches after only two applications.

A paste made of the leaf, crushed and pounded with water, is used by several tribes as a drawing poultice for boils and abscesses and infected, septic swellings. This can be applied twice daily until the area clears, often with an accompanying drink made by boiling 3 cups of chopped root in 3 litres of water for 10 minutes. This is then allowed to cool, then is strained and half a cup is taken 4 to 10 times during the day. It is particularly effective for tuberculous swelling and chest ailments, and as a gargle for scarlet fever and measles. This brew is sometimes also used with young strong shoots from the tree trunk, and is believed to be a sure cure.

An old folk remedy is a tea made of the leaves, or pieces of bark, usually by boiling up 1 cup of leaves, or 1 cup of bark, in 4 cups of water for 10 minutes. This is allowed to stand and cool and then is strained and drunk, half a cup at a time for coughs, chest ailments, swollen glands, lumbago, rheumatic aches and pains.

A root infusion made in the same way is also excellent for dysentery and diarrhoea and some tribes will chew a piece of root and swallow the juice to help stomach upsets and diarrhoea.

Height: 6 m

A leaf paste is also used to treat venereal diseases and slow-healing sores. The leaf may also be chewed and then placed over the area and bound in place with strips of bark or eucomis or agapanthus leaves.

The berry is used in porridge making and is a good famine food. In West Africa it is ground to make a coffee substitute. The African burial rites include the blinkblaar, as it is believed to be a tree of protection, in this life as well as the afterlife. Graves are fenced or covered by the thorny branches, probably more for practical purposes than magical ones.

The tree grows easily from seed as well as cuttings, and several nurseries offer well-established trees for sale. If you have the space in your garden this is a most worthwhile tree to plant as it grows into a most beautiful specimen and can be pruned, trained and trimmed to any shape.

China flower

confetti bush

Buchu

Bushman's tea

Barley sugar plant

Bush-tick
berry

Christmas berry

Bulbinella

Bracken

Bobbejaanstert

Xerophyta retinervis
(formerly *Vellozia retinervis*)

ENGLISH	Monkey's tail, black stick lily
AFRIKAANS	Olifantstert, besembos
SOUTHERN SOTHO	Mafiroane
TSWANA	Sefundi

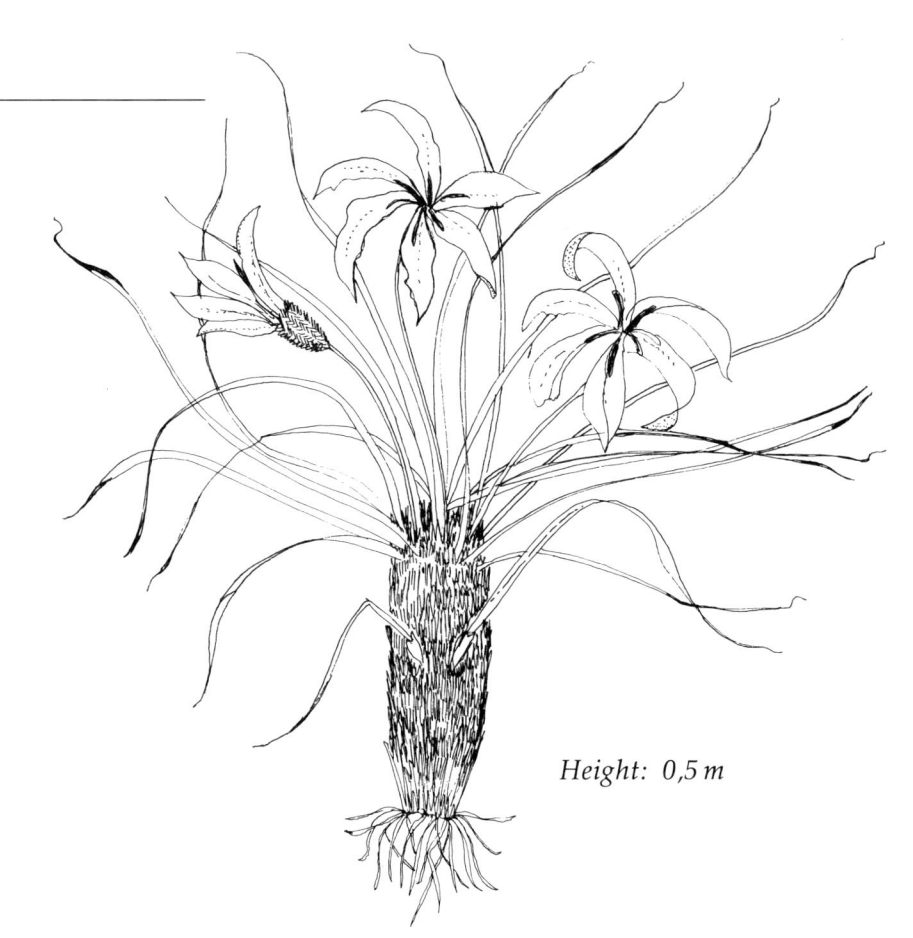

Height: 0,5 m

THE exquisite, mauvy blue funnel-shaped flowers of this strange perennial come almost as a shock in spring, when the black tufted dry plant looking like a relic from a veld fire suddenly comes to glorious life.

The dense, woody, stumpy stem — which is really a covering of roots and leaf bases around a fibrous inner stem — may be found here and there in the Transvaal and sometimes in the Orange Free State on hillsides and summit rocks, often on the southern side of hills and koppies. In spring and summer fresh tufts of rigid grass-like leaves appear and with them the beautiful, faintly scented, delicate flowers.

The plants can withstand long periods of drought owing to their strange, thick, fibrous stems which are able to soak up water rapidly and retain it for long periods. The Velloziaceae are a small family of perennials indigenous to Africa, Brazil, Madagascar and the southernmost parts of Arabia. In South Africa *Xerophyta retinervis* and *X. equisetoides* are common, both of which are used by some African tribes as a treatment for asthma.

The stem and root is dried, and this is then smoked for relieving a tight chest. The Tswana boil up the stem in water and use this as a wash for muscle tension, tight chest and over-excited children.

The bole or thick stem and root is cut and used as an effective scrubbing brush by blacks and some white farmers, and a bristly piece of stem is commonly used by the Tswana and the Sotho as a toothbrush.

I have always been fascinated by the almost prehistoric appearance of this remarkable plant, and have at times been lucky enough to find seeds, which germinate very slowly. If you are able to find seeds, remember the plant needs to be on well-drained, fairly dry soil — it does not transplant so once you put in your seed, there it must remain.

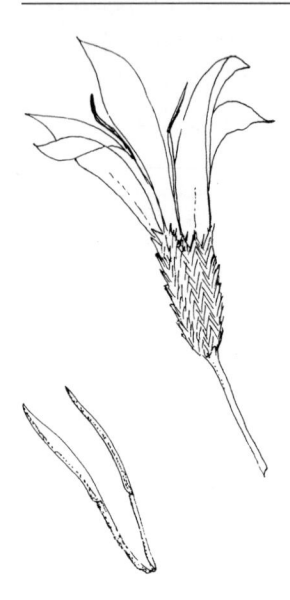

The fragile pale blueish mauve flowers appear very suddenly from September onwards, but sadly are very short-lived as they are easily destroyed by wind and rain. Try to make the plants that bloom and collect the ripening seed from these for propagation. This is a special plant that is fast disappearing, and one that we ought to propagate in earnest.

Bracken

Pteridium aquilinum

ENGLISH	Bracken fern, eagle fern
AFRIKAANS	Adelaarsvaring, brake, adelaarsvaren, arendvaring
TSWANA	Hombewe
ZULU	Umbewe

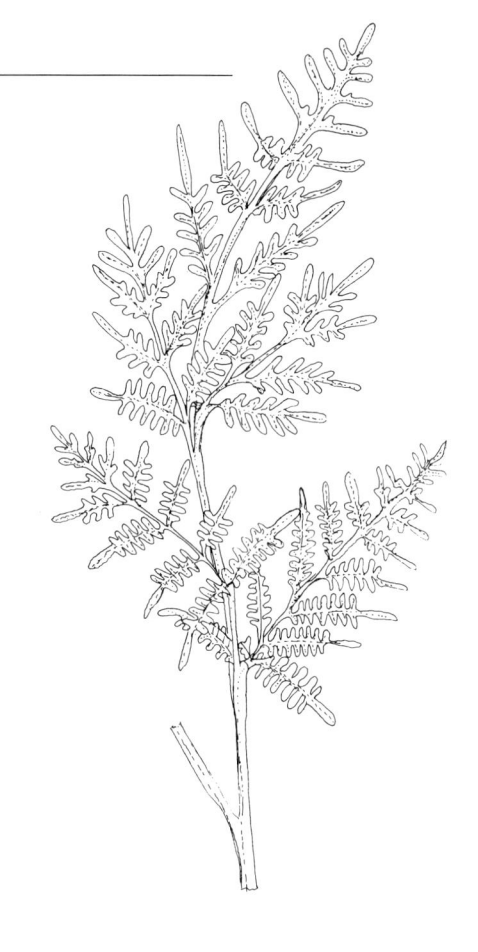

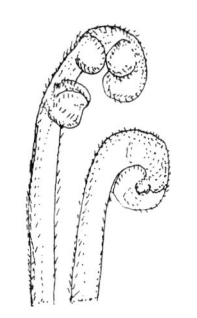

BRACKEN, the coarse, leathery, wild fern, occurs in many places throughout South Africa and throughout the world.

It covers large areas and tends to increase on badly overgrazed veld. As it is poisonous to stock, farmers try continuously to eradicate it, but its deep seated, multibranched, creeping rhizomes which shoot freely and break off easily can never really be completely dug out whole, so it is an endless problem.

Bracken occurs among the grasses on sunny mountain slopes, in ravines, on the fringes of bush, in marshy areas, on river banks, in culverts and along the edges of swamps.

The leaves are attractive, fern-like fronds, and are lovely in dried flower arrangements, particularly last year's growth as it is cinnamon-coloured with a silvery backing. Fresh green fronds dry well, and when pressed between sheets of newspaper will retain a pretty pale green colour for years.

The young curled leaf buds or shoots, commonly known as fiddleheads, can be cooked and eaten like asparagus, although the mature fern is poisonous to man and animals. The young shoots are often used as a survival food, and can be eaten raw or cooked. The food value in the ferns is relatively low, but they will help sustain life when nothing else is available. Some of the fiddleheads are more bitter than others, and some are stringy, usually depending on rainfall and age. Young fronds can also be cooked like a spinach and although these are somewhat slimy and sharp tasting, they are quite palatable when mixed with other vegetables.

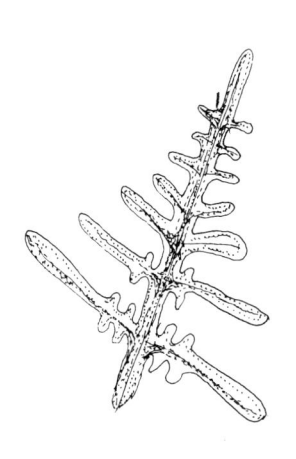

The hairy rhizomes washed well and ground up and roasted can be eaten as a vegetable or mixed into breads and porridge. The Tswana use this 'meal' to treat diarrhoea and stomach aches — a little of the cooked ground meal is eaten with water at intervals of 2 to 3 hours. The 'meal' is also used to rid the body of worms, and it is fed to pigs, dogs and goats as a food, and to clear them of intestinal infestation.

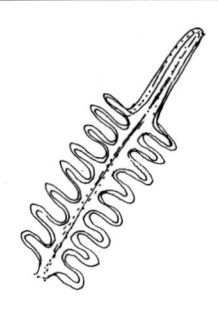

The young stem is juicy, and this well worked and softened is used by farm children — black and white — for insect bites and stings, and rashes. The juice is squeezed onto the area from the softened stem. The Zulu use the juice from the young rhizomes to treat veld sores and have found the young fronds to be of important nutritional value. They are usually boiled up in water and then eaten like asparagus. Sometimes the fiddlehead is boiled up twice, and the first water is thrown off to remove any bitterness.

Both black and white farmers have used the ash from the burned leaves of the bracken in the making of soap; and the Tswana on my farm rub the ash, which is slippery, smooth and almost oily, on to their legs to give them 'fleetness of foot' and strength. The ash has also been used in the making of glass because of its high silica content and the root, frond and underground rhizome are excellent for tanning in the leather-making process. The fronds are excellent as a long-lasting thatching for roofs.

Some nurseries offer bracken plants for sale, but before you plant it in your garden do mark off an area and set in a deep collar of plastic into the ground to restrain the evercreeping rhizomes.

A young enterprising farmer near Pretoria has turned his once much-cursed 'bracken veld' into a profitable enterprise. He sells great bunches of the fresh leaves to florists, dries the leaves for winter bouquets which he sells to florists when flowers are scarce, and he neatly packages the fiddleheads for gourmet cooks and restaurants on cardboard trays covered in cellophane — with cooking instructions! He says he does well all year round, and is considering planting more bracken!

I would love to see old-fashioned soap making start up again — and perhaps bracken could be built up into an important ingredient, making use of our wasted areas of grazing and farmland.

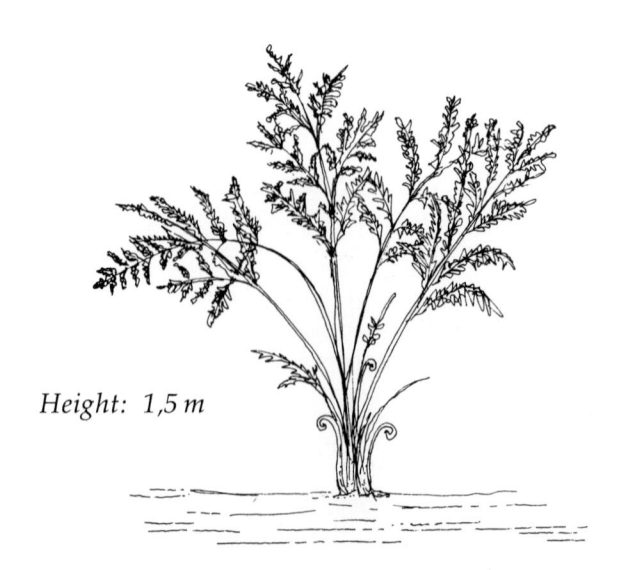

Height: 1,5 m

Bracken in mustard sauce — Serves 4

20 bracken fiddleheads
3 eggs
1 cup sugar
100 ml grape vinegar
10 ml mustard powder
little salt to taste

Wash the fiddleheads well under running water. Boil in salted water for 20 minutes. Drain.

Whisk the eggs, sugar, vinegar and mustard powder until frothy. Boil in a double boiler until the sauce starts to thicken. Stir with a wooden spoon every now and then. Add salt to taste.

When the sauce is thick, pour it over the fiddleheads which you have arranged in a dish, and sprinkle with a little chopped garlic chives, or wild garlic. Serve hot.

Brandblaar

Knowltonia vesicatora, K. transvaalensis

ENGLISH	Blister leaf
AFRIKAANS	Brandblad, tandpynblaar, katjiedrieblaar
TSWANA	Phusa
ZULU	Uxaphuza

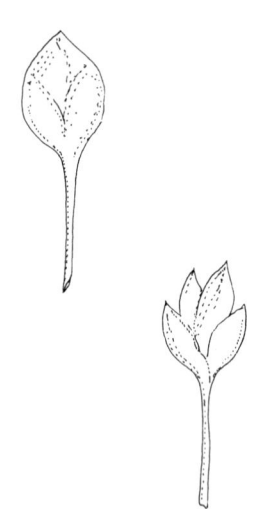

THE genus *Knowltonia* belongs to the family Ranunculaceae. Its many species were recognised by the early settlers as being similar to the hellebores found in Europe and they experimented with the leaves, using them externally as a type of mustard plaster over aching backs, joints and rheumatic areas. The burning sensation (often causing blisters) was and still is considered to be beneficial in the treatment of arthritis and rheumatism. I find the leaves extremely uncomfortable as a poultice — so I urge you to be cautious, although country people sing the brandblaar's praises!

The root stem and leaves, and to a smaller degree the flowers, contain a strong, acrid juice which reacts on the skin, raising a blister or versicant (hence the species name *vesicatora*) which is often slow to heal. In the time of Simon van der Stel it was a popular medicine, and was recorded in the *Medicinal Plant Index* by Thunberg in 1772.

The coloured people have a remedy for flu and colds which is still popular today. They make a mixture or tea of brandblaar and a pelargonium species and these dried leaves are often for sale on the Parade in Cape Town. It seems that the dose varies from family to family, but usually it is ¼ cup of mixed fresh leaves, with 1 cup of boiling water poured over it. This is left to draw for 5 minutes, then strained and drunk, a little at frequent intervals.

In the Transvaal the *Knowltonia transvaalensis* leaves are crushed and the vapour inhaled to clear a blocked nose and to ease a headache. The Tswana boil up 1 cup of leaves in 6 cups of water for 5 minutes, then strain and use this as a lotion and a wound wash; it is also dabbed onto sores and scratches.

The beautiful white anemone-like flowers of the brandblaar make a lovely garden subject and I find that ripe seed germinates easily, although it takes about 2 months to establish properly. It seems to like fairly moist soil — in the wild you'll find the brandblaar growing in the more marshy places, often in partial shade.

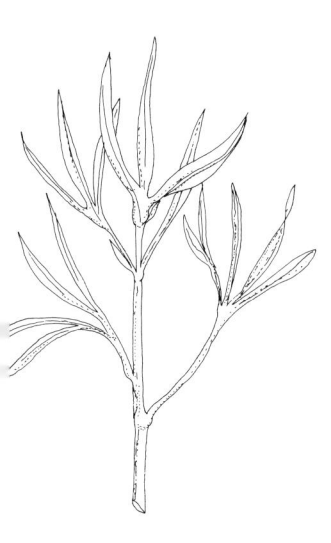

The leaves are fine and 'three-fingered', hence the name katjiedrieblaar, and the flowering head is at its best in early summer, although I have had autumn flowers depending on where I have planted it.

The spent head should be cut back after flowering as it can look untidy. If you have sensitive skin be sure to wear gloves when picking or pruning the plant.

The sangomas who visit my herb garden always request leaves of the brandblaar, which they use ground in all sorts of medicines which sound terrifyingly strong to me — but they assure me it is that strength that drives out the sickness! They also use the fresh leaves with one or two other plants as plasters or poultices for painful joints, and treat the resulting blisters with aloe juice or bulbinella juice. I was interested to see that the 'dressing' was often the castor oil shrub leaf, which is a declared noxious weed, but which grows prolifically all over South Africa. This, the sangomas inform me, is the best for drawing out pain.

Height: 0,5–0,75 m

Buchu

Agathosma betulina (formerly *Barosma betulina*), *A. crenulata*

ENGLISH	Mountain buchu, round-leaf buchu
AFRIKAANS	Boegoe, bergboegoe
KHOIKHOI	Pnkaou

A. betulina

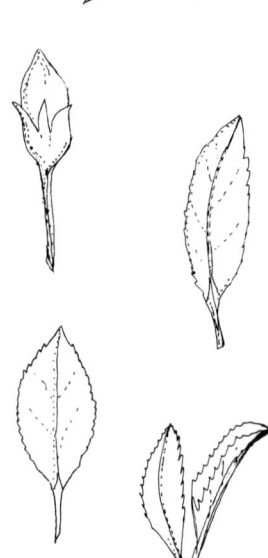

A. betulina

THE famous buchu is probably South Africa's best-known medicinal plant. Confusingly, the name buchu is given to several strongly scented plants that were introduced by the Khoikhoi to the first settlers in the Cape. The Khoikhoi dried and powdered the leaves and used this as a dusting powder for skin treatments and wounds, using the word buchu for any fragrant plant that could be used in this way — and this is probably where some of the confusion has arisen. The leaves were used for a variety of ailments — for treating wounds, stomach complaints, rheumatism, indigestion, kidney and bladder ailments — and are still used today in country districts.

Buchu leaves steeped in vinegar were once an essential part of the medicine chest. This was used as an embrocation for fractures, swellings and slow-healing wounds and country people in the south-western Cape still use this today, as do the coloured people.

The buchus belong to the family Rutaceae, to which the genus *Agathosma* belongs; and under this genus many species fall. Here I write of the well-known *Agathosma betulina* and *A. crenulata*, which are both known as the true buchu and are both grown widely for their oil and their leaves, which are dried and used medicinally.

The buchus occur naturally mainly in the south-western Cape, but have been introduced into other areas of South Africa and even overseas.

The shiny, dark green leaves are small and oval, rich in oil glands and strong smelling. The small star-like flowers ranging in colour from pinks to white are beautiful in pot-pourris and bath preparations and the shrub is an attractive addition to the garden.

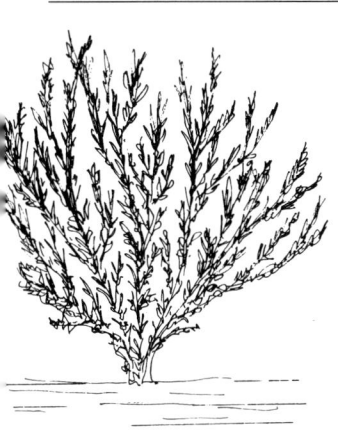

A. betulina
Height: 0,5–2 m

Buchu was first exported as a medicine in the early 1800s and it is still exported today in the form of packaged dry leaves and oil. Probably the best-known way of using it is in the famous buchu brandy, and a South African medicinal buchu wine is fast gaining popularity here and overseas as a digestive tonic that benefits rheumatic and urinary tract ailments.

Buchu brandy is made by steeping a few thumblength sprigs of fresh buchu in a bottle of brandy. Sometimes 3 or 4 cloves are added. This is shaken daily for a week then stored, and a tablespoon is taken twice a day for stomach aches, nausea, vomiting, rheumatism, bladder and kidney infections and coughs and colds.

I have used buchu brandy as a small liqueur after a heavy meal. Overseas visitors to whom I introduced it were enchanted by it, saying later that they slept astoundingly well. I later remembered my grandmother's maid telling me years ago that a teaspoon of buchu brandy would help you sleep and keep nightmares away!

Buchu is a strong herb, so use it with caution. Make a tea by pouring 1 cup of boiling water over 1 teaspoon of fresh buchu leaves. Leave to draw for 5 minutes, then strain and drink to ease cramp, colic, indigestion, chills, coughs, colds and anxiety. Often only half a cup is needed before you feel the benefit. Sip it slowly and keep the rest in the refrigerator, warming it up when required.

Tie a bunch of buchu leaves in some old pantihose and drop it under the hot tap as you fill your bath. Then relax in the hot water to ease backache and rheumatic aches and pains.

The leaves warmed in water can be used as a poultice or embrocation over a painful joint or back. Hold it in place with a crepe bandage.

The active ingredient in buchu is diosphenol (once known as barosma camphor), to which the antiseptic and diuretic effects of buchu have been ascribed. This would probably also account for the stimulation of perspiration that an infuson of buchu brings on, as well as its remarkable flushing action of the kidneys. Buchu is one of the ancient treatments for cholera and for infections of the prostate gland. It is also a remarkable treatment for gout taken as a tea twice daily — the increased perspiration is greatly beneficial to this painful affliction, and acts this way on rheumatism as well.

A number of richly scented buchus make the most beautiful addition to potpourris. Do use these very sparingly, however, as the scent is quite overpowering: *Agathosma serpyllacea* (intensely lemon-scented), *A. ciliaris*, *A. dielsana* and *A. cerefolium*.

These are also being tested for use in cosmetics, soaps, food colouring and perfumes, and although the research is only in its infancy, with this wealth of natural oils and fragrance it is certain to be important in the future. Some species have an agent that blocks ultraviolet light and research is being conducted at present for their inclusion in cosmetics, an important aspect to consider in South Africa, where our skins need special protection in the intense sunlight.

Also belonging to the buchu family is the fragrant confetti bush (*Coleonema album*, *C. pulchellum*). It is a small, upright, heath-like shrub with tiny white or pink star-like flowers. It is evergreen, easy to grow from seeds or cuttings, and its masses of flowers in spring and summer are a pure delight. The leaves are

Confetti bush
Coleonema album

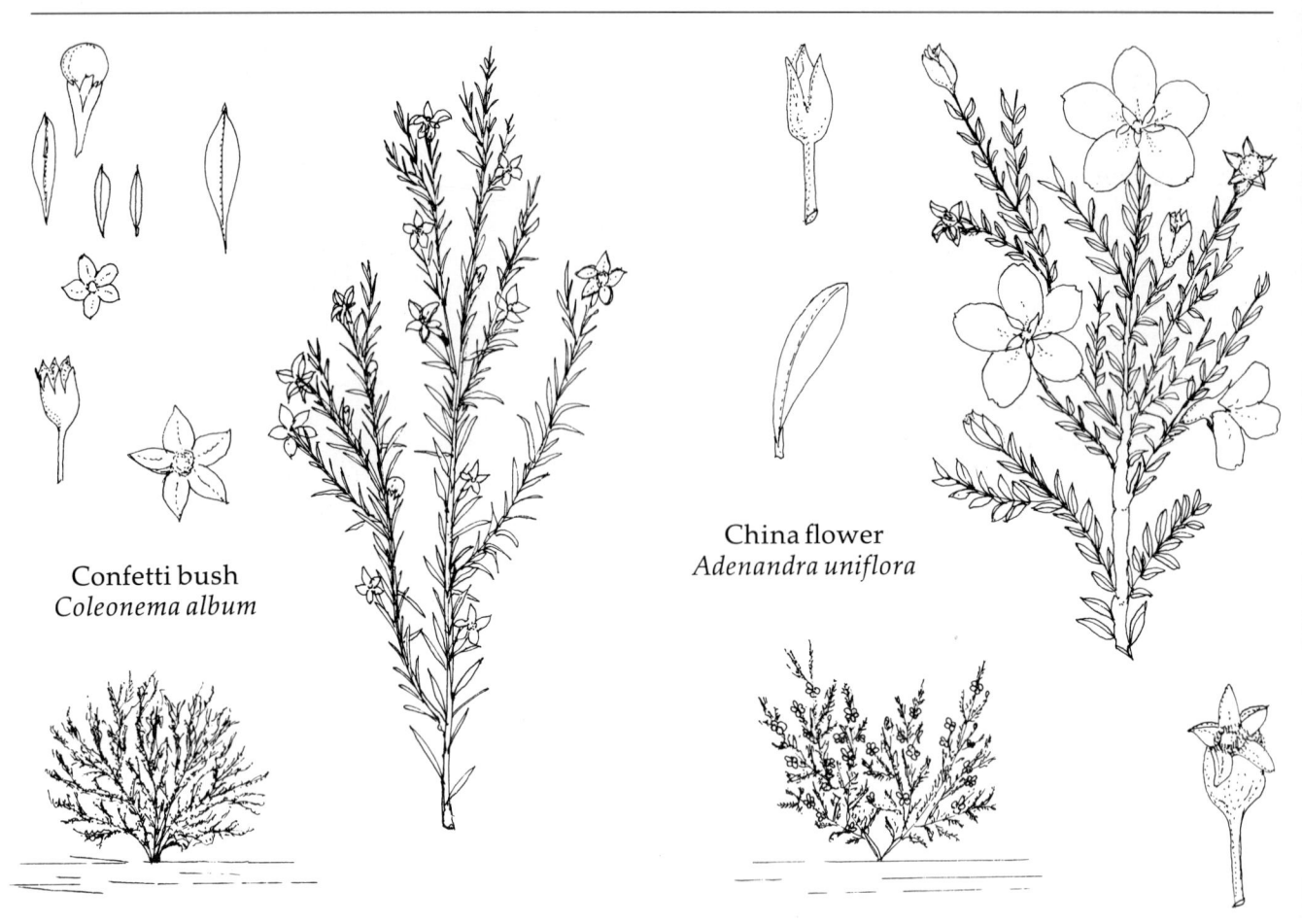

Height: 1 m

China flower
Adenandra uniflora

Height: 40 cm

used in pot-pourris, as deodorisers (a strong infusion is made and used as a wash) and fishermen rub the twigs between their hands to remove the fishy smell. They also act as insect repellents, and campers rub their bedding with the twigs to keep ants and mosquitoes away.

The exquisite china flower, also known as basterbergboegoe, Kommetjie teewater and anysboegoe (*Adenandra uniflora*), is another buchu found on the mountain slopes of the south-western Cape. With its small, porcelain-like, five-petalled flowers in pink and white marked with the finest lines of maroon, mauve or white, this plant is a most beautiful addition to a rockery. It needs well-drained soil, and as with all the buchus it needs winter watering as it originates in the winter rainfall area.

The china flower also makes a wonderful pot-pourri addition, and can be tied into a piece of pantihose and used in the bath to ease aches and pains, as a deodoriser and as a muscle relaxant.

The buchus are a most fascinating group of plants, and one could start an absorbing hobby by studying them. Kirstenbosch and various other nurseries offer the plants and the seeds for sale, so this is within everyone's reach. It is interesting to think that the buchus were one of South Africa's first medicines, and that now, so many decades later, research is still being conducted into the untapped properties of this fascinating plant.

Buddleja
Buddleja salviifolia

ENGLISH	Sagewood, wild lilac, butterfly bush
AFRIKAANS	Vaalbos, saliehout, wildesalie
SHONA	Chipambati, mupambati
SOUTHERN SOTHO	Lelothwane
SWATI	Umbataewepe
TSWANA	Mupambati
XHOSA	Cwangi, cwanci, ilotana
ZULU	Mupambati

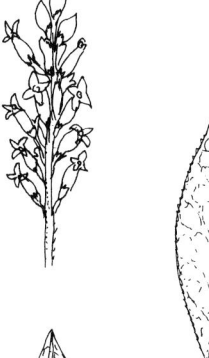

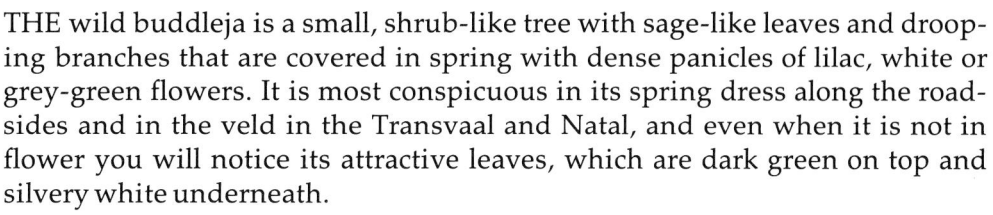

THE wild buddleja is a small, shrub-like tree with sage-like leaves and drooping branches that are covered in spring with dense panicles of lilac, white or grey-green flowers. It is most conspicuous in its spring dress along the roadsides and in the veld in the Transvaal and Natal, and even when it is not in flower you will notice its attractive leaves, which are dark green on top and silvery white underneath.

At the very first sign of spring — often in early August — the buddleja bursts forth its tight buds followed by headily fragrant flowers, and in the dry winter veld when all else is dormant, the sight is uplifting. In the evening the fragrance becomes even more pronounced, and I gather basketsful of the flowers every spring for a wild flower pot-pourri.

Butterflies love the flowers and are drawn to the bush in their dozens by its sweet honey fragrance and nectar-filled flowers — each tiny flower is tubular with a yellow centre and throat and must have the sweetest taste for the butterflies and bees. The shrub is sometimes aptly known as the butterfly bush.

Buddleja is a joy to grow in that it is so quick and so rewarding in its size and shape, is perfect for a small garden, and draws butterflies and birds. It can be pruned, trained and clipped, and I have been amazed by an avenue of buddleja lining a farm road in the eastern Transvaal that has been made into a neatly clipped hedge, an attractive 3 m high and an exquisite sight in spring and early summer. Buddleja propagates easily from cuttings, and several nurseries offer plants for sale in various colours — even a deep purple variety — all of which look attractive in the garden.

The Khoikhoi were the first to make a medicine from the leaves — probably

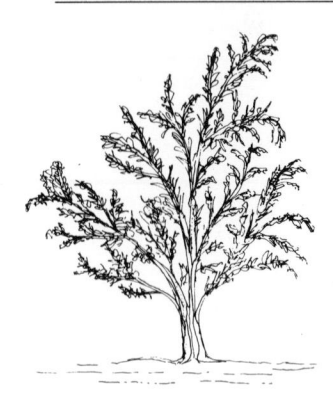

Height: 2,5–3 m

as a lotion for sores, and as an effective remedy for coughs and colds and colic.

The Tswana and the Zulu make a root decoction for stomach upsets, flatulent colic and diarrhoea. The Tswana on my farm collect the bark and steep it overnight in hot water as a treatment for sores, scratches and as a weak eye lotion for sore, red, tired eyes. The dosages vary from one family to another, but seem to be half a cup of bark or small twigs in 2 cups of hot water, left overnight and then strained.

The leaf made into a tea — 4 leaves in 1 cup of hot water — is used by the Zulu as an eyewash and a colic remedy, and the scraped root is used by the Zulu and the Sotho as a cough remedy, by steeping it in hot water and then using the 'tea' as a gargle and a medicinal brew — a little sipped at frequent intervals.

The Tswana make a springtime tea of the flowers — ¼ cup of fresh flowers steeped in 2 cups of boiling water — to use as a wash or lotion for sores, and sip a little for strength for the new season's crop planting! I love that idea — but when I tried drinking a little of the tea I found it made no difference to my springtime energy! Even when the crop planting was insurmountable, I found a good breakfast did more for my strength and stamina than the wild sweetness of that tea!

Springtime pot-pourri

1 basketful wild buddleja flowers (about 10 cups)
5 cups wild jasmine (Jasminum multipartitum)
5 cups Tulbaghia fragrans *flowers*
3 cups minced, dried orange, lemon and naartjie peel
1 cup mixed cinnamon, nutmeg and allspice, roughly crushed
jasmine or honeysuckle oil
1 cup coarse salt

Dry the flowers on newspaper in the shade, turning them every day. When completely dry, mix all the ingredients. Store in a large jar or bucket that can be sealed well. Shake daily. After 3 weeks add more oil or spices until the pot-pourri smells to your liking. Fill bowls, sachets or jars with it and place them in cupboards.

Bulbinella
Bulbine frutescens

ENGLISH	Cat's tail, burn jelly plant, stalked bulbine, grass aloe
AFRIKAANS	Katstert, geelkatstert, balsem kopieva, copaiba
SOTHO	Khomo-ya-ntsukammele, sehlare-sa-pekane, sehlare-sa-mollo
TSWANA	Ibucu
XHOSA	Intelezi, ingelwane
ZULU	Ibhucu, intelezi

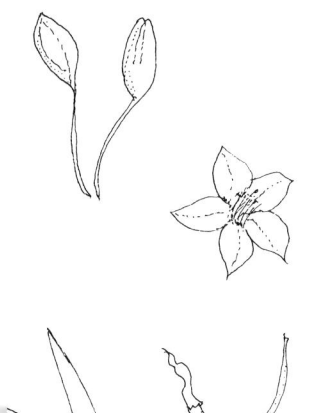

THE name bulbinella has confusingly been commonly used (often incorrectly) to name so many of this large group of plants. There are several varieties of *Bulbine frutescens*. Some have long, thin, dark green leaves while others have pale, squat leaves that grow in a neat compact plant, but the most common one, found in so many South African gardens as a popular rockery plant, is the yellow-flowered, juicy-leaved bulbinella much loved for its soothing jelly-like juice that can be so conveniently applied with a mere squeeze of the leaf.

I use bulbinella almost daily for all sorts of things. I have it growing outside the kitchen door for quick application for burns, on the front stoep as a potplant for mosquito bites on summer evenings, and in the far corners of the garden for scrapes, cuts, grazes and sunburn. All the farm workers have taken plants home to grow at their houses too!

Bulbinella must be one of nature's most remarkable medicine chests all in one, and as it grows so quickly, so easily and so abundantly, no one need ever be without it. A piece pulled off an established plant with a bit of stem will root quickly, and in no time form a cushion of succulent leaves with pretty yellow or orange flowers. It thrives in any soil and is used extensively in landscaping, on road islands, rocky hillsides, and in places where little else grows. It likes full sun, and seems to need very little water — but with the odd spadeful of compost and a good watering once a week, it makes a most handsome garden subject.

The medicinal uses of bulbinella are amazing, and I keep finding new ones! Liberally apply the freshly squeezed juice frequently to burns, blisters, rashes, insect bites, itchy places, cracked lips, fever blisters, cold sores (even up inside the nose), pimples, mouth ulcers, cracked fingers, nails and heels, bee and wasp stings, and sores and rashes on animals (I have used it on the very sensitive tummies of little puppies for rashes or eczema and it immediately soothed). No home with children should ever be without bulbinella as it is an instant first aid remedy for those daily tumbles and scrapes and it stops the bleeding!

Flat dwellers can grow bulbinella in a pot on a sunny windowsill or in a large tub on a balcony — it is pretty and decorative and an absolute joy to grow and use.

Height: 15 cm

Bulrush
Typha capensis

ENGLISH — Catstail
AFRIKAANS — Matjiesgoed, papkuil, palmiet
SOTHO — Motsitla
XHOSA — inGcongolo, umkhanzi
ZULU — Ibhuma

Height: 1,5 m

THE bulrush *Typha latifolia* is a well-known plant occurring in many places in the world along rivers and streams, beside dams, and in marshy, wet areas. In South Africa *T. capensis* grows wild throughout the country and although extensive research has been done on the overseas variety, not much has been done here on our indigenous variety.

The long, strap-like leaves are used for thatching and mat and basket making by many African people, but the plant also has medicinal values.

A decoction of the root and basal stem is used externally by the Zulu as a wash, and as a medicine taken internally for the treatment of venereal diseases. Fresh roots are dug out of the river banks, peeled and scraped, and are then boiled in water — usually 1 measure of roots to 2 or 3 measures of water. The decoction is then left to stand overnight, and is strained the next morning and used. This strong tea is also used by the Zulu and the Xhosa to assist the delivery of the afterbirth — externally as a wash, and again drunk as a medicine — for both man and animal. The Sotho drink a tea made from the roots during pregnancy, believing it will strengthen the uterine contractions.

The Tswana on the farm have shown me how they use the soft, woolly inflorescence to staunch bleeding, packing it around wounds (a method also used by the Chinese and the American Red Indians!) and the Zulu, Sotho, Xhosa and Tswana women use the soft, woolly seed heads packed into rags or pieces of material as an absorbent pad during menstruation.

A tea made of the root and lower stem — half a cup of chopped pieces to 2 cups of boiling water — has been used in South Africa and overseas for the treatment of dysentery, bowel haemorrhages, diarrhoea and enteritis. The usual dose is 1 to 2 tablespoons taken every half hour until the condition clears. As this tea is considered to be diuretic and cleansing, it is sometimes used to treat kidney and bladder problems.

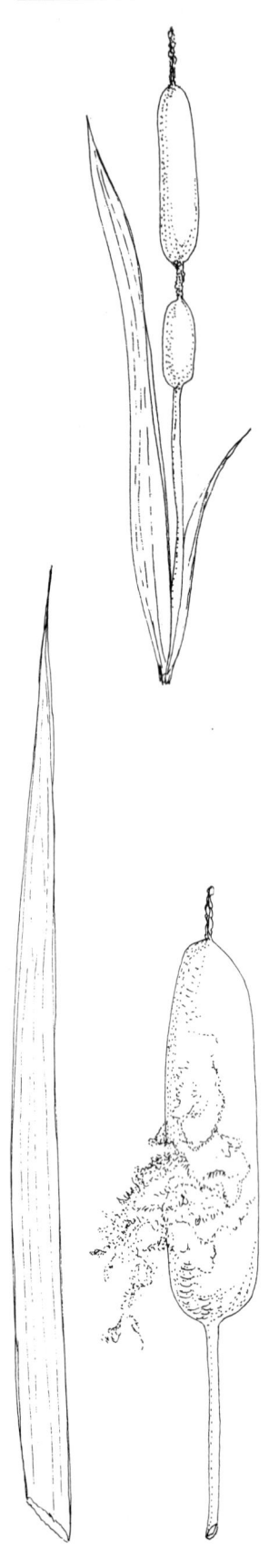

A Shangaan sangoma told me that she uses a tea of the bulrush as a blood cleanser in treating bladder and urethra infections and as an external wash. She collects fresh bulrushes from the river daily and carries out the treatment for 4 or 5 days until the patient feels better. I found this interesting, as so often the treatment is for only 1 or 2 days. She also showed me how she twisted and softened the leaves and used these as bandages to hold a dressing in place. She took a large castor oil leaf (*Ricinus communis*), warmed it in hot water, and placed it over a sprained and swollen ankle of one of the farm workers. Then, deftly twisting a bunch of leaves she softened them to a flexible bandage which she wound round the ankle over the castor oil leaf to hold it neatly in place and knotted the ends. This dressing was left on overnight. The leaf never moved, and the next morning I was surprised to see the man walking easily on the foot with no apparent discomfort. Her belief was that as the bulrush grows in wet paces, it will therefore draw off the extra 'water' that causes the swelling in a sprained ankle, and the castor oil leaf is soothing and drawing for easing the pain.

The hollow bulrush stem can be made into a flute, and the Xhosa have a ritual when a water diviner is initiated in which this flute is used. Because the plant grows in water, they believe the sound of the flute will guide them to the underground water.

The root and lower part of the stem are edible. The new young shoots that appear above the mud in spring can be cooked like asparagus, and when eaten with butter and a little salt are quite tasty. The mature root and lower stem can be ground into a meal and made into flat cakes which can be cooked over the open fire. Recent studies have found that the food value of this plant is equal to that of rice or mealies. It just doesn't taste as good!

I have minced the new shoots and used them in soup, and found them pleasant. An old Tswana man told me to pick the young flowering heads (which eventually become the dark brown velvety seedheads that make interesting dried flower arrangements), boil these in water, then drain them and fry them gently in a little fat. For this the bulrushes need careful watching to see when the bud forms in early summer.

I grow bulrushes easily near a tap, and although they like their feet in wet ground they do fairly well this way. But do take care if you plant them near a dam that you restrict the area where you plant them as they are terribly invasive.

Some farmers chop out the roots and stems and feed them to pigs and cattle as a famine food.

Cape Willow

Cape Honeysuckle

Cancer bush

Carpet geranium

Clivia

Crinum lily

Bushman's tea
Athrixia phylicoides

ENGLISH	Kaffir tea
AFRIKAANS	Boesmanstee, kaffertee
LOVEDU	Mothathaila
SOUTHERN SOTHO	Sephomolo
XHOSA	Itshelo
ZULU	Umtshanela, itiye-la-bantu, itshalo, umtshanela

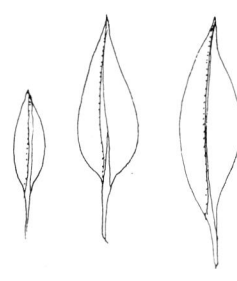

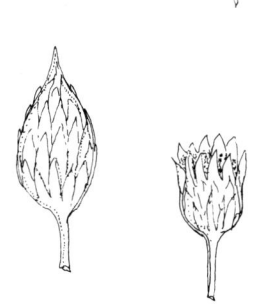

THIS is a small, pretty shrub, much branched, with thin, white, woolly stems, small dark green pointed leaves with white woolly backs and small pink or mauvy pink daisy flowers with a bright yellow centre. It grows to about 50 cm in height and is common in the eastern parts of the country in the open veld. In the coastal areas it flowers from May to July and further inland flowers appear from mid to late summer. The flowers vary from the palest pink to all shades of pink and mauve to deep purple, depending on the soil and area.

The plant has been used for many decades — by both black and white — as a medicinal tea. It was probably introduced to the colonists by the Khoikhoi, and then as the Voortrekkers moved further inland and up the coast, the San and the other indigenous people would probably have taught the travellers the uses of the plant.

As a medicinal tea it is used for cleansing or purifying the blood. It is helpful in treating boils, bad acne, infected wounds and cuts, and may also be used as a wash (¼ cup of leaves in 1 cup of boiling water — leave for 5 minutes to draw and then strain). The same brew can be used as a lotion dabbed on to the boil, skin eruption or cut.

The tea is also excellent for coughs and colds, and as a gargle for throat infections and loss of voice. The Zulu make a tea from the root of the plant for coughs, and use it as a purgative; and the Xhosa and the Sotho chew the leaf for sore throats and coughs. Several tribes and many whites drink an infusion of the leaf as a health-giving tea as the taste is pleasant. In some country districts it is actually planted in farmhouse gardens to be near at hand for tea.

A strong tea is also much valued by the Sotho as a soothing wash for sore feet, and after washing the feet they bind them in castor oil leaves *(Ricinis communis)* and often sleep in these green bandages ! The Bushman's tea and the castor oil leaves have a deep-acting effect on the hard, horny skin of the feet and on the muscles, and this is a much valued treatment.

The plant propagates from ripe seed collected at the end of summer and needs space, full sun and well-drained soil. It benefits from light pruning from time to time to keep it tidy in the garden.

Some nurseries in the eastern Cape offer Bushman's tea plants for sale, so do keep a look out for them: this is a lovely plant to grow.

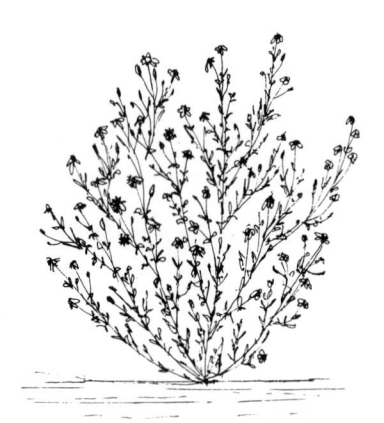

Height: 50 cm

Bush-tick berry
Chrysanthemoides monilifera

AFRIKAANS	Boetabessie, bietou, bokbessie
SOUTHERN SOTHO	Ntloyalekwaba
ZULU	Itholonja

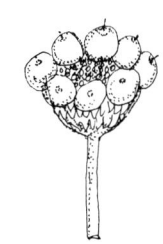

A spreading, dense, evergreen shrub, the bush-tick berry is an important wind breaker and soil binder, and a valuable source of food for the indigenous people.

During spring and early summer it is covered with masses of bright yellow eight-petalled daisy-like flowers about 2 cm in diameter. These are followed by luscious, purple berries which have given the plant its common name — the berries look exactly like bloated ticks! The species name *monilifera* means 'bearing a necklace', and this refers to the beautifully arranged berries that cover the plant necklace-like in their growth habit.

Hardy and undemanding, this plant's natural habitat is along the coast from the Cape Peninsula through to the Humansdorp area and on to Natal. Its remarkable wind and drought-resistant qualities have made the plant popular with conservationists and municipalities, as it grows well on sand dunes and on cleared arid areas like building and factory sites, thus helping to re-establish the natural plant life.

The fruits formed an important part of the Khoikhoi diet, and the Sotho, Zulu and Xhosa believe the fruit to contain blood-strengthening and purifying qualities. The juice is given in small, frequent doses — usually mixed with water — to a man suffering from impotence and to those recovering from a weakening illness like a stomach ailment or gastritis. The ripe berries are added to their porridge or the juice taken in water or tea, sometimes combined with other herbs that also give strength.

For adolescents the bush-tick berry is much sought after, as apart from its blood-strengthening and purifying qualities, it is also believed to clear up adolescent acne and skin problems and give a brave heart! The leaves contain a lot of alkaline substances and the ash from the burnt leaves and stems was used by the colonists in the making of soap. An excellent spray for mildew on plants can be made from this ash: ¼ bucket of pure bush-tick berry leaves and stems burned into an ash to 1 bucket of hot water. Leave overnight, then splash onto plants every day for 4 days.

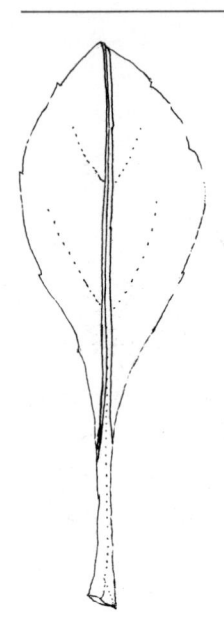

The ripe berries can be made into a delicious jam, but the fresh berry is so loved by children that there never seem to be enough left to make into jam! The birds, monkeys and baboons also relish the fruit, but the leaves are toxic to livestock, so farmers should take care.

I have tasted a nourishing syrup made from the berries, that forms a base to which iced water or soda water is added, and this makes a delicious summer cordial and keeps well undiluted in the fridge.

The plant grows easily from ripe seed, thrives in adverse conditions, and is spectacular, attractive and a remarkable garden subject when pruned into a neat bush. It needs full sun and winter watering to ensure a good crop of berries in summer.

Height: 1,5 m

Bush-tick berry cordial

4 cups ripe bush-tick berries
2 cups water
2 cups sugar
6 cloves
1 thumblength piece ginger root

Boil all the ingredients together gently for about half an hour in a closed pot. Stir from time to time to prevent sticking or burning. Allow to cool. Strain through a fine sieve. Bottle the syrup in a well-corked bottle. Serve diluted with iced water — usually 1 part syrup to 8 parts water.

Cancer bush
Sutherlandia frutescens

AFRIKAANS	Kankerbos, gansies, kalkoenbos, bitterblaar
SOUTHERN SOTHO	Musapelo
ZULU	Unwele

THE three species of *Sutherlandia*, all known as cancer bush, are *frutescens*, *microphylla* and *tomentosa*, all of which are cultivated as garden plants. They range in height from 0,5 m to 1,5 m and have red flowers and balloon-like seed heads. Long respected and used in medicine, the cancer bush was introduced to the colonists in the early days by the Khoikhoi, and it has been used ever since as a wonderful remedy for a variety of ailments.

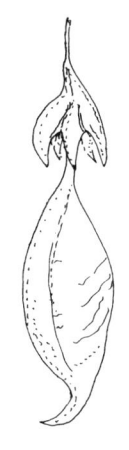

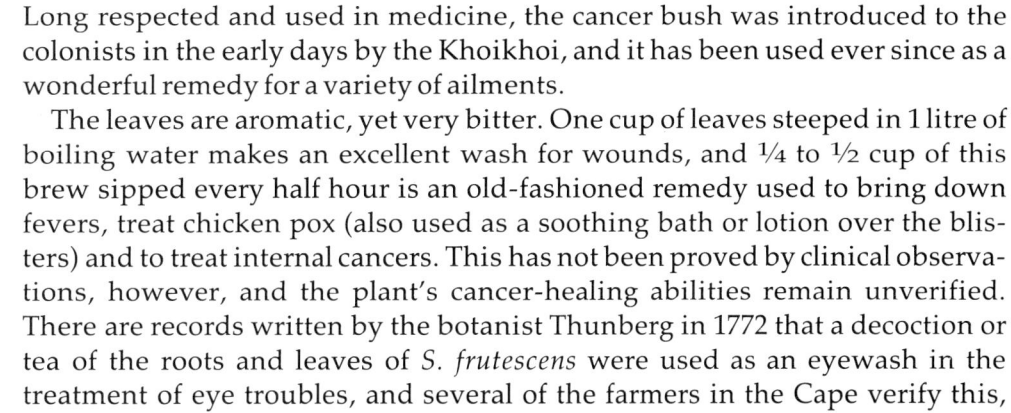

The leaves are aromatic, yet very bitter. One cup of leaves steeped in 1 litre of boiling water makes an excellent wash for wounds, and ¼ to ½ cup of this brew sipped every half hour is an old-fashioned remedy used to bring down fevers, treat chicken pox (also used as a soothing bath or lotion over the blisters) and to treat internal cancers. This has not been proved by clinical observations, however, and the plant's cancer-healing abilities remain unverified. There are records written by the botanist Thunberg in 1772 that a decoction or tea of the roots and leaves of *S. frutescens* were used as an eyewash in the treatment of eye troubles, and several of the farmers in the Cape verify this, saying that their workers still use cancer bush to treat eye and other ailments today.

In spite of the bitterness of the leaves, *Sutherlandia* is relished by browsing sheep and cattle, though cows eating a lot of it bring through that bitterness in their milk.

A weak infusion of the leaf (2 leaves in 1 cup of boiling water) can be taken for influenza, rheumatism, liver ailments, haemorrhoids (use the stronger lotion here as an external application), bladder, uterus and 'female' complaints, for diarrhoea, stomach ailments and for backache. Many people use *Suther-*

landia as a tonic and believe that a little taken before meals will aid digestion and improve the appetite.

I have used cancer bush in the bath to bring down a fever and ease the aches and pains of flu, and found it very soothing and refreshing. I used half a bucket of leaves, flowers and seed pods covered in water, and brought this to the boil. This made about 6 litres of strong tea, which I boiled for 15 minutes and then stood with the herb still in it until it was pleasantly warm. I then poured this tea into the bath water and relaxed in it for 15 minutes. It was necessary to use it only once, as by the following morning the fever was gone, and I felt much better.

Interestingly, the sangomas who come to get plants from my herb garden (I encourage this to spare the digging up of the wild plants) invariably ask for *Sutherlandia*. They use it to treat internal cancers and for fevers and flu, and mix the powdered leaf with sugar and water for coughs and colds. It is a much respected plant and I try to have seedlings going all the time.

Try growing cancer bush in your garden — it's a lovely talking point, and the seed balloons make attractive dried flower arrangements as well as being good tossed into a soothing bath after a strenuous day.

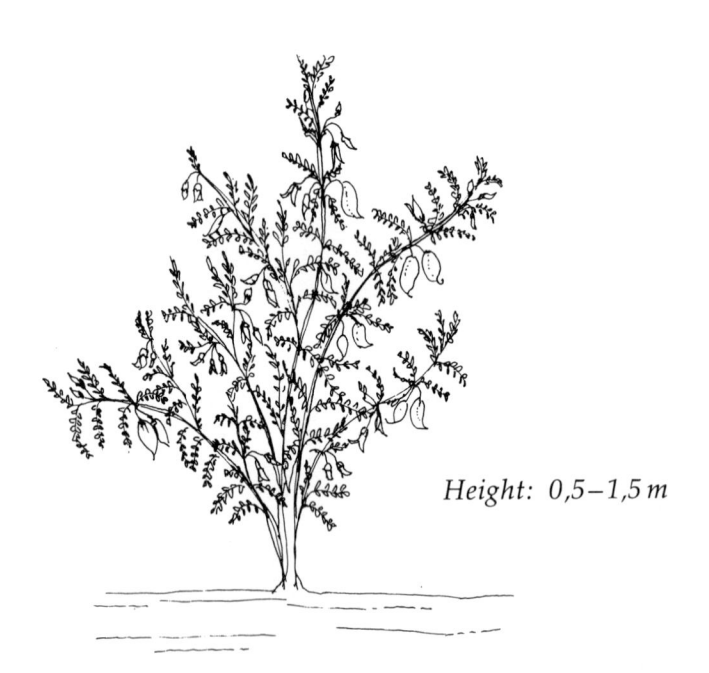

Height: 0,5–1,5 m

Cape honeysuckle
Tecomaria capensis

ENGLISH	Tecoma
AFRIKAANS	Handskoentjie, trompetters
SOTHO	Molaka
XHOSA	iCakatha
ZULU	Lungana

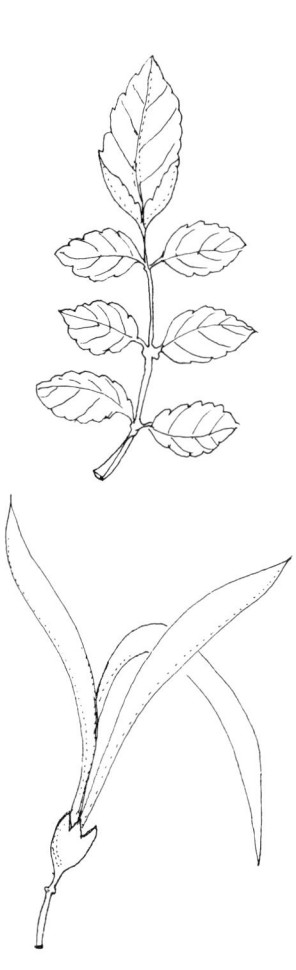

THE Cape honeysuckle is a colourful, climbing shrub that occurs mainly on the coast from Uitenhage to East London, Transkei through to Natal and up as far as the Transvaal bushveld. It is a much-loved garden plant with many uses. It can be pruned into hedges or shaped shrubs, it can be trained up trellises and pergolas, over fences and arches, and forms a striking groundcover over rocky slopes and difficult gardening areas. With its evergreen, lush, dark, shiny leaves and its brilliant orange flowers, it is a tremendous asset in the garden. My favourite way of growing it is over a pergola. Trained up a frame it makes the perfect green and living 'gazebo', and it remains lush and green all year round, with dense, inviting shade. The long, trailing stems root where they touch the ground, so once you have it established in your garden you'll have many plants that can be taken up and planted elsewhere. It benefits from pruning, and is a quick grower. The sun birds attracted by its bright flowers are a joy to watch.

Medicinally, Cape honeysuckle is much used and respected. The Sotho and the Xhosa use the dried, powdered bark in a tea for bringing down fevers, and for relieving pain and sleeplessness. The Zulu use it in the same way, but also for chest ailments like bronchitis and pneumonia, and diarrhoea, dysentery and stomach pains. The powdered bark is used to rub into bleeding gums and several people in the eastern Cape still use it in a relaxant tea, particularly to bring down fevers and ease the aches and pains of flu. The bark is scraped from the stems and dried in the shade for 3 or 4 days, after which it is crushed and powdered and stored in an airtight jar for later use.

The Xhosa cut pieces of stem and thread them into a necklace for nursing mothers, believing it to encourage milk flow and to make the baby strong and vigorous.

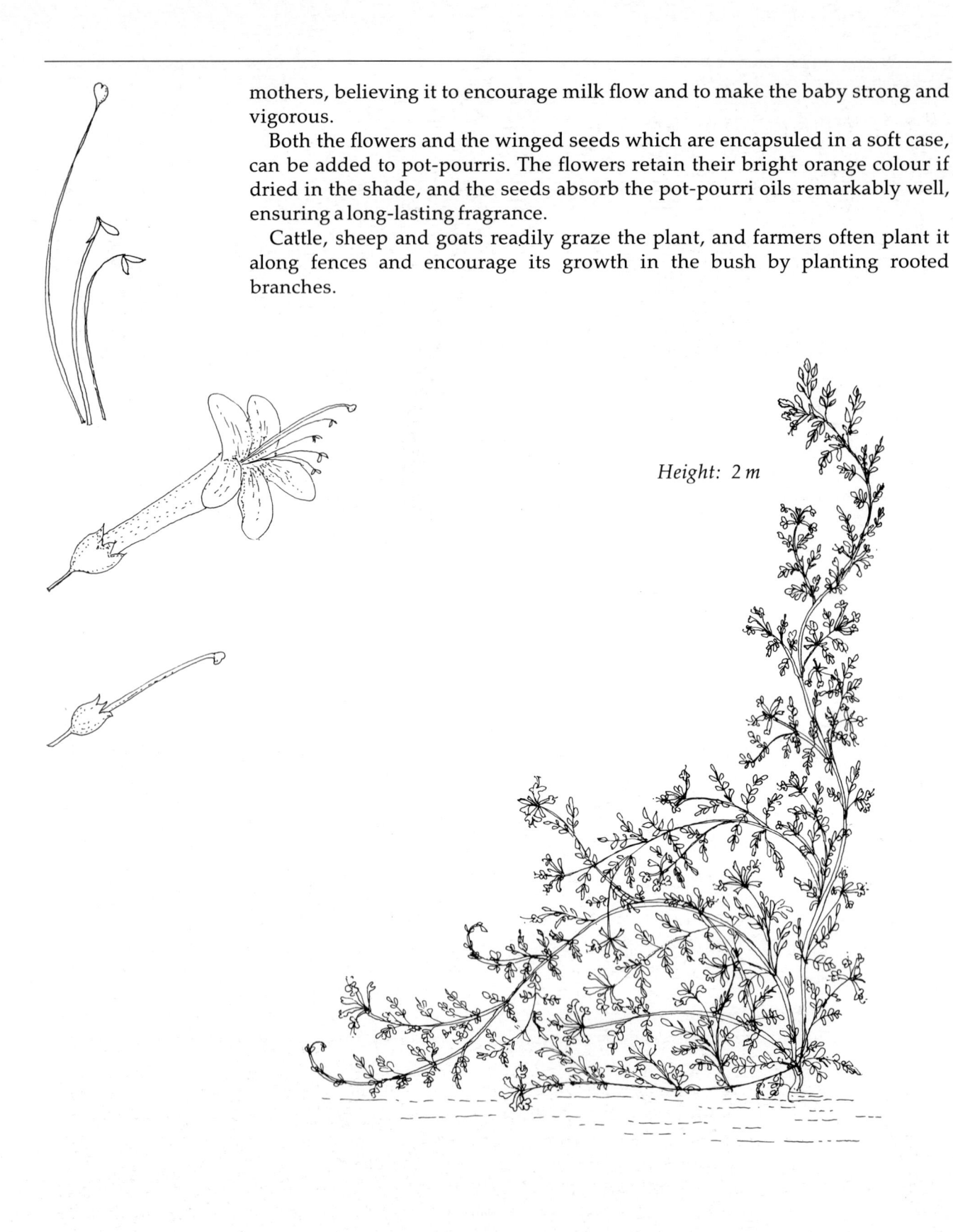

Both the flowers and the winged seeds which are encapsuled in a soft case, can be added to pot-pourris. The flowers retain their bright orange colour if dried in the shade, and the seeds absorb the pot-pourri oils remarkably well, ensuring a long-lasting fragrance.

Cattle, sheep and goats readily graze the plant, and farmers often plant it along fences and encourage its growth in the bush by planting rooted branches.

Height: 2 m

Cape willow

Salix mucronata
(formerly *S. capensis*, *S. gariepina*)

ENGLISH	River willow
AFRIKAANS	Rivierwilg, wilde wilgerboom, Kaapse wilger
SOTHO	Moluane
SOUTHERN SOTHO	Moduwane
TSWANA	Modibonoka
XHOSA	Umgcunube, mqcwimbe
ZULU	Damara more-bonoku, umyezana

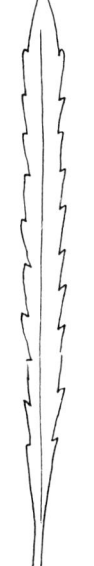

THE willow is one of the oldest sources of medicine, including aspirin, and is still used today for a number of ailments. In its long medicinal history it has been a much-respected tree and people in South Africa have made use of it since earliest times.

The Cape willow is indigenous to South Africa and is still used today as a treatment for rheumatism. The Khoikhoi were probably the first to use the tree, both as a medicinal tea and as a wash for skin ailments, and in the treatment of fevers, particularly rheumatic fever, malaria and heat exhaustion.

The Xhosa and the Tswana make a strong tea by boiling 1 cup of leaves in 4 cups of water for 10 minutes. This is then strained when pleasantly warm and half a cup is drunk twice a day, or in feverish conditions 4 to 8 times during the day. A large quantity of leaves and soft branches boiled in enough water to cover them for 20 minutes, then cooled and strained, is used as a skin lotion, and as a wash for heat rash, skin rashes and inflammation, and for scalp itches, sores and inflammations. This lotion is believed by many rural people to be an excellent hair treatment and they use it to stimulate hair growth.

The leaf is bitter, but nevertheless an infusion is drunk as an appetiser before dinner (usually a 6 cm twig is drawn in 1 cup of boiling water for 5 minutes, then strained and sweetened with sugar and flavoured with a slice of lemon).

A tea made of the bark and the soft twigs stripped of their leaves is used for headaches and neck stiffness (usually ¼ cup of bark to 2 cups of boiling water, draw for 10 minutes, then strain and drink). If it is sweetened with a little honey it is a surprisingly pleasant and soothing drink, and it seems to quell nausea as well, despite the bitterness.

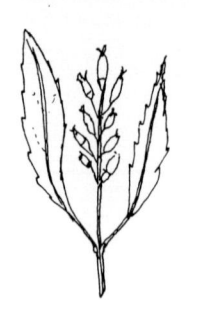

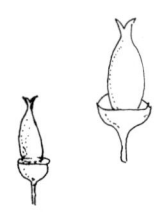

The Zulu bind the long, soft branches around their waists for abdominal and kidney pains, and to give them strength, and legend has it that a returning warrior was garlanded with twisted branches of willow and if he brought home proof of the slaughter of the enemy, he was given a necklace of small pieces of willow wood strung together as the mark of a hero.

Goats, sheep, cattle and fowls relish the leaves when grass is scarce, and the wood is light, soft and workable and excellent for roof rafters and for carving into bowls, spoons, grain mortars and flat dishes.

The Cape willow grows naturally along river banks, vleis and dams all over South Africa except in the Transvaal and in some parts of Natal, but it has been planted along river banks in these provinces by farmers. The Cape willow does not have the long, weeping branches of its close cousin the weeping willow, although in older specimens the branches do hang low.

The trees are easily propagated by cuttings, or thick truncheons placed into moist ground that has been well dug and composted. The piece of branch will soon root, provided it is kept wet, and will quickly grow into an attractive, pale green tree of about 12 m in height. It is exquisite in its pale spring and light summer foliage and is excellent in the garden as its roots do not have the voracious water-seeking habits of those of the weeping willow.

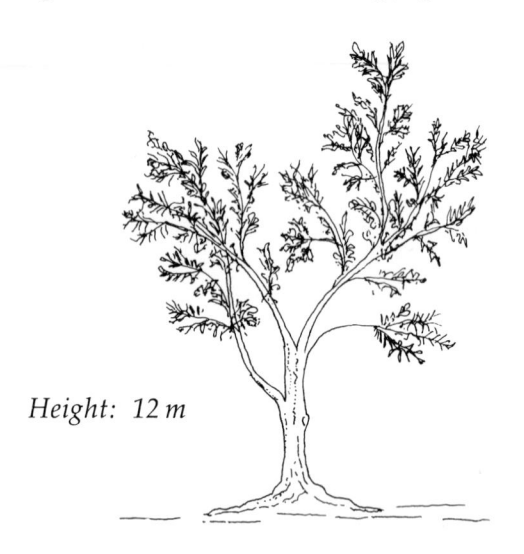

Height: 12 m

Cape willow hair growing oil

1 cup Cape willow leaves
1 cup maidenhair fern leaves
10 cloves
1 whole nutmeg
½ cup brandy
2 cups maize or sunflower cooking oil (maize oil is best)

This is an old recipe used by the early settlers, and one my grandmother knew of as a young girl in Stellenbosch.

Mix all the ingredients in a pot. Stand the pot in a large pot of boiling water and heat for 2 to 2½ hours. Leave to cool overnight. Next morning strain and bottle. Rub this oil into the hair and scalp twice a week before shampooing or, if you are bald, rub a little into the scalp every evening before bathing.

Carpet geranium
Geranium incanum

ENGLISH	Wild geranium
AFRIKAANS	Bergtee, vrouebossie, vrouetee, mannetjie rabassam, tee, meidjiejanwillemse
SOTHO	Ngope-sethsoha, ngope-setshowa, tlhako
TSWANA	Mlako
XHOSA	Tlako

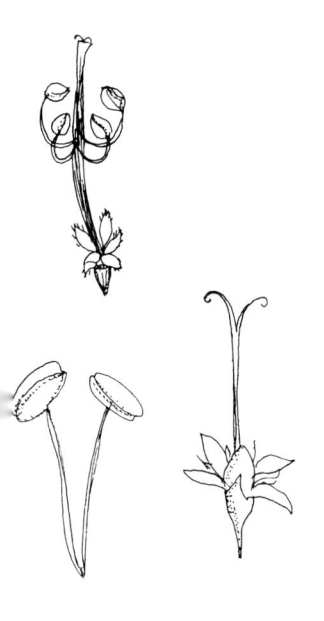

THIS delicate, pretty, soft groundcover has become a much-loved garden plant both here and overseas, and its showy magenta flowers all through the summer make it rewarding to grow in all sorts of places.

I was first introduced to it in my childhood while staying with my grandmother in Gordon's Bay. She grew it in her terraced garden and her coloured maid made a pleasant-tasting tea of the leaves every morning which I shared with her, never then realising how important this herb is for all sorts of ailments.

It is excellent as a strong tea for expelling worms: ¼ cup of fresh leaves to 1 cup of boiling water — allow to draw for 10 minutes, then strain and drink warm first thing in the morning every morning for 10 days. This is also used for dogs and cats, mixed into their food, and bowls of it placed for the dogs to drink throughout the day.

The Khoikhoi used the wild geranium as a tea for irregular or excessive menstruation, expelling the afterbirth and for bladder infections in women. The colonists made it in the same way, and so the wild geranium became an important garden subject through the years.

The Basuto use it as a tea for diarrhoea, and as a digestive, and the Tswana drink it in a mixture with wilde als and one or two of the scented geraniums for colic, diarrhoea and dysentery.

I have drunk a tea (¼ cup of fresh leaves in 1 cup of boiling water, steeped for 3 or 4 minutes, then strained and sweetened with a little honey) for pre-menstrual tension — 1 cup first thing every morning for 4 or 5 days — and found it very comforting and helpful.

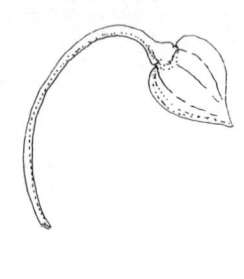

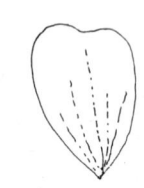

It grows prolifically in the Cape, but can also be found in the other provinces. In the heat of the Transvaal it seems to do best in partial shade, although I have it growing abundantly and attractively in full sun in my herb garden where it gets a good watering twice a week. It spreads easily from both seed and runners, so do give this lovely soft and delicate groundcover space in your garden.

The flowers are edible and can be used to decorate salads and fruit salads. I have crystallised them like violets and used them to decorate cakes and puddings, which always causes much interest.

Many nurseries sell the plants, and the seeds are available through special outlets. Do try to search this lovely plant out — it is charming and will give many years of pleasure, as it reseeds itself and spreads easily by runners.

Height: 15 cm

Chamomile
Matricaria glabrata, M. africana

ENGLISH	Wild chamomile
AFRIKAANS	Wilde kamille
XHOSA	Umsolo
ZULU	Umsolo

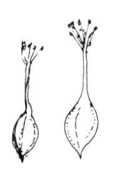

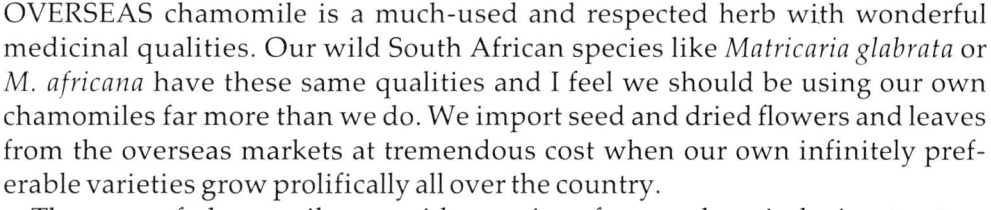

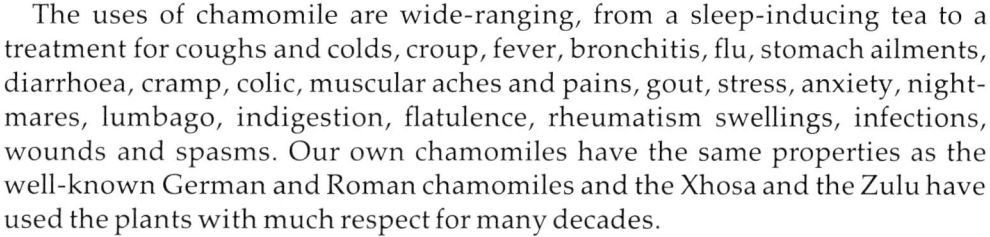

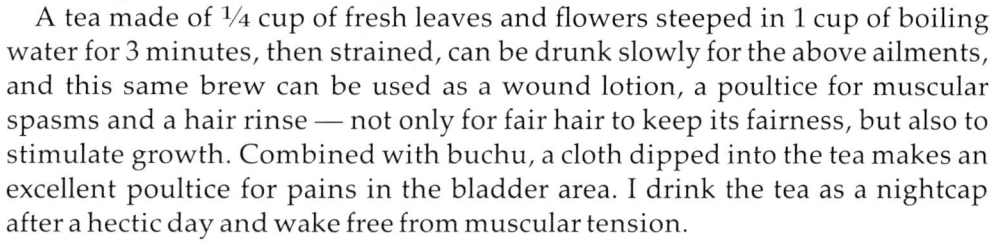

OVERSEAS chamomile is a much-used and respected herb with wonderful medicinal qualities. Our wild South African species like *Matricaria glabrata* or *M. africana* have these same qualities and I feel we should be using our own chamomiles far more than we do. We import seed and dried flowers and leaves from the overseas markets at tremendous cost when our own infinitely preferable varieties grow prolifically all over the country.

The uses of chamomile are wide-ranging, from a sleep-inducing tea to a treatment for coughs and colds, croup, fever, bronchitis, flu, stomach ailments, diarrhoea, cramp, colic, muscular aches and pains, gout, stress, anxiety, nightmares, lumbago, indigestion, flatulence, rheumatism swellings, infections, wounds and spasms. Our own chamomiles have the same properties as the well-known German and Roman chamomiles and the Xhosa and the Zulu have used the plants with much respect for many decades.

A tea made of ¼ cup of fresh leaves and flowers steeped in 1 cup of boiling water for 3 minutes, then strained, can be drunk slowly for the above ailments, and this same brew can be used as a wound lotion, a poultice for muscular spasms and a hair rinse — not only for fair hair to keep its fairness, but also to stimulate growth. Combined with buchu, a cloth dipped into the tea makes an excellent poultice for pains in the bladder area. I drink the tea as a nightcap after a hectic day and wake free from muscular tension.

All chamomiles grow easily from seed, and once you have them in the garden they seed themselves in all sorts of places at all times of the year.

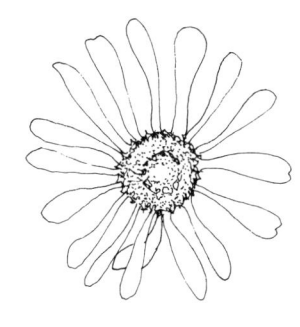

Height: 15 cm

Wild chamomile massage oil

1 cup sweet oil*
½ cup dried wild chamomile flowers and leaves
3 or 4 drops lavender oil
1 stick cinnamon

Gently warm the oil, chamomile and cinnamon in a double boiler for 15 minutes. Take off the stove and allow to cool while still standing the mixture over the boiling water. Strain out the chamomile and cinnamon, pour into a bottle, add the lavender oil and shake. Use as a massage oil.

* Available from chemists

Wild chamomile and Hottentotskooigoed pot-pourri

few drops lavender oil
1 cup minced dried lemon peel
2 cups dried chamomile flowers
2 cups dried Hottentotskooigoed flowers
½ cup coriander seeds
½ cup cloves

Add the lavender oil to the lemon peel. Seal in a jar and shake daily for a week, while you dry the herbs.

Mix all the other ingredients. Add the lemon peel and lavender oil. Keep in a jar and shake daily for a week. Add more oil if you wish. Fill sachets with this and use as a cupboard freshener.

Woolly chamomile (Lasiospermum bipinnatum)

I was introduced to woolly chamomile by a farmer near Bethal who grew it as a fumigant. He burned the leaves and flowers to fumigate his poultry houses and store rooms, having been taught this as a boy by the Sotho on the farm, who used it to fumigate their houses.

The seeds germinate easily, and the plant forms a frothy green mat in both sun and shade. The flower makes a soft, woolly seed head, hence the name, and the birds love to use this as a lining for their nests, scattering the seeds all over the garden. Sparrows and cisticolas seem to love it the most!

I use woolly chamomile in my bath, pouring a kettle of boiling water over 2 or 3 handfuls of leaves, stems and flowers, leeting it steep, then straining and adding this water to the bath. It is wonderfully relaxing. I also use the flowers and the woolly seed heads in wild flower pot-pourris, and find they hold the perfume of added essential oils well.

Added to a massage oil dried leaves and flowers of both chamomiles make a very soothing rub for muscle spasm, particularly on the feet and behind the heels.

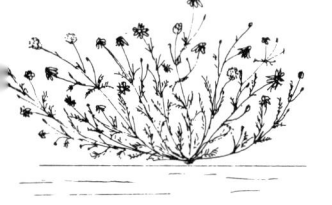

Height: 20–30 cm

Christmas berry
Chironia baccifera

ENGLISH	Wild gentian, piles bush, toothache berry
AFRIKAANS	Bitterbos, aambeibossie, meidjie willemse, tandpynbossie, perdebos

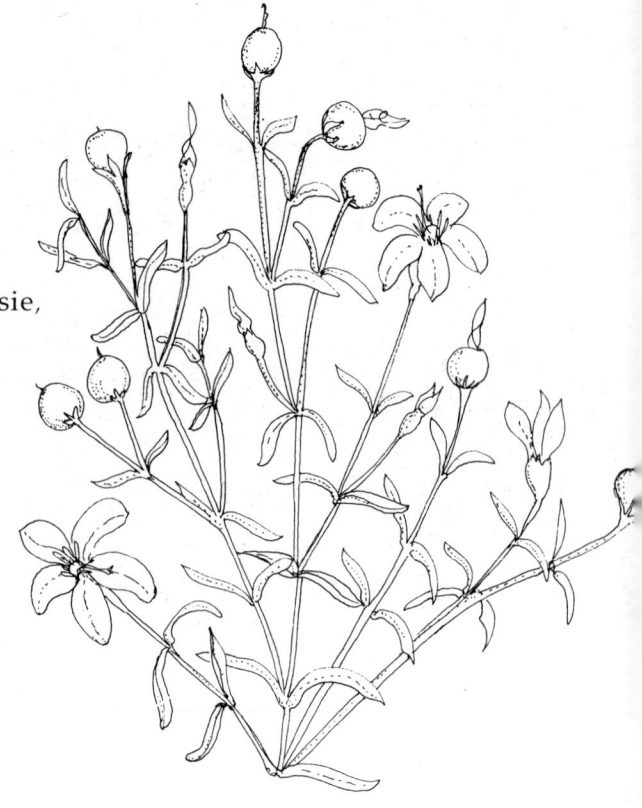

EVERGREEN, shrubby and attractive, the popular Christmas berry is a much sought after garden plant which propagates easily from seed, and is a pure delight to find on the sand dunes in the Cape and Natal. It grows to about 30 cm in height and in summer the entire plant is covered with pink, star-shaped flowers followed by pea-sized shiny red fruits.

Although it is a moisture-loving plant it does remarkably well in the wild in dry areas, even on sand dunes. It needs good compost and a deep weekly watering to do well in your garden inland, and can withstand a certain amount of cold and a little frost.

It is a famous medicinal plant and is much used throughout South Africa by whites, coloureds and certain African tribes. An infusion made of the leaves, stems, roots and flowers is famous as a blood purifier to cleanse the body in the treatment of acne, skin diseases, heat rash, veld sores, boils, abscesses and venereal sores. A tea or decoction made of the whole plant is drunk as a cleanser for skin problems and haemorrhoids (¼ cup of fresh herb to 1 cup of boiling water — draw for 3 minutes, then strain) and this same brew is used as a wash and as a lotion applied to the area at frequent intervals (4 to 8 times per day).

It has a purgative action, and the Khoikhoi used it as a fast-acting purgative by making a strong brew from it and drinking it first thing in the morning. A word of caution here, however, as it is a strong plant and can produce sleepiness and increased perspiration.

Some African tribes make a strong tea from the plant as a remedy for leprosy, which they drink and use as an external application.

Perhaps its best-known use is as a purgative for those suffering from

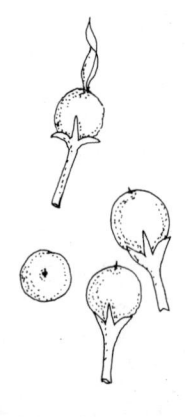

Ginger bush

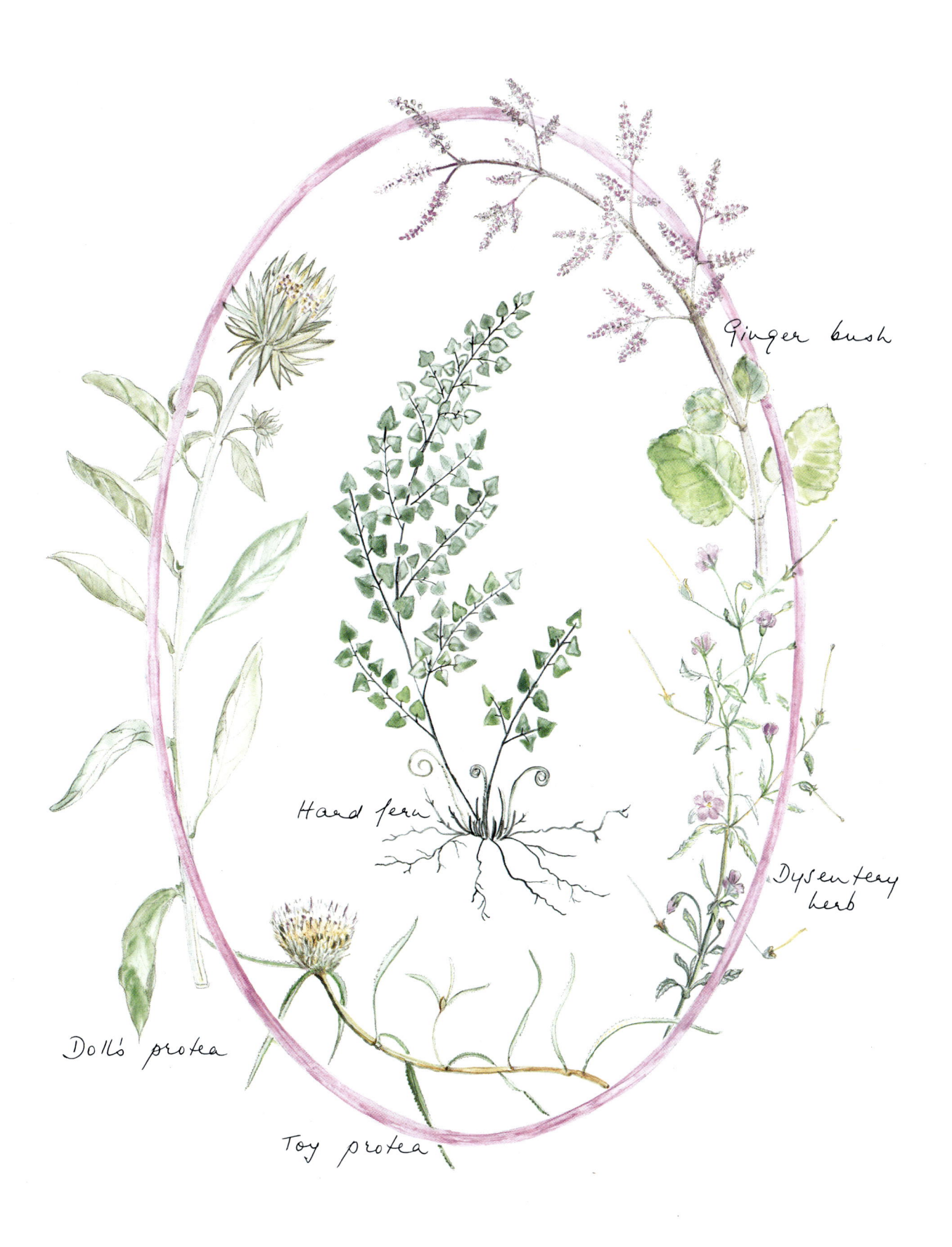

Hand fern

Dysentery herb

Doll's protea

Toy protea

Curry bush
Helichrysum
foetidum

Curry bush
Helichrysum petoelatum

Woolly chamomile
Lasiospermun bipinnatum

Wild chamomile
Matricaria africana

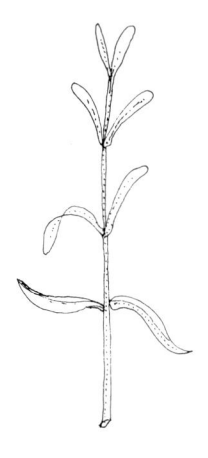

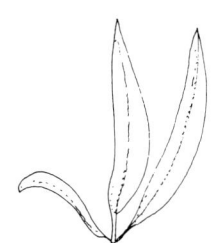

haemorrhoids (hence the name 'aambeibossie'). A pad of cottonwool soaked in the tea and applied at night to the piles is said to be soothing, comforting and shrinks the piles. I was told by a coloured herb seller on the Parade in Cape Town that she pounds the leaves, stems and roots, adds warm water and spreads this on a piece of cloth which she binds in place over boils, abscesses, or over piles, and leaves it on overnight, and by the next morning there is marked improvement. She also makes a tea (1 cup of leaves to 1 litre of boiling water, stand for 3 minutes, then strain and drink half a cup at a time) to ease and cure diarrhoea, which she said she'd used for children and old people as it is safe. When I queried its purgative action she said it cleansed the body of impurities, 'not blocking it'! She also said the old people used it both internally and externally to treat syphilis and that they made 'mixtures' which they sold to the sailors for a good profit!

I delighted in the colourful description and as she spoke many interested onlookers gathered round, verifying, laughing, adding anecdotes, vividly describing how they too were dosed as children by their grandmothers on Saturdays and how they had 'the trots' all Sunday!

The sunlit marketplace, the people's laughter, the green baskets of fresh herbs, the umbrellas, the sea gulls wheeling overhead, the plants uniting us all — it was a moment I'll not easily forget. And now, whenever I see the exquisite Christmas berry in its brightness and beauty, I am reminded of the faith that the people had in this plant, and still have today, I hear the laughter of those in the marketplace, and I salute that remarkable little plant and thank God for it.

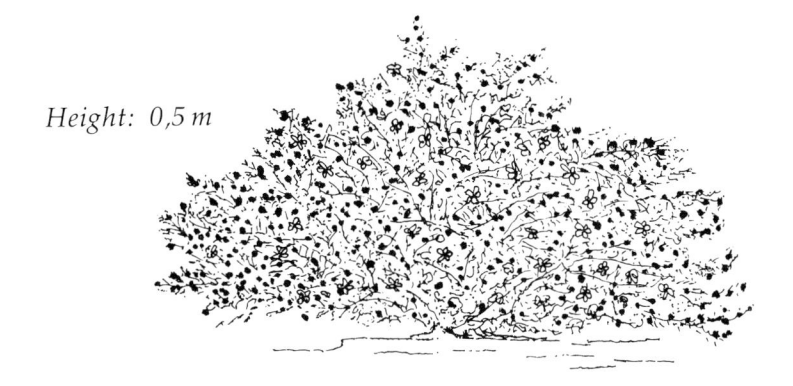

Height: 0,5 m

Clivia

Clivia miniata

ENGLISH	Bush lily, St John's lily
AFRIKAANS	Boslelie
XHOSA	Umayima
ZULU	Ubuhlungu beyima

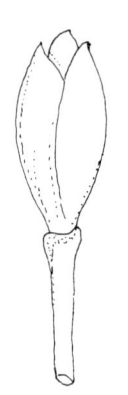

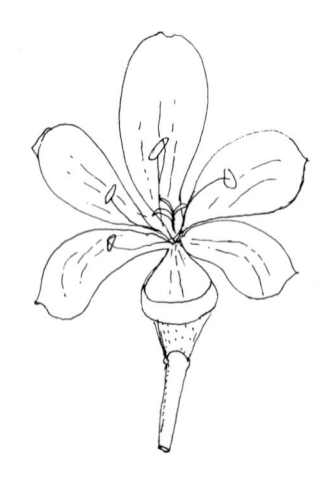

THE beautiful clivia grows wild in shady ravines from Haga-Haga in the eastern Cape, along the coastal forests of the Transkei and Natal and in moist, shady places in the eastern Transvaal, and has been a much sought after and much admired pot plant both here and abroad for over a century.

The flower is lily-like, with glorious umbels of dark orangy pink flowers which are most abundant during spring, but in hothouse conditions can flower all through the summer. They are followed by bright green and red berries. The straplike leaves are thick and dark green and the roots are fleshy, juicy and almost corm-like. Often suckers appear around the parent plant which have their own roots, and these can be cut away and set in moist soil in pots to start new plants. Propagation by the dried ripe berries is also fairly easy, but do remember the seeds only ripen approximately 10 months after the flowering season.

As a pot plant the clivia is exceptional, but do choose a big pot with rich composted soil, and place it where it will get morning sun. Do not divide the plant unnecessarily as it does not like being disturbed.

Clivia are much sought after by the Xhosa and the Zulu as a treatment for snakebite. I was told by an old Zulu that the poison is sucked out and a tourniquet of vines applied to the affected limb, then the clivia root is crushed and the juice applied to the area.*

The root juice diluted in hot water is also administered to the victim of

---

* The treatment of snakebite has become a controversial issue and this treatment is *not* recommended. Tourniquets and sucking the poison out are no longer advocated.

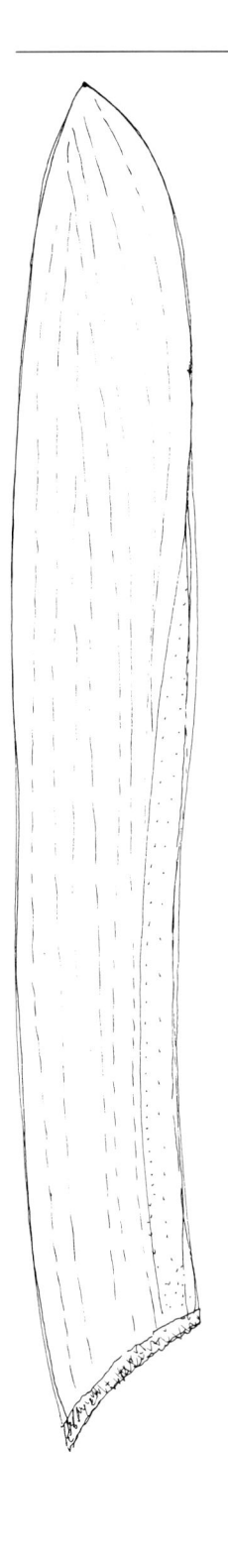

snakebite — but quite how much I was unable to ascertain, as a heated argument with the younger members of the family ensued!

A weak tea made of the chopped roots has been used for decades by African tribes to treat febrile conditions and to ease childbirth. Once the baby is safely born, the young mother is given a drink of this tea to help her milk flow. Should she have difficulty in feeding the newborn, this treatment is continued until the lactation process is functioning well.

Interestingly, recent laboratory tests have shown that the clivia contains certain alkaloids, including cliviine, which consists essentially of a substance called lycorine, a lactone-containing alkaloid which would account for the clivia's remarkable ability to facilitate childbirth. The fascinating question is how those old Zulu folk-lore 'doctors' knew about it so long ago. I was told by a Zulu family that only the women knew of the plant's uses with regard to childbirth and that they taught the men how to use it for snakebite. I was also warned that the plant is poisonous and therefore not to experiment with it.

Many blacks — both rural and urban — regard the clivia as a sign of extreme good fortune and to have one growing near the home is considered to bring wealth in children, cattle and crops, good rains and good health.

Commercially clivia are a thriving business both in South Africa, their country of origin, and overseas. Their flowering period is long and even when the plant is not in flower its dark, glossy leaves are attractive in glades under trees or in pots. A friend of mine who grows clivia commercially has found after sessions of sorting and peeling the seeds that have been soaked in water to facilitate the removal of the skins prior to sowing, that her nails are strengthened and toughened and remain so for some time afterwards. Working as she does with clivia seeds she hardly ever has a broken nail. I often wonder what the cosmetic industry would make of that — perhaps a new nail cream called Clivia Claws!

Height: 0,5–0,75 m

Coral tree
Erythrina lysistemon

ENGLISH	Common coral tree, lucky-bean tree
AFRIKAANS	Gewone koraalboom, kafferboom
NDEBELE	Umkawane
SHANGAAN	Nsisimbane
TSWANA	Muvale
ZULU	Umpumbuluku

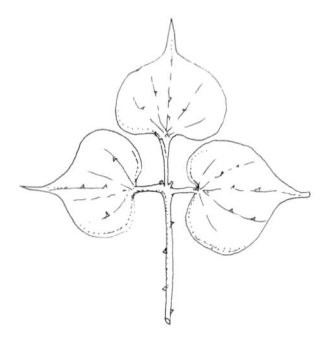

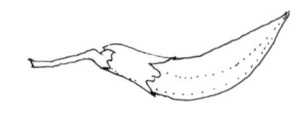

THERE are many beautiful species of *Erythrina*, and all are easily grown, drought resistant and rewarding. Every garden should have at least one!

The commonest of all the coral trees, *E. lysistemon*, is a medium to large, attractively shaped tree that is starkly beautiful in winter when its bare pale grey branches are seen at their best. In spring its abundant clusters of brilliant scarlet flowers are in striking contrast to the dullness of the veld. The flowers are borne on the bare branches in large clusters. They consist of long petals enclosing other petals and even the stamens and sepals are a brilliant red. The flowers appear from July to October, and once they are over clusters of long, slender segmented black pods that are sharply waisted between each seed, appear. These split open to reveal the brilliant red lucky bean seeds which are avidly collected by many tribes to be made into trinkets and necklaces and sold as lucky charms. At present medical tests are being done on the crushed seeds for a variety of alkaloids and components, possibly pain killing, which may find their place in curative medicines.

The leaves are three-foliate, large, and with a tapering apex. These appear only after the spring flowers are over, and are deciduous.

The coral tree occurs naturally in a wide range of altitudes and habitats, growing happily thrust up against other trees in forest and scrub, in dry woodland, in the open veld, on bare hillsides, high rainfall areas and even in coastal dune bush. It varies in size and height according to its habitat and extends from the Transkei, Natal, eastern and western Transvaal and into Zimbabwe. Cultivated specimens are common all over the country and are often planted along streets.

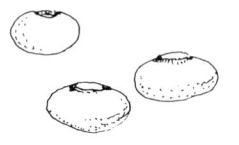

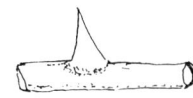

Although the coral tree prefers a frost-free climate it is able to withstand frost. I have seen beautiful huge specimens thriving in the Orange Free State and the Witwatersrand.

Some years ago I was fascinated to find a farmer in the far north-eastern Transvaal who had marked the boundaries of his farm with coral trees. In midsummer when the seeds were mature he and his staff collected great truckloads of them. The pods he sold to florists and markets for dried flower decorations and the lucky bean seeds to curio shops and roadside stalls. He enjoyed a brisk trade year after year from a crop that was drought resistant, needed no maintenance, no watering, no fertilising or care and gave him in return a substantial income, with only one month's work per year — the actual collecting of the pods — and in spring people came from far and wide to see the glory of those several kilometres of brilliant flame-coloured boundary. He also did a brisk trade with the witchdoctors, and as a bonus he said in all his years of farming nothing was ever stolen from him, as the coral trees protected him!

In African folklore the coral tree is much respected. One custom is that when a man dies his relatives will take a truncheon from a tree growing near his home and plant it on his grave, which will protect him in the afterlife. Some tribes cut strips of bark from all four sides of the trunk and use these to bind a bunch of specific wild herbs together (one is possibly silverleafed vernonia). These are boiled in water and the resultant tea is taken by women in labour to ease the pains.

The leaves are crushed and applied to sores and suppurating wounds, and open wounds are disinfected and treated with the ash of burnt bark. The Tswana and the Zulu make a strong tea by boiling a cupful of leaves in 2 cups of water for half an hour, then straining it and allowing it to cool, and they use the warm liquid as an earache remedy. A little is dropped into the ear and the rest is used as a poultice — a cloth is wrung out in it and held behind the ear. The root can also be boiled in the same way and used as a lotion or poultice over sprains and bruises, and sore, aching feet. Crushed leaves placed in the shoes (be careful to remove any of the small thorns!) also ease sore feet and cracked heels.

I have noticed that some of my gardeners wrap strips of bark which they peel from the branches — carefully removing any thorns first — around the handles of their spades and forks when the midday sun takes its toll. They tell me it gives them strength and soothes sore hands! All these uses indicate painkilling effects so perhaps there is a future for the coral tree in medicine.

Best of all, this is one of the few trees that grows from a whole branch cut from the tree and thrust into the ground. It roots quickly, as long as it is kept damp for at least two months, and I have seen a whole tree cut down and replanted some distance away with the help of a front end loader. It went on growing as if nothing had ever happened!

A beautiful sight is the coral tree and the wild pear *(Dombeya rotundifolia)* grown together — the contrast of their scarlet and white flowers in spring is spectacular.

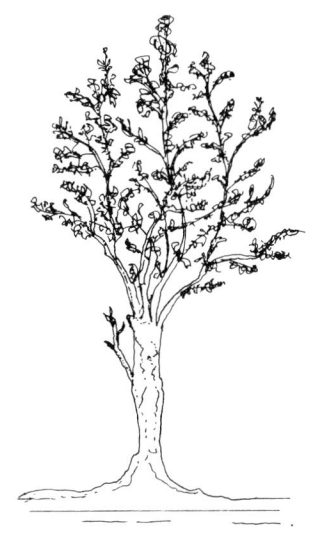

Height: 8 m

Crinum lily

Crinum bulbispermum

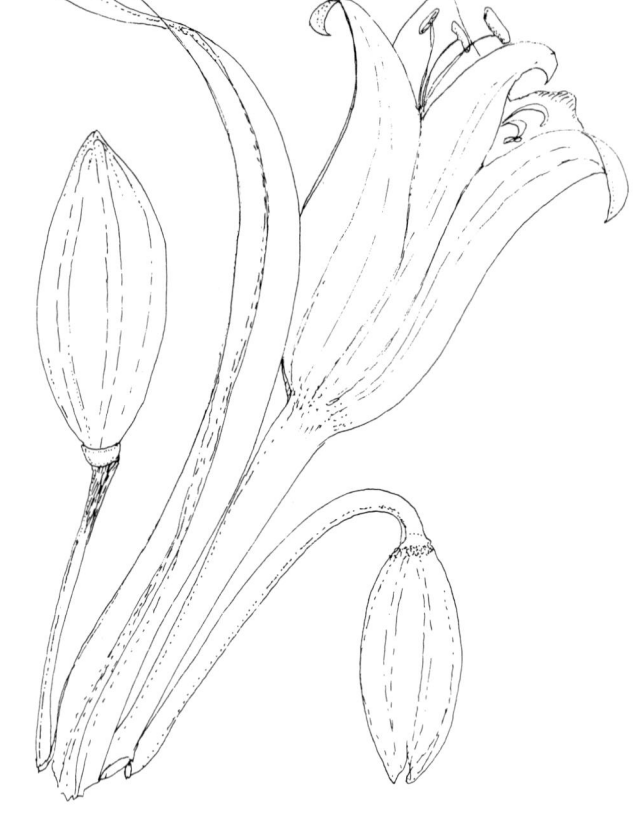

ENGLISH	Orange River lily, river crinum, wild amaryllis
AFRIKAANS	Boslelie, seeroogblom, rivierlelie
SOUTHERN SOTHO	Lelutla
TSWANA	Mototse
ZULU	Umduze

THIS beautiful and eye-catching lily with its pale pink and white trumpet-shaped flowers can be seen in midsummer on the flat plains of the Transvaal, the Orange Free State, where it is the floral emblem, through into Natal and the Cape Province.

Most lily species prefer moist, marshy places, but this resilient variety seems to be happy in all sorts of soils, and can go for long periods without water. In long drought periods I have been pleasantly surprised to find the unexpected and enormous inflorescence of exquisite flowers bravely standing in the dry veld grass when all else was arid and shrivelled.

The enormous bulb produces a thick column of arching, sheathing leaves that spread out to about 60 cm. Cattle browse on the leaves, so it often looks untidy and stringy in the veld, but in the garden these tough, strap-like leaves are attractive and unusual in their grey-green colouring.

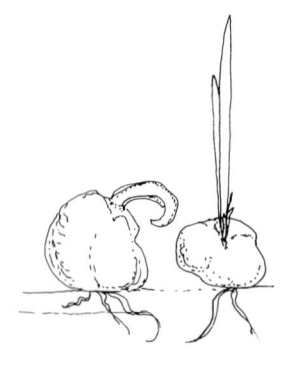

A fleshy, flowering stalk about 50 cm tall emerges through the leaves to the one side, and the terminal umbels of drooping pink and white flowers, sweetly scented and prolific, are surrounded by papery bracts. A dark pink line running down the petals, which curl back exposing the dark stamens and styles, gives the flowers a most unusual beauty, and after the flowers are spent a thick fleshy marble-like fruit with bulging seeds emerges. These seeds propagate easily, and can be sown immediately they ripen in well-composted soil. Do allow a metre between each one as they spread their leaves widely and they do best if they are undisturbed for some years. In winter some of the leaves will die back, particularly in the frost areas, but new leaves appear in August, so the plant is attractive all year round.

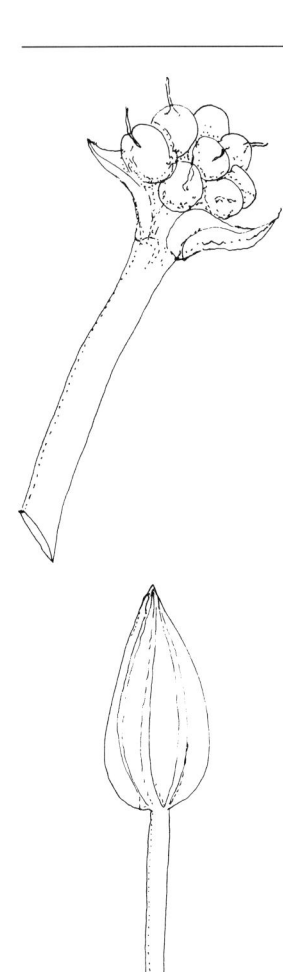

The plant is a favourite medicinal herb, and the leaves, flowers and bulb are all used. The Sotho make a strong brew of the leaves and sliced bulb for treating colds, coughs, as an external application or wash for wounds, scrofula and haemorrhoids.

The Zulu and the Tswana apply the roasted bulb to aching joints, rheumatism and backache, and as a drawing poultice for abscesses and suppurating sores. The strap-like leaves are used to bind dressings in place, and the flower bound over a swollen joint or sprain soothes and helps reduce the swelling.

Juice squeezed from the leaf base is used by several tribes for earache and roasted pieces of the bulb are placed behind the ear or over the ear to ease the pain. Some tribes also use the crushed bulb as an application for haemorrhoids, held in place at night by tight-fitting panties or bound nappy-like with strips of material. The Zulu also apply the roasted bulb to varicosities, often using the leaves to hold the pieces of the bulb in position overnight.

Some tribes believe a brew made of the leaves of the crinum lily to be remarkably effective in the treatment of malaria and the same brew is taken by the Zulu as a treatment for rheumatic fever (usually ¼ cup of chopped leaves to 1 cup of boiling water, stand for 5 minutes then strain). The Tswana on my farm drink a brew of crushed leaf bases and stalks to increase the flow of urine in bladder and kidney infections, and apply the warmed, sliced bulb over the kidneys to ease discomfort.

I was once glad to see that some herd boys who found the lily in the veld collected the seeds and planted them near their homes. The plant is so prized and so used that I always fear it will become even rarer, and so I grow the seeds in bags for the sangomas who come from far and wide to get them from me, and I encourage them to save the prolific seeds and grow them in turn.

Each summer during the flowering period — from October to sometimes as late as February — I pick off a few buds for my wild flower pot-pourris, but as the valuable seeds are my pride, I only pick off one or two buds and watch eagerly as the seeds swell and ripen after the beautiful flowers fade.

This special plant is much prized in my herb garden, and if the plants are attacked by caterpillars, I make a strong khakibos brew and pour this over the plants, or I sprinkle dried khakibos all around them to protect them.

Height: 23–30 cm

Cross-berry
Grewia occidentalis

ENGLISH	Assegai wood, bow wood
AFRIKAANS	Kruisbessie, knoppieshout
NORTHERN SOTHO	Mogwana
SOUTHERN SOTHO	Lesika
TSWANA	Mokukutu
XHOSA	Umnaqabaza
ZULU	Iklolo

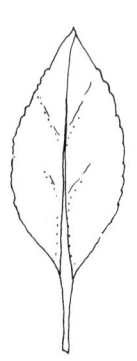

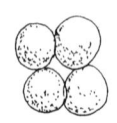

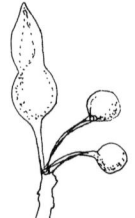

THE cross-berry is a small shrub or tree that is widely distributed throughout the Transvaal, Natal and the eastern Cape. It is found in a variety of habitats from mountain slopes to thornveld, from wooded areas to the more arid highveld.

The branches are pliant and tough, and the bark, which contains gum and tannin, is strong and fibrous and is used to make string or rope. The evergreen shiny leaves, and mauve, ten-petalled, star-shaped flowers make it a pretty and worthwhile shrub to grow in the garden and the edible, square-shaped fruits which give the name cross-berry are much loved by birds. In the veld cattle graze the leaves and monkeys and baboons enjoy the ripe yellow berries in midsummer and autumn, competing with the birds for the tasty fruits. I have seen the shrub trimmed into a perfect ball shape and found that it benefits from pruning, which is useful to know if you have limited space in your garden. Left to its wild somewhat sprawling shape it is equally attractive and interesting as a garden subject.

The cross-berry is a much-used medicinal plant. The Zulu, Tswana and Xhosa soak the bark and small twigs in hot water and use this as a wash and lotion for wounds. Several tribes make a tea of the leaves and twigs and use this for impotency, barrenness and to ease childbirth or hasten the onset of labour if it is retarded, and a wash is made by diluting the tea for both mother and child after the birth.

The plant is probably best known for its strong yet pliable wood. The San and Khoikhoi used the wood for many generations to make bows, and even today farm children often seek out the good branches for this purpose. The bark is stripped from the branch (and dried and stored for medicinal use) and the pliable, sweet-smelling stick is then bound and tied into the bow. It is left to mature for a day or two before being used. It is strong and light in weight, and much prized.

The wood does not splinter and is used by the Zulu and Xhosa for assegaai handles, axe handles, and for re-inforcing roofs and barricades. The Southern Sotho use thin pieces of the wood for scarifiers and lances and household utensils. Being pliable, it can be woven into basket handles and gates, and it is long lasting. Small shrubs dug from the wooded areas are often planted and nurtured near their homes as the wood is so valuable to them.

To start your cross-berry off, it needs a large hole, 1 m by 1 m in depth and width. Fill the hole with compost and rich soil and water it generously. Plant the cross-berry in it, trampling the soil firmly around it. Water twice or three times weekly until it is well established. Thereafter I have found a good weekly watering suffices.

Several nurseries offer the plants for sale so do look out for this unusual shrub, and if you can't find an established plant for sale, hardwood cuttings from a strong mother plant are easy to root in wet sand.

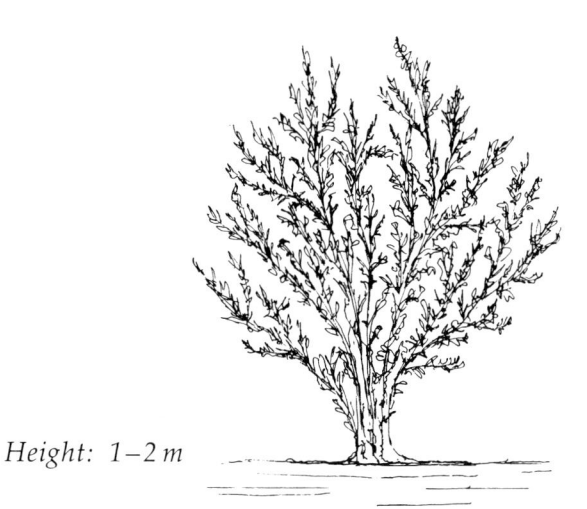

Height: 1–2 m

Curry bush
Helichrysum species

ENGLISH	Everlastings
AFRIKAANS	Hottentotskooigoed, Hotnotskooigoed, Hottentotsbedding, sewejaartjies, kerriebos

THE name Hottentotskooigoed covers several species under the genus *Helichrysum*. The leaves are usually a soft grey and curry-scented, the stems are soft and the shrub scrambles or makes a dense groundcover, spreading attractively over rocks or down hillsides. The vagrant Khoikhoi, or Hottentots, used the soft leaves and flowers as bedding, hence the name, and campers still do this today. The pleasant smell the leaves give off aids sleep, and the profusion of growth in the coastal areas makes it an easily accessible plant for hikers and campers. It grows at its best indigenously usually within 14 km of the sea, but through the years some nurseries have started propagating it and it grows well even in the Transvaal with its dry winters and summer rainfall.

There are several medicinal uses for the *Helichrysum* species and those living in the Cape could start a fascinating collection, as they all make lovely garden subjects. Here are a few to get you started.

Helichrysum auriculatum

A pleasant and health-giving tea is made from this plant (¼ cup of fresh leaves and stems steeped for 3 to 5 minutes in 1 cup of boiling water, then strained and sweetened with a little honey if desired, is the standard brew) as a treatment for backache, kidney diseases and kidney stones and heart ailments. It was once used as a daily tea in the Cape by the colonists and the Khoikhoi, and was a very much respected drink.

Helichrysum crispum

A tea made of the leaves (standard brew) is drunk for heart conditions, kidney ailments and backache. It calms a racing heart, and is said to be effective for coronary thrombosis and hypertension. The Khoikhoi used it as a calming tea and introduced it to the colonists in this way. It is also an effective heart treatment for animals.

Helichrysum foetidum

The leaves warmed and washed in water make an excellent poultice for a festering, infected sore, as they have astringent qualities which draw out the infection. The leaves and flowers are aromatic and are used in pot-pourris and in bedding and pillows.

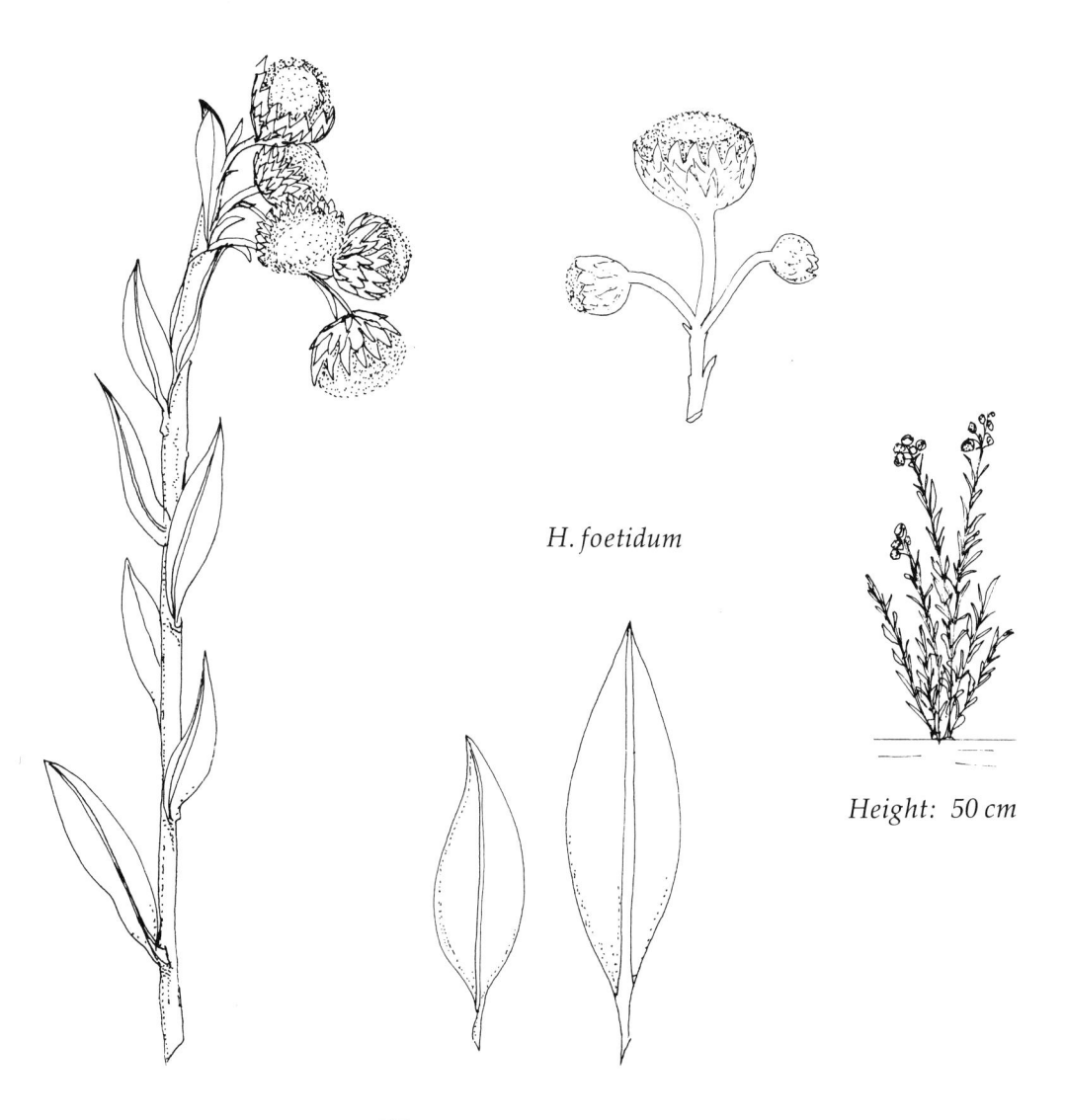

H. foetidum

Height: 50 cm

Helichrysum peteolatum

I grow this species very successfully in my herb garden, where its small, furry, grey leaves make a lovely groundcover, at times climbing up into other plants. A tea of these leaves (standard brew) is taken for heart conditions, stress, hypertension, anxiety and over-excitement. The flowers are much loved in dried arrangements and are exported to many countries, and the plant is also cultivated overseas.

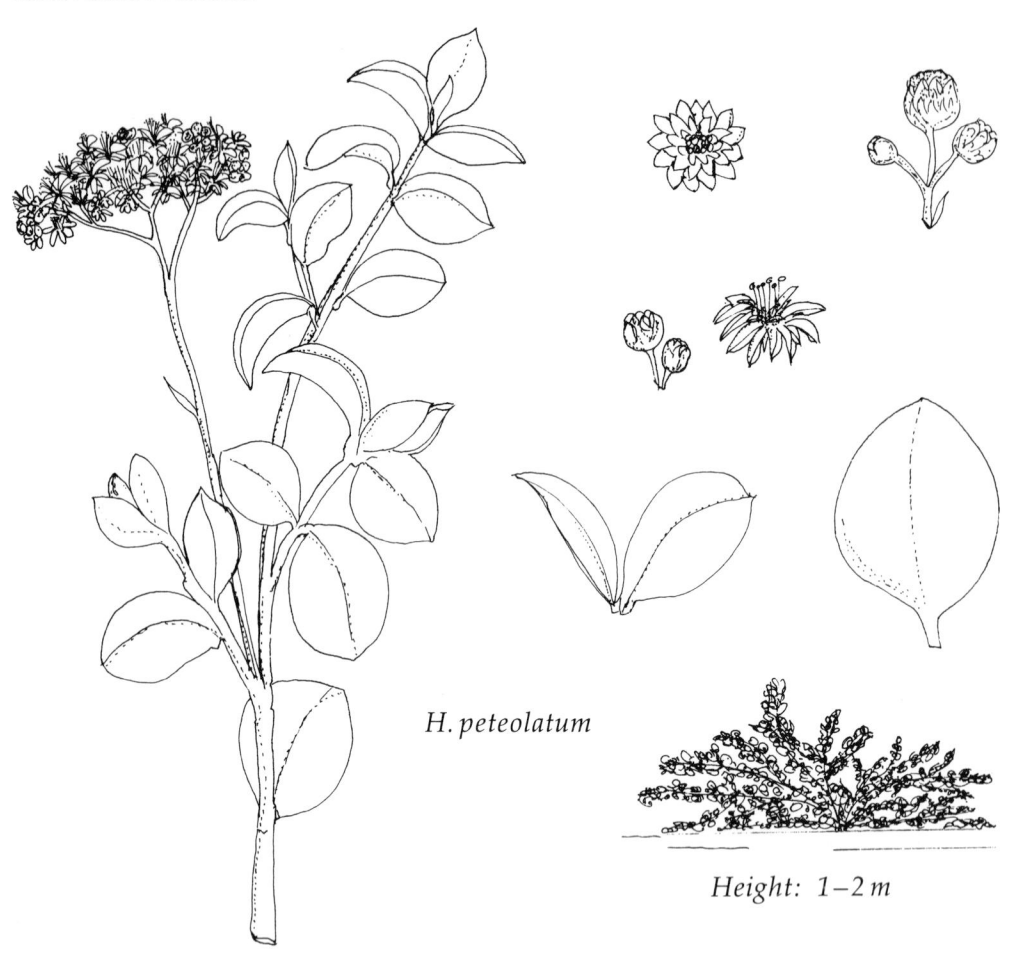

H. peteolatum

Height: 1–2 m

Helichrysum odoratissimum

The dried leaves and stems of this fragrant plant are used by the Sotho to fumigate their huts. Added to fat (usually sheep fat) and boiled in it, the leaves give off their scent, and make a soothing ointment.

A tea (standard brew) helps colic and stitch, and aids sleep, muscle tension and cramps as well as coughs and colds.

A leaf can be chewed to give relief to heartburn and flatulence and is also a soothing wound dressing applied directly to the area and bound in place and replaced by fresh leaves twice daily.

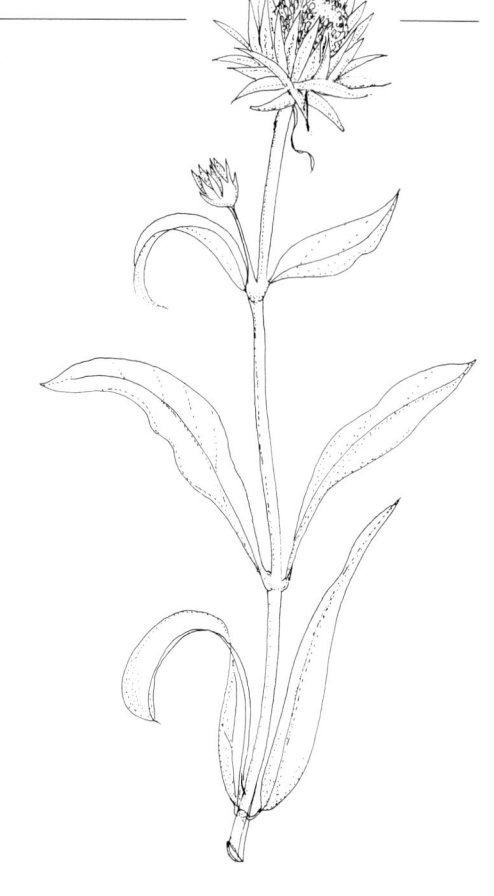

Doll's protea
Dicoma zeyheri

ENGLISH　Toy sugar bush, silver everlasting
AFRIKAANS　Jakkalsbos
TSWANA　Thlonga
XHOSA　Umgele

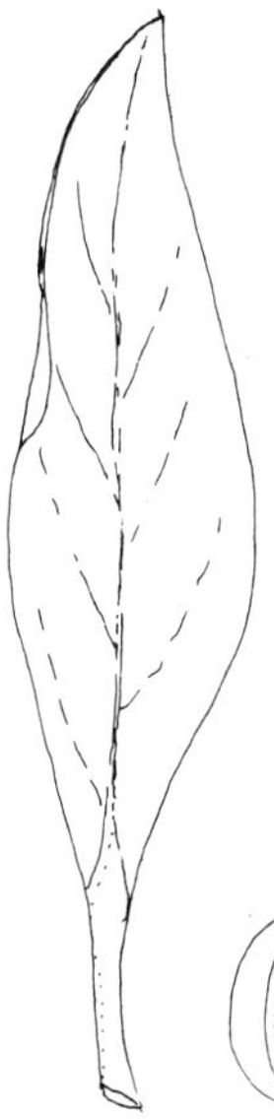

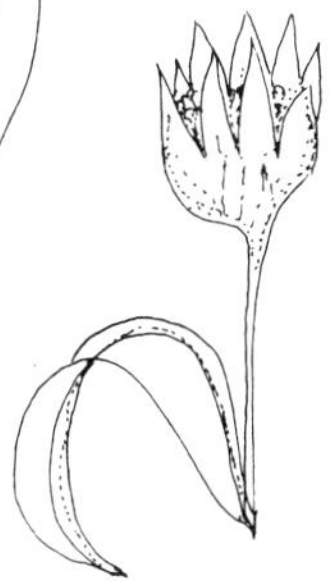

THE doll's protea is a small, herbaceous perennial with prickly, thistle-like flower heads that have stiff silvery green and mauve bracts which are tough and long-lasting. The flower head measures 4 to 5 cm in diameter and looks like a little protea, hence the common name, and the leaves are stiff and green on the top surface and silvery felted underneath. Because of the long-lasting qualities of the most unusual flowers, this plant is much sought after by florists and cut flower growers.

It grows wild in the grasslands all over the country, flowering between January and May, and propagates only by seed. It needs full sun and water during the summer months, and a well-drained soil — a rockery would be the ideal place to grow it. In high frost areas it seems to be fairly frost resistant.

A decoction is made of the doll's protea root for treating back pains, anaemia and illnesses that cause weakness of the limbs. The Tswana use a tea of the plant as a wash over the legs after a long illness to give the patient new strength, and the back is washed with the tea as well. The Zulu give a decoction or tea to a mother after the birth of her baby if it has been a long and difficult birth to build up her strength and give her 'good red strong blood'.

Height: 20 cm

Toy protea (Dicoma anomala)

AFRIKAANS	Aambeibos, gryshout, koorsbossie, korsbossie, maagbossie, maagwortel, swartstorm, vyfaartjies, wurmbos
MANYIKA	Chiparurangomo
SHONA	Chifumuro
SOUTHERN SOTHO	Hlwejane
SOTHO	Kloenya
TSWANA	Thlongati, thlonya
XHOSA	Inyongwane

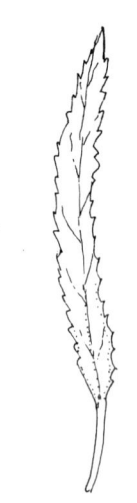

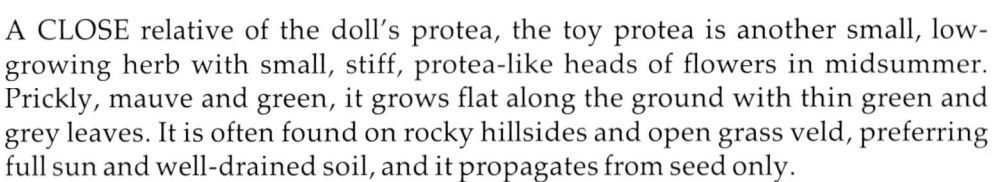

A CLOSE relative of the doll's protea, the toy protea is another small, low-growing herb with small, stiff, protea-like heads of flowers in midsummer. Prickly, mauve and green, it grows flat along the ground with thin green and grey leaves. It is often found on rocky hillsides and open grass veld, preferring full sun and well-drained soil, and it propagates from seed only.

As a medicinal herb the toy protea is famous and much-respected, and is used for a wide variety of ailments by several African tribes and many whites.

A decocotion or tea made of the plant (usually 1 cup of scraped and chopped root and 6 or 7 leaves are boiled in 2 litres of water for 20 minutes and then strained) is taken 2 tablespoons at a time for the treatment of diarrhoea, dysentery and intestinal worms. In this last instance the decoction is taken on an empty stomach first thing in the morning and then twice more in the day for 3 days until the worms are cleared from the system. The same brew is used for treating gall sickness in cattle and sheep.

The Southern Sotho use this decoction for the treatment of venereal diseases and as a purgative, and for the treatment of colic, stomach upsets and toothache.

The Xhosa and the Sotho use the dried powdered stem, leaves and especially the root, for treating sores, veld sores, scabs on the heads of children, wounds and ringworm, and also use it on horses and cattle to treat sores and ulcers.

Some white farmers cultivate the plant in their gardens as a quick remedy for several ailments. The root mixed with gin is used as an application for haemorrhoids (the common name aambeibos comes from this) and a tea or decoction is drunk to bring down fevers.

In general, however, the plant is used by African tribes for coughs and colds, to clear blood disorders, as a tonic and heart herb, to assist the circulation, particularly for varicose veins, to relieve colic and stomach upsets, to treat worms, and as a charm to protect against poisonous food!

The Tswana on my farm chew the root and swallow the saliva to clear sore throats, coughs and colds, and for diarrhoea and dysentery. The Sotho use the root in this way too, and believe the plant combined with one or two other plants will cure sterility. Best of all, by drinking a tea of the root, the singing voice can become clear and high!

What an amazing plant! Do look after it if you have one in your garden — it is precious and becoming rare.

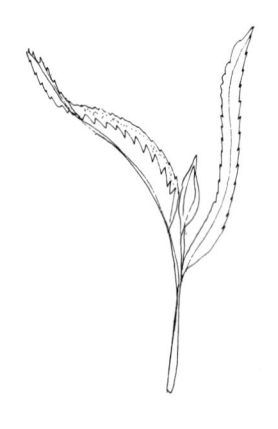

Height: 10–15 cm

Dysentery herb
Monsonia angustifolia

ENGLISH	Crane's bill
AFRIKAANS	Alsbossie, naaldbossie, teebossie, assegaaibos, rabas
SOTHO	Makorotswana
TSWANA	Remarungana
XHOSA	Igqitha

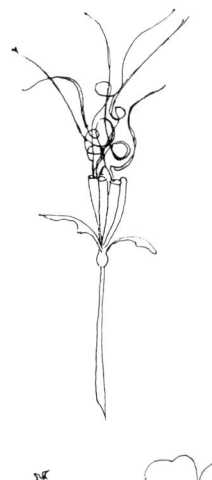

SEVERAL species of *Monsonia* have been used in the past in the treatment of dysentery, typhoid fever, intestinal haemorrhage and diarrhoea and this pungent smelling, unobtrusive little herb with its tiny, crinkled, five-petalled pale mauve flowers is still much sought after.

A tea is made of the herb by pouring 1 cup of boiling water over ¼ cup of leaves, stems and fruits. Stand, steep for 5 minutes, then strain. Sip ½ cup at a time at intervals of 2 hours throughout the day until the condition clears. This tea, a tablespoon at a time, is also popular as a treatment for heartburn, dyspepsia, flatulence and digestive disturbances. As the plant is annual and not always available, the dried herb may be stored. In this case use only 1 tablespoon of dried leaves to 1 cup of boiling water. Steep for 5 to 7 minutes, strain and use as above.

A stronger tea (½ cup of herb in 1 cup of boiling water steeped for 5 minutes) may be used as a lotion for snakebite and a little sipped at the same time. It is also used to treat anthrax in cattle, and to dose calves and lambs for diarrhoea. The juice of the stem and leaf is also squeezed onto anthrax pustules and slow-healing sores.

Because the plant is so astringent some tribes drink the tea and use cooled tea as a local application to varicose veins and ulcers, and as a wash for haemorrhoids. The Tswana also use it as a wash during menstruation and after childbirth.

The Sotho use a weak tea as a treatment for eye infections, and as an eyebath for opthalmic disorders in cattle. Some white farmers still feel this is the best treatment to date, and treat their cattle with the herb frequently.

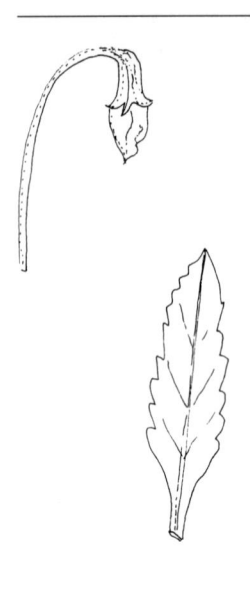

M. angustifolia is annual and can be found in the open grassland throughout South Africa. It is most easily recognised by its long fruits and tiny, bell-shaped flowers. Although the seeds germinate erratically, I have been fairly successful in growing it from seed, but find it grows easily if left to go quite wild. Once you become familiar with it you will find it in the veld and along the roadsides. A slightly different variety, *M. burkeana*, has flatter leaves, is perennial and is found from the northern Cape through the Orange Free State into the Transvaal and further up into Zimbabwe and Angola. It too can be used as effectively as its cousin, *M. angustifolia*.

The *Monsonia* species are named after Lady Anne Monson, who was very well-known in botanic circles and worked with the Swedish naturalist Linnaeus in the eighteenth century. She became very interested in the dysentery herbs and they now bear her name.

Height: 15 cm

Flame lily

Maidenhair fern

Heather

Harebell

Hottentot's tea Hotnotskool Horsetail Huilboerboon

Elephant's foot
Dioscorea elephantipes

ENGLISH	Dioscorea, cortizone plant
AFRIKAANS	Hotnotsbrood, olifantsvoet, skilpadknol
XHOSA	Nakaa
ZULU	Ingweva

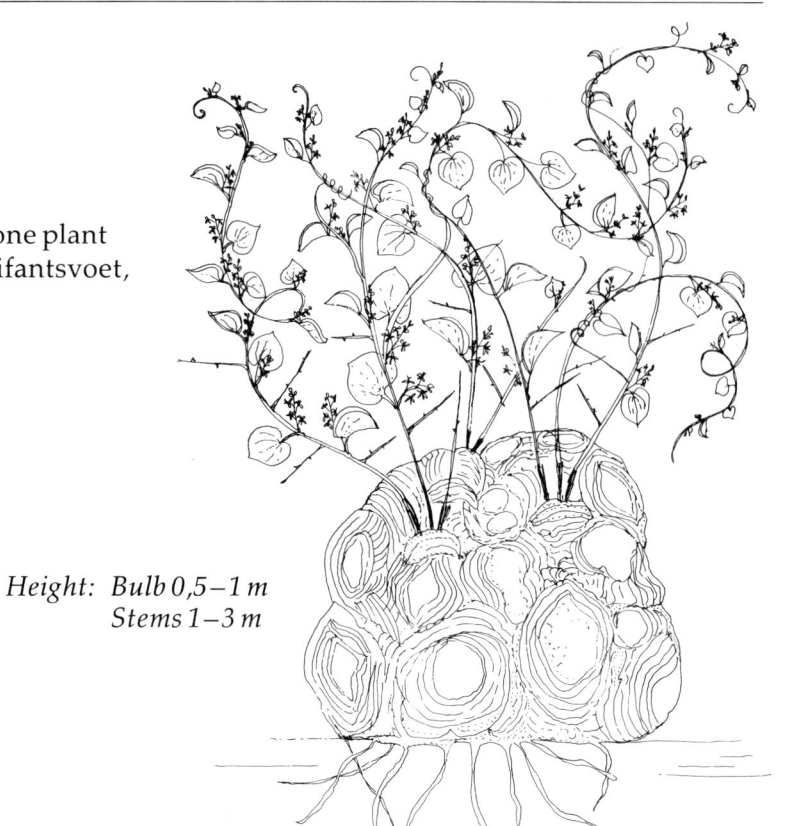

Height: Bulb 0,5–1 m
Stems 1–3 m

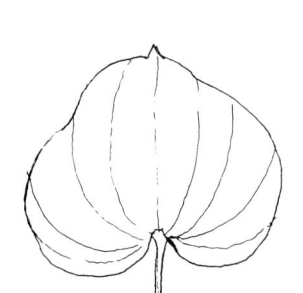

THE elephant's foot is one of South Africa's strangest and rarest plants. It is really a huge tuber that rests on the ground, with fibrous roots on the under-surface. The annual stems which grow out of the top of the tuber are widely climbing, twisting vigorously into the surrounding bush. The smooth stems have alternate, small, almost heart-shaped leaves and in summer small racemes of greeny yellow flowers can be seen, followed by winged, three-angled, thumbnail-sized fruits.

The plant grows predominantly in the Humansdorp-Uitenhage-Patensie area, and one hot February morning I was taken by friends to a farm in that area where the farmer had given us permission to climb into the mountains to look at his elephant's foot plants which grew in a wide area there. We climbed almost perpendicular slopes, searching amongst the rocks and brush, but found nothing. We had given up and sat drinking some refreshments, when we realised that the strange, gnarled, dark rocks we were leaning against were in fact the elephant's foot tubers! Some were over half a metre in height, some a metre in diameter, and some were so embedded in the rocks as to look exactly like a part of the rock. The tangle of stems winding metres from the mother plant hid more tubers. It was a most amazing sight. The gnarled, dark tortoise shell-like markings of the tuber (hence the name skilpadknol) looked like some petrified prehistoric monster, and the great size of the plant in some places could not be fully seen.

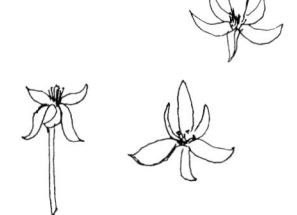

The farmer told us that the Khoikhoi hollowed out the biggest plants for dwellings and ate the solid flesh of the plant which contains starch and sago (hence the name Hottentotsbrood). In about 1778 Patterson, a botanist at the

Cape, recorded that the Khoikhoi ate the plant as a valuable strength-giving food. The plant was first cultivated in England in about 1775 and is still grown in European hothouses. The hothouse at Kew has a good specimen still flourishing.

What is most amazing about this plant is that it contains a natural cortizone, dioscorine, which is the characteristic alkaloid of the genus *Dioscera*. This is used in homeopathic medicine and will surely be one of our most important sources of natural cortizone in the future.

The plant is extremely slow-growing and is protected, so very few nurseries have it for sale. I managed to procure a small plant from an indigenous nursery some years ago, and find it does well if left undisturbed in a partially shaded wild area. The long summer stems indicate its adaptation to the Transvaal climate but the tuber seems to have grown not a fraction since I planted it.

The Karoo botanic gardens near Worcester have good specimens, which are well worth a visit.

Flame lily
Gloriosa superba forma *superba*

ENGLISH	Gloriosa lily, Turk's cap, dragon lily, superb lily
AFRIKAANS	Vlamlelie, geelboslelie, rooiboslelie
NDEBELE	Amakukulume
ZULU	Ihlamvu

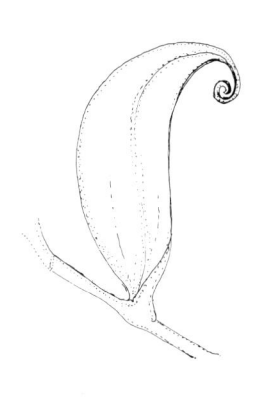

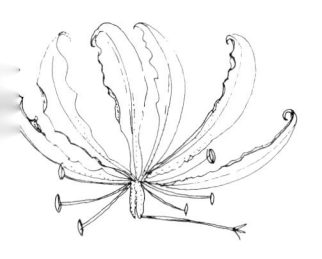

THE flame lily is a tuberous climber that uses the tendrils formed by the tips of the leaves as a support to hook itself into trees and shrubs. It occurs in masses along the coastal bush in the eastern Cape, Natal and up into Zimbabwe and East Africa.

The plant grows from a finger-like flat tuber which sends up one stem. By midsummer several flowers have opened up along that stem and sometimes the stem bears side branches which also bear flowers. In autumn the stems die back, leaving the tuber to go into its dormant period. The flowers have beautiful red and yellow wavy flame-like petals, hence the name.

Although the plant is poisonous, the tuber is much prized by blacks and whites as a medicinal application for haemorrhoids, sprains, strains and bruises. Slices of fresh tuber folded into a linen or cotton cloth are bound over the area, and some tribes use the juice as an external application for snakebite and scorpion stings. The root tuber is also used as an antiparasitic, applied externally by squeezing the juice from the tuber onto the area, and the Ndebele and Zulu use it for skin eruptions, tick infections and screw worm on cattle with remarkable success.

A word of caution here, however: the plant is extremely toxic and even if it is externally applied the skin does absorb the poison to an extent, so be guided by your doctor or qualified herbalist before applying the plant in any way. Keep the tubers away from pets and children when you dig them up and take care when cutting the flowers that the juice from the stem does not come into contact with the mouth or eyes. The latex from the fruit can kill a dog within 15 minutes, so handle with care!

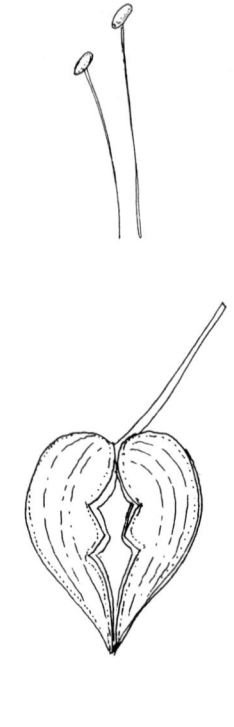

The plant is a violent emetic and if it is eaten it results in death within 4 hours. Strangely enough, the whole plant with its highly poisonous rootstock is dug up and greedily devoured by porcupines. I have had enough nocturnal visits from a pair of endlessly hungry and destructive porcupines to know they have cast iron stomachs!

The plant is easy to grow. Space the tubers approximately 100 cm deep and 30 cm apart in richly composted, well-dug soil in a semi-shaded position. Set up a trellis or support as they will climb to about a metre in height. I planted my tubers where they get morning sun and deep shade from midday on, and they thrive in that position. They need a lot of water during spring and early summer, but once the stem and leaves die, they need no more watering. Do not disturb the tubers as they will multiply in that position, and the following spring the new growth will become evident as soon as the weather warms up. After about 4 years, dig up the clump in July and divide the tubers, replanting them in their new position as soon as possible.

If the flowers are left to fade on the stem they form pretty seeds, and some tribes thread these into colourful necklaces which are considered to be a lucky charm, giving the wearer protection and strength.

At present pharmaceutical tests are being done overseas on the plant, but at the time of writing the results of these tests are still unknown. Perhaps one day the flame lily will be an important medicine much needed to treat the masses in Central Africa and India.

The Tswana on my farm use the juice of a leaf tip as a treatment for pimples and skin eruptions, and said that a Masai farm worker had taught them this when they were children. I find this fascinating, as it is not often that one tribe will educate another on the important medicinal uses of a plant.

The flame lily is Zimbabwe's national flower and is being grown commercially there now. Many ex-Rhodesians in South Africa grow their little bit of green heritage in their gardens, far away from their homeland, and it is with much tenderness that they nurture their precious plants.

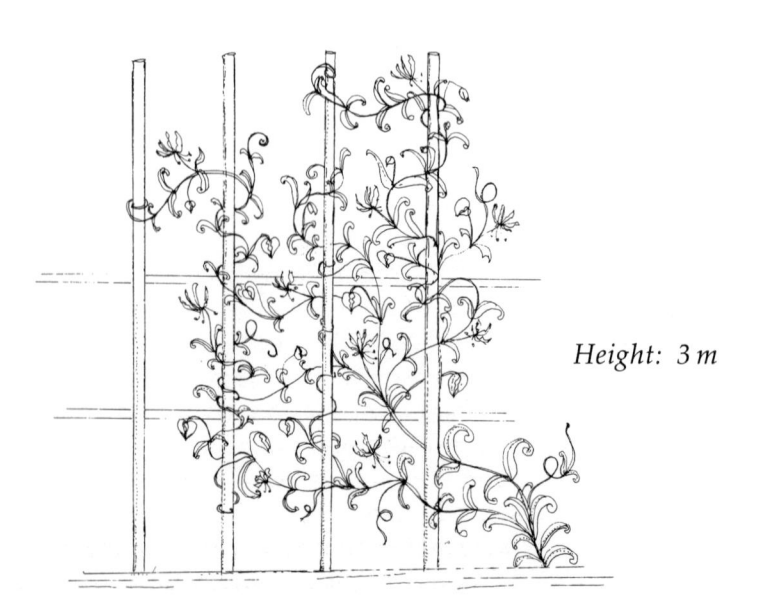

Height: 3 m

Ginger bush
Tetradenia riparia (formerly Iboza riparia)

ENGLISH Wild ginger, iboza, misty plume bush
AFRIKAANS Watersalie
TSWANA   Bosso, ibosso
ZULU   Ibozane, iboza

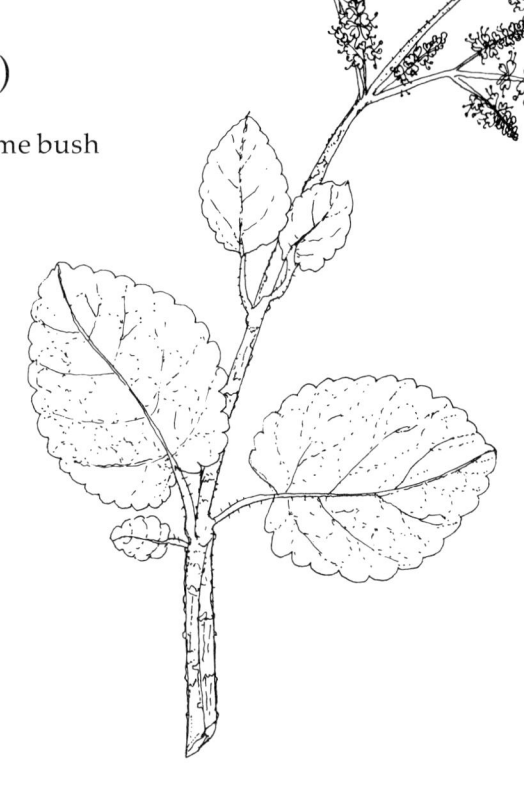

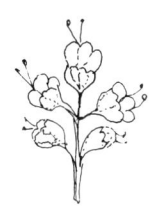

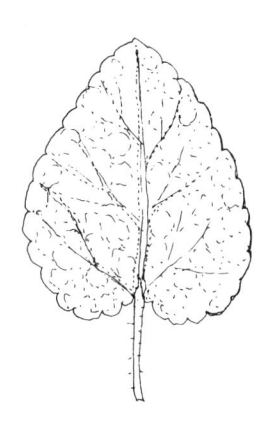

THE beautiful ginger bush belongs to the great Lamiaceae or salvia family. It has the aromatic leaves characteristic of this family, and is native to Natal and the eastern Transvaal, and to some extent the northern Transvaal. It cannot withstand frost but I have been very successful in growing it as a large container plant, which I move into my greenhouse as soon as the first frosts arrive and the cold winds of May blow, and it winters there well. It is deciduous and is probably one of the easiest of all plants to propagate. Merely cut off a branch and stick it into the ground! I have never had a failure, and I have grown a whole long fence of wild ginger by just continuously taking cuttings and sticking them into the ground alongside the mother plant.

The exquisite, tiny pink, white or pale mauve flowers borne on large, attractive panicles are the earliest of all the spring flowers, and they appear in great profusion — soft, feathery, almost misty in appearance — when little else is in flower. I pick great sprays of flowers for spring arrangements, and they last beautifully in water. I also pick the flowers, and later the leaves, as they appear after the flowering period is over, for pot-pourris.

The leaf made into a tea is an old remedy first used by the Zulu for coughs, colds, chest and respiratory ailments. Usually the dosage is 2 leaves to 1 cup of boiling water, stand for 5 minutes, then strain and drink. Sometimes the young shoot of the plant is used for the same purpose, and this tea is also taken for colic, stomach upsets, stomach ache, flatulence, diarrhoea and nausea. The Tswana take a small dose at a time to soothe fevers and calm the patient, and believe it to be excellent for old people.

An old malaria remedy used in the eastern Transvaal is to take 1 cup of the tea made from the leaves at bedtime. Just one dose is apparently enough to cure the patient by the following day. In the treatment of malaria this could be a remarkable natural medicine, and one hopes that medical tests will be done on it in the future.

A sangoma from the Louis Trichardt area who comes to my herb garden for plants told me of her forefathers' treatment of malaria, dropsy and diarrhoea with iboza. She said two leaves crushed and briefly chewed would help bring down a fever, and the chewed leaves bound on the inner area of the wrists for 3 or 4 hours would calm and ease the fever. She also said that she bathed patients suffering from malaria in a weak, tepid infusion of the leaves and bound their wrists with softened leaves. She was taught as a child to grow the plant near at hand for its amazing medicinal properties, and was given iboza tea as a winter remedy for colds and bronchitis. The dried leaves were stored at the end of summer and kept as a winter medicine, or young shoots were used when no leaves were available.

The Tswana, Zulu and the Venda all use iboza for treating animals, often combining it with *Fagara capensis* (small knobwood tree) in the treatment of gall sickness in cattle.

The plant grows well over a metre in width and height, and in the wild it is often found along streams and in moist places. However, even with not much water it adapts to all sorts of soils and conditions. It ideally needs light shade at midday or afternoon shade, but I have grown it successfully in full sun with not much watering and although the bushes are not as big as their shaded, well-watered neighbours, they still grow well.

Several nurseries offer the plants for sale, and once your bush is well established, take summer cuttings and propagate it lavishly. It is a most beautiful and rewarding plant or small shrub to grow.

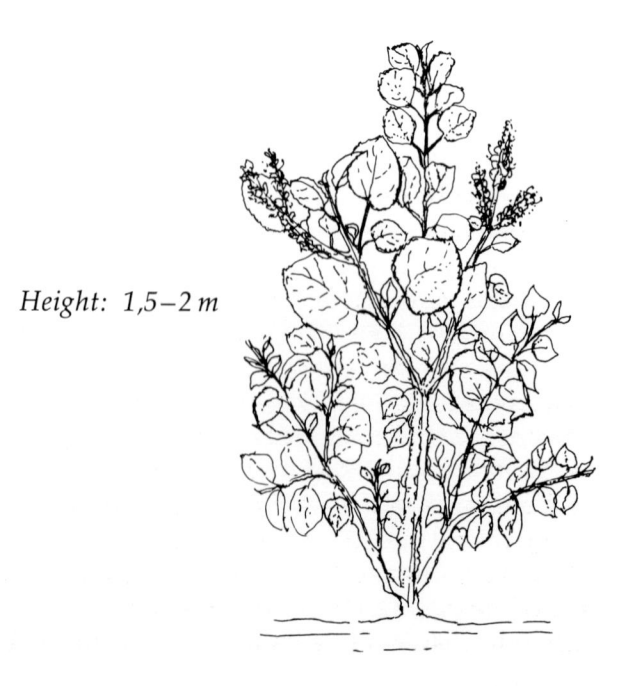

Height: 1,5–2 m

Hard fern
Pellaea calomelanos

ENGLISH   Wild fern
SOTHO   Lethsitha, pata-lewana
TSWANA   Legogoana
ZULU   Phaladza

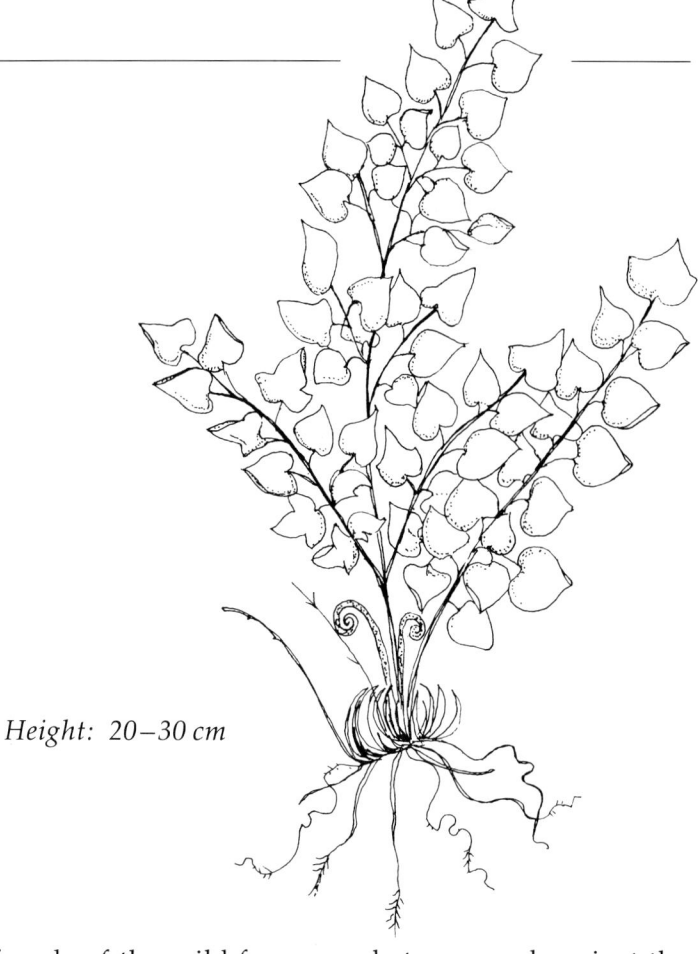

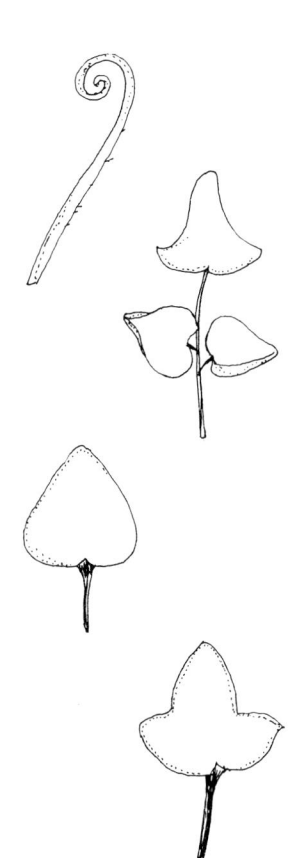

Height: 20–30 cm

THE stiff, grey green fronds of the wild fern grow between and against the rocks in all sorts of soilless places, in the blazing sun on bare hillsides, as well as in cool forest areas. It is a special medicinal plant that I have loved since my childhood when I first found it in the hills around Pretoria and pressed it in my drawing book. Only many years later did I learn from a Tswana gardener, who had been taught by his grandmother, to make a tea of the leaves for coughs and colds. (About ¼ cup of fresh leaves drawn in 1 cup of boiling water for 5 minutes and sweetened with honey if desired.) He also used the root or rhizome boiled and pounded into a soft mass as a dressing or poultice to draw an abscess or boil or to clear an infected sore.

Just as the beautiful cultivated maidenhair fern is used by many whites for chest conditions, so the indigenous variety is used by the indigenous people for the same purposes.

After the rains the fern is at its best, and it is then that I carefully pick the fronds and put them into vinegar for the bath (see recipe overleaf) or press them for dried flower arrangements — they stay pale green for ever! I use a small frond boiled up in milk (1 thumblength piece to 1 cup of hot milk) and sweetened with honey and drink this at bedtime to calm myself after a hectic day. Some tribes give this drink to a child frightened or upset, and smoke the leaf for asthma and chest colds. The Zulu often burn a few green leaves on the fire and inhale the smoke to clear a head cold.

Best of all is to use it in the bath to soften the skin. Tie 2 or 3 fresh fronds in a piece of muslin, toss this under the running hot water tap and then use this as a wash ball (I often add a handful of oats to the bag to really cleanse the skin).

You can rub soap over the ball as well and use it as a skin-softening and cleansing scrub.

Cultivation of the hard fern is difficult as only the tiniest of plants will transplant. Never try to move it once it is established in your garden, and keep an eye open for some nurseries have plants established in bags every now and then.

Hard fern tea — Serves 1

2 thumblength sprigs hard fern
1 sprig sage
slice fresh lemon
1 cup boiling water
2 teaspoons honey
dash of brandy

Pour the boiling water over the hard fern and the sage. Stand, steep for 3 to 5 minutes, then strain, add the lemon, honey and lastly the brandy. Sip slowly.

Hard fern bath vinegar

1 bottle white grape vinegar
6–10 mature fronds hard fern

Push the fern into the vinegar and stand the bottle in the sun for 10 days. Shake and turn it daily (you can add more fronds after 5 days, removing the old should you wish to make the vinegar extra strong). Use a little in your bath to soften the skin.

I make a large quantity in midsummer to last me through winter, and I find this vinegar replaces the need for a bath oil, which can clog the pores. It keeps the skin soft and moist, preventing it from drying out and becoming scaly and flaky in winter. You can also use apple cider vinegar for a very special bath treatment, made in the same way as above.

Diluted 1 to 10 with tepid water this also makes an excellent astringent rinse or lotion for oily skin on the face, neck, chest and shoulders. Dab on 2 to 3 times daily, or after your bath.

Harebell
Dierama pendulum

ENGLISH	Zuurberg harebell, angel's fishing rod
AFRIKAANS	Grasklokkies
SOUTHERN SOTHO	Lethepu
TSWANA	Sepu

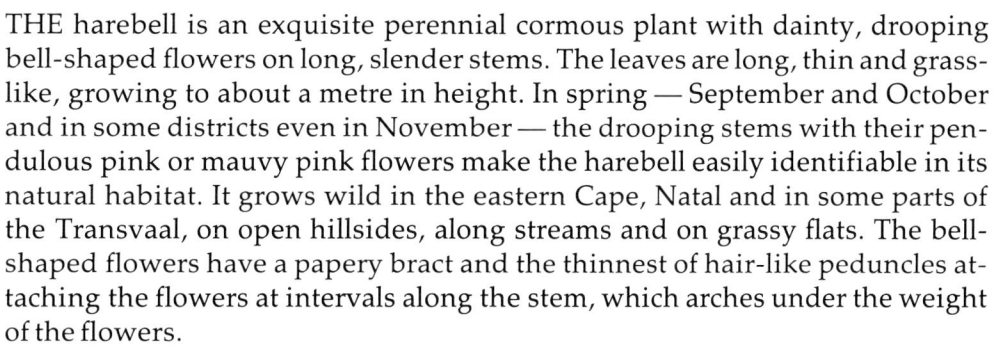

THE harebell is an exquisite perennial cormous plant with dainty, drooping bell-shaped flowers on long, slender stems. The leaves are long, thin and grass-like, growing to about a metre in height. In spring — September and October and in some districts even in November — the drooping stems with their pendulous pink or mauvy pink flowers make the harebell easily identifiable in its natural habitat. It grows wild in the eastern Cape, Natal and in some parts of the Transvaal, on open hillsides, along streams and on grassy flats. The bell-shaped flowers have a papery bract and the thinnest of hair-like peduncles attaching the flowers at intervals along the stem, which arches under the weight of the flowers.

The grass-like clump is evergreen, preferring a moist, semi-shaded position, but I have grown it successfully in full sun in a rockery, where it does well with a good weekly watering.

The corm is crushed, boiled up in water, and used by the Southern Sotho as a purgative enema for severe constipation and abdominal flatulence, though it is a somewhat drastic enema that should be given with caution. A tea of wild verbena *(Pentanisia prunelloides)* is drunk to weaken the powerfully strong effect of a harebell enema, or given to someone who reacts badly to the enema.

The Zulu and the Tswana crush the corm and apply it to bruises or contusions and use the crushed flowers mixed with hot water and sour fig leaf juice for bites, stings and infected skin rashes.

I grow harebells abundantly as they demand nothing other than a good weekly watering, and in spring I gather armfuls of the flowers which I carefully layer in silica gel (making sure the gel fills each bell flower). When I shake them out 4 days later I am left with a simply exquisite dried harebell that lasts indefinitely in shape and colour. Masses of these in a tall glass vase are the envy of all who gaze on them, for the dainty bells on their slender stems are rare and exquisite and yet so easy to preserve in all their beauty.

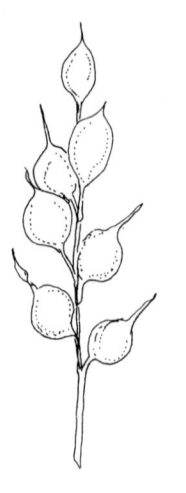

The corms can be bought through various wild flower nurseries or mail orders, and will grow with ease. They are small, round and covered with fibre, and should be planted in well-composted soil at least a metre apart as they grow rapidly into large clumps. The big clumps can be divided in winter and the corms replanted. In some areas the leaves will dry and become unsightly, and these can then be clipped or burned off, and the new growth in spring will be lush and thick. Clip back the leaves to 3 or 4 cm in length on the corm and plant in well-composted, moist soil.

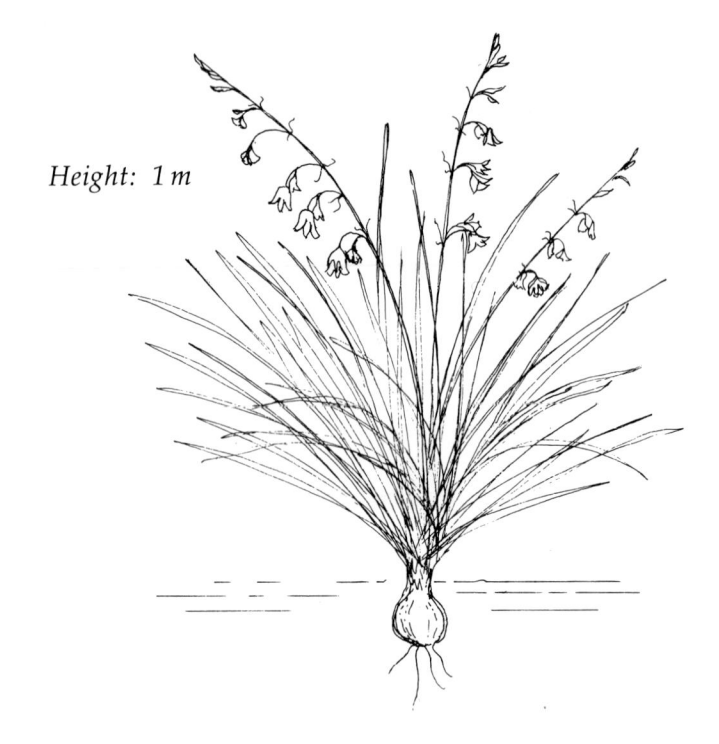

Height: 1 m

Heather
Ericaceae family

ENGLISH   Ling, heath, erica
AFRIKAANS   Heide

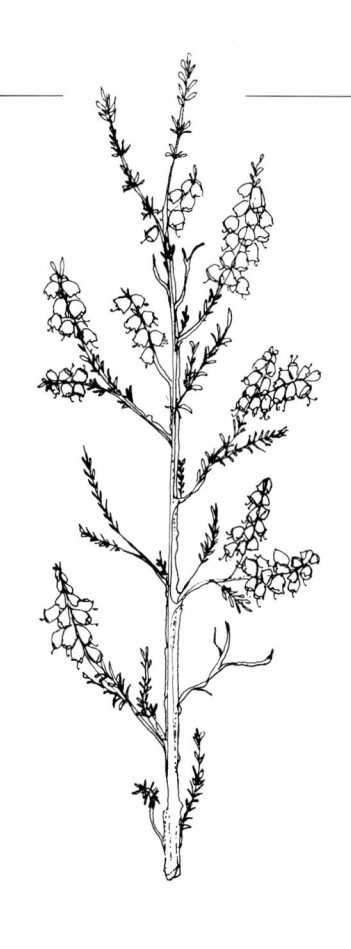

THERE are a great many varieties of exquisite South African heathers, all belonging to the large Ericaceae or heath family. They all do best in light, sandy, acid soils, rich in compost and with good drainage. This is important for gardeners wishing to introduce these special plants into their gardens for, as a rule, the heathers are difficult to cultivate and haphazard preparation will result in poor specimens. Many nurseries offer the plants for sale, so do not be tempted to dig up or take cuttings of the heathers in the wild — apart from it being against the law, they just will not grow.

In general the ericas are erect, much-branched shrubs with small, bell-like flowers which are borne in profusion. The leaves are small and finely formed and the branches are dark and woody, and make excellent kindling. They occur naturally in coastal areas, particularly the Cape, sometimes Natal, and occasionally inland on hillsides and roadsides. Most like winter rainfall, but cultivated specimens do well in summer rainfall areas as well.

As a child I was intrigued by my Scottish grandmother's belief that heather tea would ease stomach aches, give energy, soothe depression and nervousness, help insomnia, anxiety, tension and fear (infuse one small sprig in a cup of boiling water, steep for 5 minutes, then strain and drink). She also showed me how the Scottish Highlanders made a wonderful cough syrup from heather. She lived at Gordon's Bay and her favourite heather was the common one, *Erica hirtifolia*, whose beautiful tiny mauve flowers delight the eye when travelling in the Cape. She would gather sprigs and boil these in water to which honey was added, slowly simmering it on her old coal stove while the South Easter whined and howled around the house. My sister and I were given small cups of this sweet, wild tea to sip as we sat in front of the fire, and it did much to soothe our winter sniffles and coughs and, with hot buttered toast, was our afternoon treat. My great grandmother said it was good for her rheumatism, and as a girl in Scotland she had used a strong brew of heather flowers for coughs and colds, as well as a tonic tea and a soother of nightmares.

Some farmers place hives on heather-covered hillsides. The honey from the heather is one of the most flavoursome of all honeys — farm stalls and country markets sometimes offer it for sale.

Dried heather flowers are beautiful in pot-pourris as they retain their colour and seem to absorb the essential oil well. I have found that clipping my bushes after the summer flowering time not only neatens but seems to encourage new growth the following season. I use every leaf, twig and branch finely chopped in pot-pourri, as they make an excellent base.

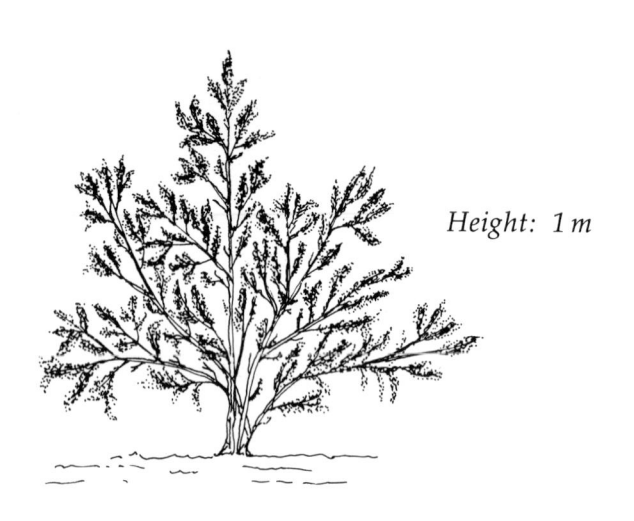

Height: 1 m

Heather cough syrup

3 cups heather flower sprigs
6 cups boiling water
1–2 cups honey
10 cloves

Boil the ingredients gently together for half an hour in a double boiler or covered stainless steel pot. Remove from the stove and allow to cool. Strain and drink half a cup at a time, warmed, 3 or 4 times a day, for coughs and colds or as a nightcap; or morning and evening for anxiety and depression, lack of energy or lack of appetite. Keep any excess in the fridge.

Heather pot-pourri

4 cups heather flowers
4 cups finely chopped heather twigs, branches and leaves
4 cups currybush flowers and leaves
2 cups dried minced lemon peel
4 cups wild lavender tree leaves
1 cup crushed cloves, cinnamon and nutmeg
½–1 cup coarse sea salt*
essential oil

* Omit the salt if you live near the sea, as the salt is hydroscoptic and serves the purpose of keeping the pot-pourri from becoming dusty, brittle and dry if you live in a dry atmosphere inland.

Dry all the ingredients in the shade on newspaper. Turn daily. Combine them once they are well dried, and shake together in a large tin or jar. Add the essential oil (I like lavender or honeysuckle oil in this mixture). Shake daily for 2 weeks, then fill sachets or bowls with it. Revive with a little essential oil from time to time.

Honeybush tea
Cyclopia genistoides

ENGLISH   Honey tea
AFRIKAANS   Heuningtee, bossietee, suikertee, heuningbostee, heuningblomtee

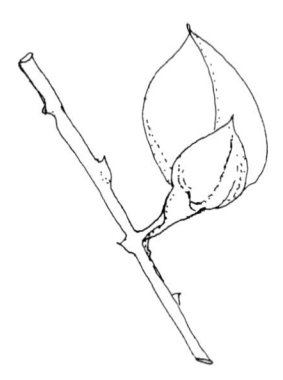

THIS somewhat unobtrusive bushy shrub grows on the hillsides in the southern, south-western and south-eastern Cape. It has been used in the Cape since around 1750 as a tonic tea, excellent for soothing coughs and colds, and Carl Thunberg recorded it in 1772 as 'honingtee'. There are several other species of *Cyclopia* which are also called 'heuningtee' but here I write of *Cyclopia genistoides*.

The yellow-green bush grows up to about 60 cm in height, the leaves are narrow and needle-like, growing in threes up the stems, and in spring and summer the small, bright yellow, peaflower-like flowers appear on the tips of the branches and hum with bees.

The leaves, twigs and flowers are picked and then fermented in water or cured in wet sacking, and then dried and packaged as a health-giving tea.

I use 2 tablespoons of tea in 1 litre of boiling water, which I boil for 20 minutes to give it flavour, then strain. People from the southern Cape often like it stronger, so experiment yourself until it has that sweet honey flavour that makes it so special. The tea can be reboiled and re-used although it eventually loses its flavour altogether.

The plant is astringent and caffeine free, although it contains tannin, and I am always intrigued how those early settlers knew that this was a 'tea' plant and how to use that tea. Most people who have grown up on honeybush tea say it was the Khoikhoi who taught the colonists the uses of the plant and how to make the tea.

The treatment for catarrh, pneumonia, tuberculosis, as well as coughs and colds was 1 cup of honeybush tea boiled for at least 20 minutes 4 or more times per day. It was also given after a long illness as a restorative, and to the old and weak to give them strength and vitality.

Today honeybush tea, like rooibos tea, is enjoying ever-increasing popularity as a healthy and flavoursome tea. You can buy the ready-packaged tea at farm stalls, country stores and at some health food shops in the Cape, and it can be ordered from farmers in the Langkloof. Although the propagation of the plants is from seed I have not been successful in growing the plant in the Transvaal. It needs those Cape mountain slopes and the glorious winter rains, but some nurseries in the Cape may have plants for sale so keep a look out for them. It is a most rewarding plant to have in the garden.

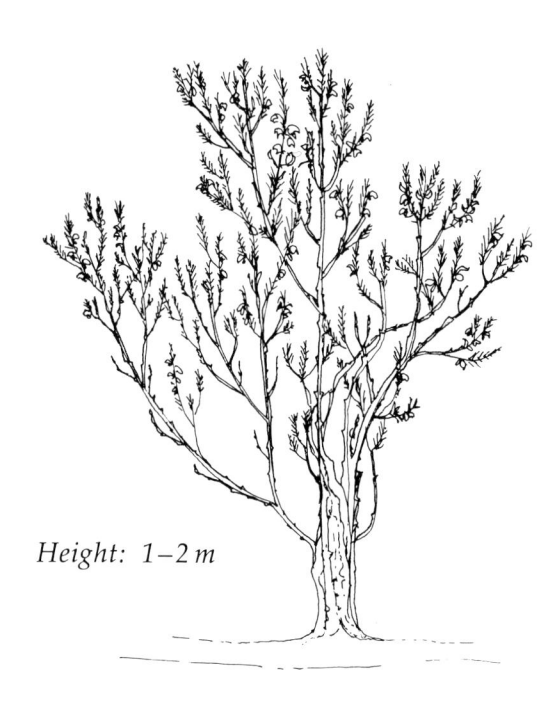

Height: 1–2 m

Honeybush tea jelly   Serves 6

3 tablespoons gelatine
1 litre honeybush tea (boil for half an hour and then strain)
1 litre fresh orange juice or 1 litre fresh peach juice and pulp
honey to sweeten

This is a healthy, refreshing jelly and is particularly good for children. I have also made it with grapes and grape juice, and grated apple and apple juice for an invalid, and it is always much enjoyed.

Soften the gelatine in a little hot water and mix into the warm tea. Add the fruit juice and fruit pulp. Stir enough honey to sweeten (I use about 4 tablespoons). Pour into a glass bowl or jelly mould. Place in the fridge overnight to set. Serve with cream or custard.

Horsetail

Equisetum ramosissimum

ENGLISH   Drill grass
AFRIKAANS   Perdestert
ZULU   Isikhumukele

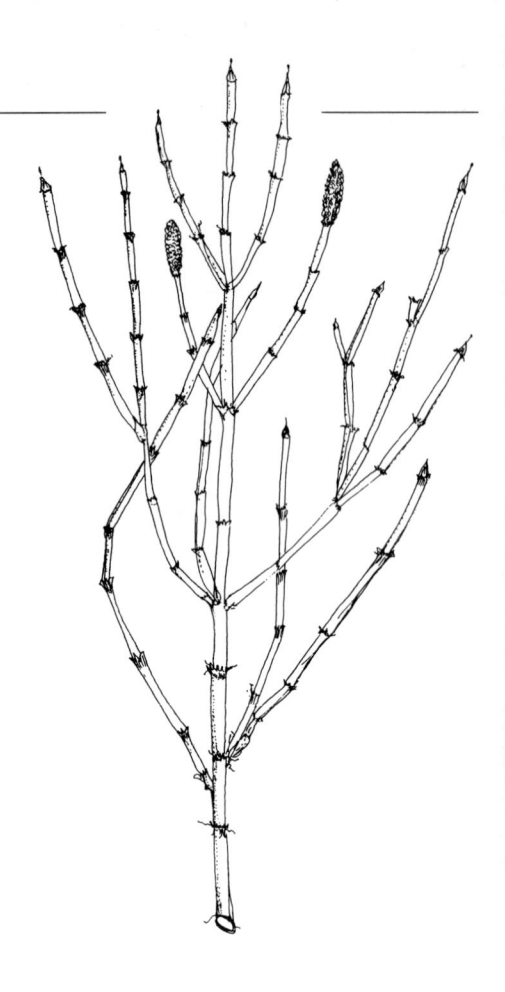

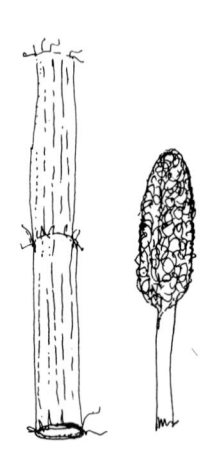

HORSETAIL is a strange, brittle grass with long, jointed, horsetail-like stems. In midsummer small, cone-like 'flowers' that carry the seeds or spores emerge and it pushes up runners in all sorts of unusual places, emerging between paving stones, under hot rocks, in deeply shaded moist areas along streams, and even in beach sand. It will seek out its favourite place and although it seems difficult to grow and to propagate, it will suddenly emerge from an unexpected place, green and vibrant.

It is much respected as a medicine and is probably best known for its diuretic qualities taken as a tea, and for prostate, bladder and urinary infections. It is useful also in treating diarrhoea and dysentery and as a wash for wounds and sores. Use ¼ cup of the fresh needle-like leaves and stems to 1 cup of boiling water — stand, steep for 5 minutes, then strain and drink about ½ cup every 2 to 3 hours until the condition eases. Use this same brew to wash wounds, dabbing it on as a lotion, which will staunch bleeding and help clear infection.

The Zulu particularly respect this magical plant and use dried powdered stems mixed into water to treat tummy upsets, particularly in children. The Sotho also use it for colic and colds in the same way, usually 2 teaspoons in 1 cup of warm water.

The European variety, *Equisetum arvense*, is also a valuable medicine and is used to treat stomach ulcers, intestinal ulcers, inflammation of the vagina (the tea is used as a douche), to reduce glandular swellings, to dissolve bladder stones, for dysentery, earache, toothache, as a wound wash, and to treat prostate problems. Our indigenous horsetail was used by the colonists for all these treatments and is still used today by many whites.

Honeybush
tea

Jasmine

Melianthus major
Kruidjie-roer-
my-nie

Kattekruie

marula

knobwood

wild medlar

lavender tree

As well as being a remarkable medicine, the horsetails are excellent pot scourers and bucket cleaners. The hard silica content makes them hard to beat as a cleaning material. The ash from the burnt plant mixed with water (1 cup of ash to 2 litres of water) is an effective spray against mildew and the boiled plant (3 cups of fresh plant boiled in 3 litres of water for 20 minutes) is also extremely effective if it is sprayed or splashed on to mildew-infected plants. It also acts as a tonic to ailing plants, probably owing to its high silica content.

It is poisonous to grazing animals so farmers have rooted it out which makes it difficult to find, but some specialist nurseries may have it established in bags.

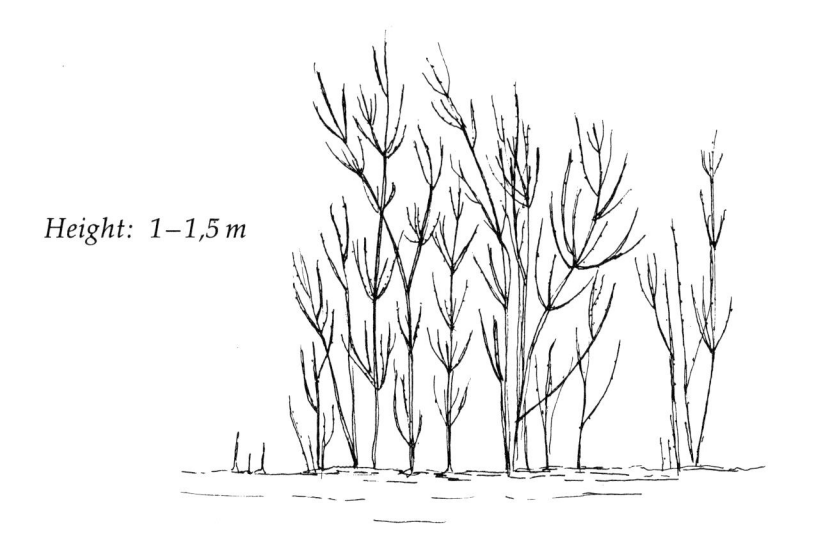

Height: 1–1,5 m

Hotnotskool

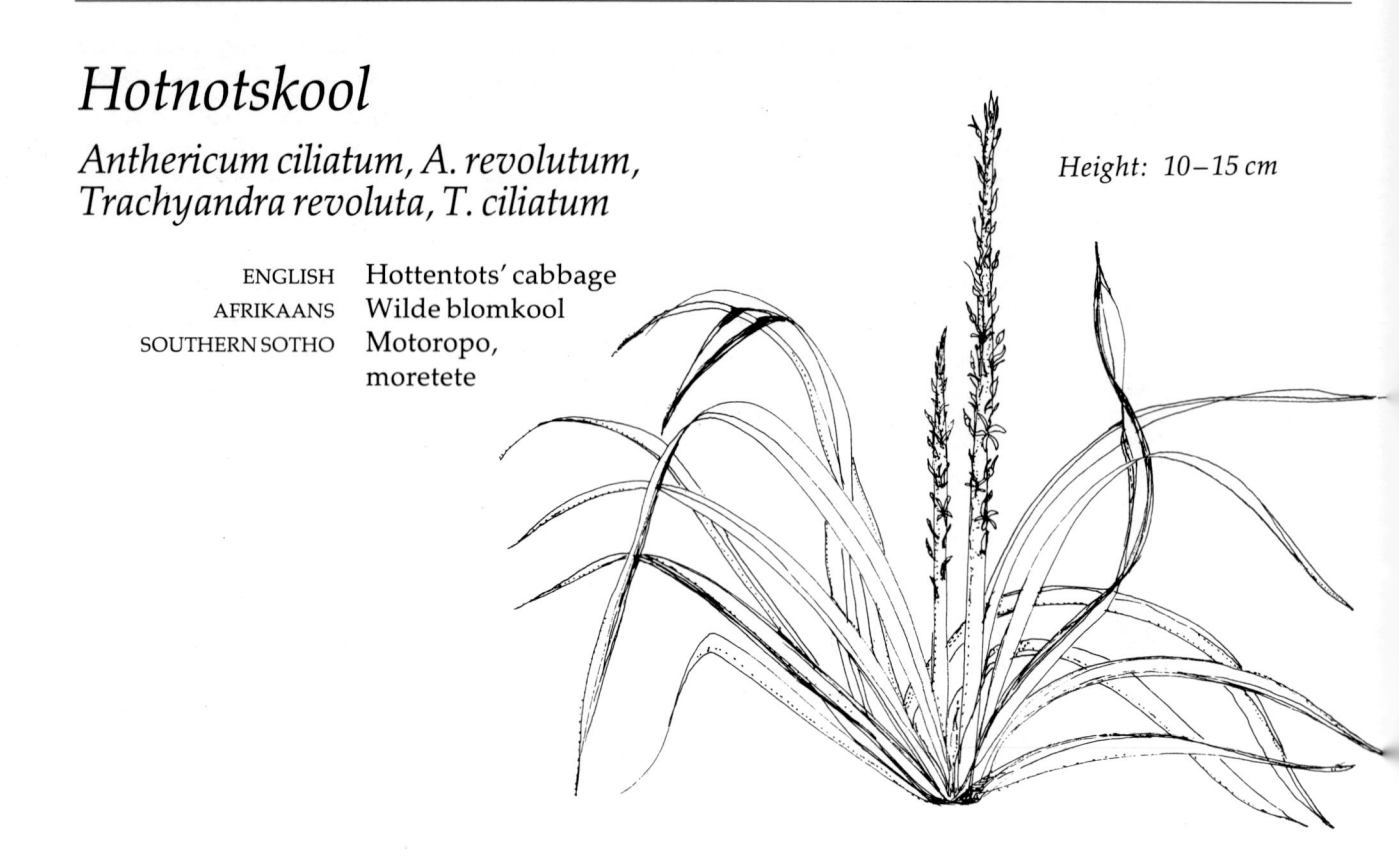

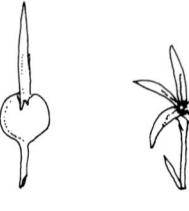

Anthericum ciliatum, A. revolutum, Trachyandra revoluta, T. ciliatum

Height: 10–15 cm

ENGLISH	Hottentots' cabbage
AFRIKAANS	Wilde blomkool
SOUTHERN SOTHO	Motoropo, moretete

THE juicy, unopened inflorescence of this small, unspectacular, grey green plant is known as the Hotnotskool. It grows in the sand veld particularly in the Cape and the Orange Free State and grows particularly abundantly in the winter rainfall areas. It comes up in lawns and roadsides, cultivated lands, open veld and in the garden, where most people consider it to be an irritating weed!

The whole plant is soft and juicy with small, thin, grey-green strap-like leaves that grow out of the slightly bulbous root stock, and the flowering head which looks like an asparagus shoot grows out of the leaves from the centre. It is not very noticeable in its unobtrusive greys and soft green, but once you get to know it you will keep a sharp look out for it, as it is one of the most delicious of all wild vegetables.

The soothing sap from the stem and bulbous part of the root is used by the Sotho as a lotion for sores on the heads of their children or for sores on the legs, rubbing the juice into the area at intervals. A coloured farm worker near Clanwilliam showed me how he crushed the stem and root and applied it as a dressing to a minor burn, and I have found it to be quickly soothing on scalds.

In the Orange Free State the Sotho make a decoction of the plant which is used to treat hysteria and shock, but the proportions used seem to differ from family to family. Often the water in which the young shoots are cooked is used as a tea for bladder infections diluted and warmed and the plant is eaten at the same time, but again dosages seem to vary.

The young shoots make a healthy and delicious dish when steamed and served with butter, salt and pepper and a squeeze of fresh lemon juice.

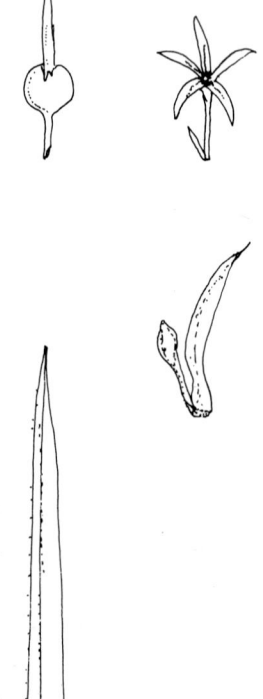

I have managed to grow Hotnotskool in the Transvaal by carefully collecting the seeds from the ripe flowering heads growing in the veld and treating them as an annual. They seed themselves all over the vegetable garden once they are established, and they are such a wonderfully tasty vegetable that I nurture them with care.

Hotnotskool bredie   Serves 6–8

8 mutton loin chops
½ cup brown flour
little sunflower cooking oil
2 large onions, peeled and chopped
4 large ripe tomatoes, skinned and chopped
2 apples, peeled and chopped
2–3 cups Hotnotskool, sliced into 3 cm lengths
2–4 potatoes, peeled and diced
2 tablespoons fresh, chopped parsley
½ tablespoon fresh thyme
salt and pepper to taste
2 tablespoons honey
1 litre water

This bredie is the nicest way to prepare Hotnotskool and it is best with Karoo lamb.

Roll the mutton chops in the flour. Heat the oil in a deep heavy pot and brown the mutton chops in the oil. Add the onions and brown lightly, then add the tomatoes.

Add all the other ingredients and half the water. Simmer with the lid on, adding a little more water from time to time, and stir to prevent burning and to keep the bredie moist and succulent.

Adjust seasoning, and sprinkle with fresh parsley. Serve with brown rice and a green salad.

This bredie keeps well in the fridge or freezer.

Hottentots' tea

Helichrysum nudifolium

ENGLISH	Wild everlasting
MPONDO	Icolacola
SOUTHERN SOTHO	Phefswana-basia
SWATI	Ludvutfane
TSWANA	Phefo, mathabelo
ZULU	Isidwaba-somkhovu

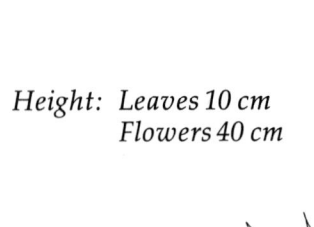

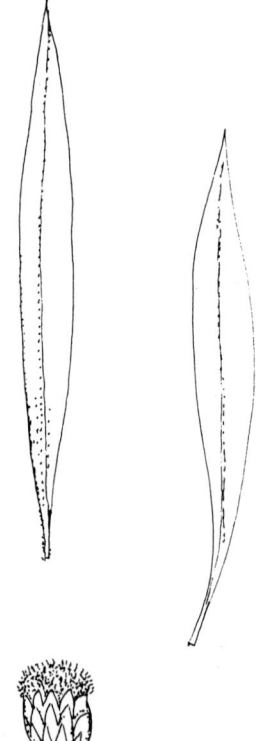

Height: Leaves 10 cm
Flowers 40 cm

THIS is one of our oldest and best known medicinal plants, first recorded in 1884 by Carl Pappe, a physician who came to South Africa in 1831 and later became the country's first professor of botany. He called the plant 'caffer tea' and recommended it in the form of an infusion as a demulcent for catarrh, phthisis and other 'pulmonary affections'.

It is common from the Cape through to Central Africa, and is used medicinally everywhere it is found by all the indigenous people. The pretty pale yellow flower heads on slender grey stalks which appear above the grass in the open veld and along the roadsides are at their best in October and November, and last long after Christmas.

Primarily the leaves of the Hottentots' tea are used to treat colds, coughs and chest complaints in the form of a tea. Usually the dose is 1 cup of boiling water poured over ¼ cup of fresh leaves, and the tea is left to stand for 5 minutes, then strained and 2 or 3 cups drunk daily to clear the condition.

In the Transvaal the Tswana also use the flowers in their tea, and often save dried flowers for winter use. Sometimes they mix wilde als leaves with the leaves and flowers of the Hottentots tea and use this as a gargle for sore throats, or drink it as a tea.

Some tribes also eat the leaves for colds, and some use the root, chopped and brewed, as a strong tea for all chest complaints, including tuberculosis.

The Zulu and the Tswana use the leaves crushed and warmed in hot water as a wound dressing or as a poultice over boils and abscesses or swellings, and change the dressing twice daily. A useful ointment is also made by burning the leaves and adding the ash to Vaseline or sheep fat, which is then used over bruises, swellings or rheumatic joints. The Southern Sotho make a soothing steam bath by pouring an infusion of the plant over stones that have been heated in the fire and placed in a tin or metal bath. The 'tea' is made by pouring 10 cups of boiling water over about 3 cups of leaves, stems and roots of the plant. The resultant steam is then inhaled and the tea is used as a wash for a feverish patient, or for one suffering from fear and bad dreams.

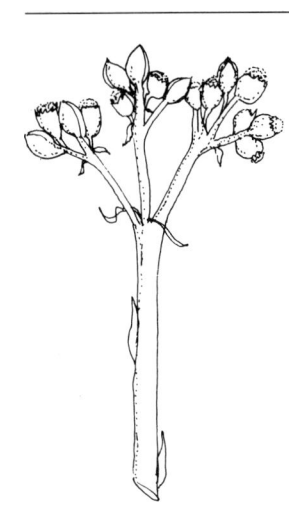

In Natal the Zulu make a strong tea of the plant and use it as a lotion over swellings, haemorrhoids and bruises. They also pulp the leaves in the tea and use this as a wound dressing.

In some country districts the farmers still use the plant as a wound dressing for cattle and sheep, and add a little of it to their food if the animal is not in good condition. The crushed, softened leaf is excellent as a treatment for infected tick bites on dogs. I have found it works best if pounded with the leaf of *Bulbine frutescens* (bulbinella) and then dabbed on to the infected area 3 to 6 times a day.

Hottentots' tea is a pleasant, health-giving tea and one that is still much favoured today in rural areas. The plant grows fairly easily from ripe seed collected in midsummer, but it does need full sun, well-drained soil and not too much water. It is a tough perennial that flowers year after year, but cannot be transplanted. You will need to start new plants from seed to increase your stock. The flat-topped, pale yellow flower heads dry beautifully for dried flower arrangements and can be sucessfully grown commercially. Pick the flowers when they are almost fully opened — some parts of the umbel must still be in tight buds — then dry by hanging the bunches upside down in a cool, airy place. After 1 week in hot weather they will be dry enough to use or to sell. They retain their lovely pale yellow colour and the flowers are also lovely in pot-pourris as they absorb the essential oil extremely well and their faint 'wild everlasting' scent enhances the pot-pourri.

Huilboerboon
Schotia brachypetala

ENGLISH	Tree fuchsia, weeping boer-bean, weeping schotia, Hottentots bean tree, African walnut
PEDI	Molope
SHANGAAN	Uvovovo
SWATI	Uvovovo
TSWANA	Umutwa
VENDA	Mutanswa
XHOSA	Mfofof, umgxam
ZULU	Uvovovo

THE beautiful tree fuchsia, commonly known all over South Africa as the huilboerboon, is a large tree with a rounded crown, dark green glossy compound leaves, and great masses of glorious crimson waxy flowers in spring and summer. The flowers have no petals and the red sepals and filaments growing closely together along the ends of the branches give the appearance of fuchsias. Nectar is produced copiously by the flowers and drips continuously from them, giving rise to the name huilboerboon or weeping boer-bean. Birds, butterflies and moths love the nectar and this tree is a wonderful way to draw them to your garden.

The tree occurs naturally in a fairly wide area from Natal and the Transkei up through the eastern parts of the Transvaal and on into Zimbabwe and Zambia. Its preferred habitat is open deciduous woodland and scrub and it needs a temperate, frost-free climate, although it can withstand a fair amount of cold. It usually grows to about 16 m in height.

Many seeds are produced after the flowers in about March/April and these are beautiful in dried flower arrangements. The flat woody pod which is 6 to 10 cm long splits open to reveal flat, oval, cinnamon-coloured seeds about 2 cm long with a pale waxy aril. These can be roasted and eaten, or roasted and ground into a powder which makes a good coffee substitute. The seeds and the pod absorb the essential oil in pot-pourri making, acting as an excellent fixative and the dried flowers also absorb the oil well and retain their colour well in pot-pourri.

The wood from the tree is tough, hard and heavy and dries to a dark walnut brown colour, making superb furniture. The bark has a high tannin content,

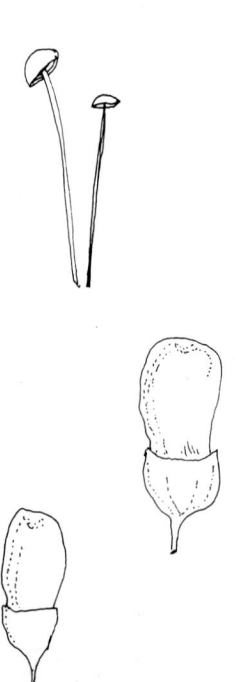

S. brachypetala

Height: 8 m

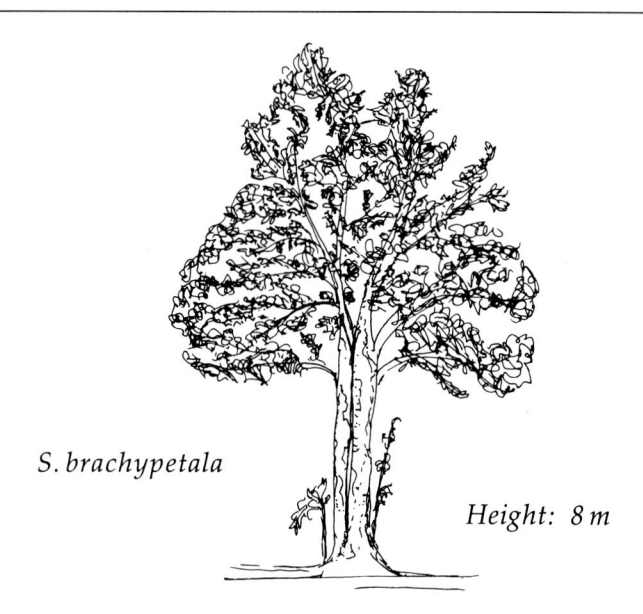

making it a useful medicine for diarrhoea. The Xhosa, Zulu and Tswana make a tea of pieces of the bark and drink it in small doses to ease a hangover or a headache, or take a tablespoon to ease heartburn (usually about 1 cup of bark pieces boiled in 3 to 4 cups of water for 20 minutes, then left to cool and strained). This same brew can be taken a tablespoon at a time at frequent intervals for diarrhoea and nausea and the Tswana give a little at a time to a child with colic. A little is sipped with cold water after vomiting.

Schotia afra, or the karoo boer-bean, is used in the same way. This is a much smaller tree, shrubby, twisted and gnarled, and it grows in more arid regions, along dry watercourses and in Karroid bush and scrub.

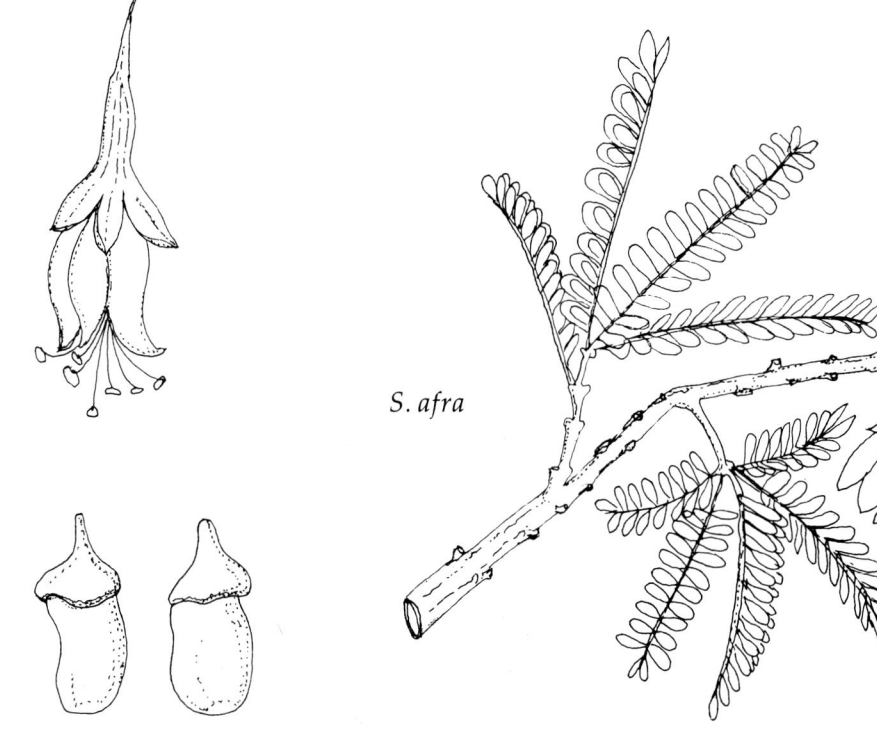

S. afra

105

S. afra

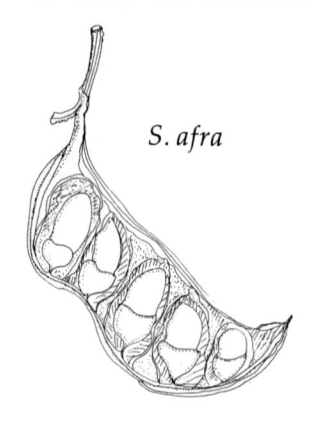

The seeds of this tree can be eaten green or roasted and ground into a meal and made into porridge and pancakes. These were important in the diet of the early settlers and the Khoikhoi. This tree occurs in the coastal districts of the eastern and southern Cape where it is a valuable windbreaker, but it grows only to about 4 m in height.

S. afra var. *angustifolia*, also known as Hottentots bean (Xhosa: 'UmGxam') is another Karoo boer-bean very similar to *S. brachypetala* which grows inland in Namaqualand and Namibia. It has finer, feathery leaves and is used in the same way as its close relatives. The crushed flowers are also used as a wound healer, by grinding them to a paste and applying it to an inflamed area like a poultice. This tree needs very little water so it is prevalent in the drier areas of the country, and its beautiful red flowers can be seen late into the summer in the arid scrub.

Nurseries offer the young trees for sale in bags. The fruit of the *Schotia* is an important famine food and the tree is a source of excellent wood, so we should be planting them everywhere!

S. afra   Height: 5 m

Jasmine

Jasminum multipartitum

ENGLISH   Many-petalled jasmine
AFRIKAANS   Wilde jasmyn

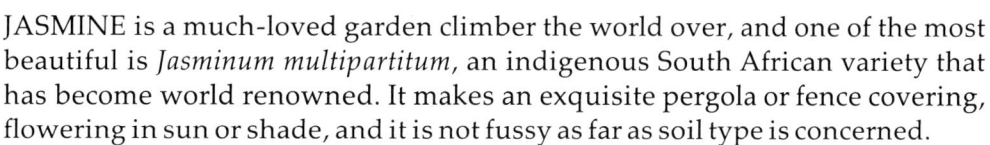

JASMINE is a much-loved garden climber the world over, and one of the most beautiful is *Jasminum multipartitum*, an indigenous South African variety that has become world renowned. It makes an exquisite pergola or fence covering, flowering in sun or shade, and it is not fussy as far as soil type is concerned.

The white, waxy flowers are about 5 cm wide and the glossy leaves are oval and pointed. The twisting grey and green stems make fascinating shapes, and the climber can be trained into any shape. One of the most beautiful is to train it up a sturdy pole onto which a square or circular windy-dry washing frame has been fixed. Within two years the whole frame can be covered if you train the stems, weaving them in and out of the wires weekly. This makes an

Height: Up to 3 m

evergreen garden umbrella and in spring its fragrant pink and white flowers are an absolute delight.

The flowers dried and added to China tea make an excellent after-dinner digestive. Pick the fresh flowers for bath vinegars or for flavouring honey — just push 6 to 10 flowers into a jar of honey, stand for 2 weeks and strain. This honey is a wonderful relaxant if taken in a little hot water at bedtime. I find 2 teaspoons in a cup of hot water the most pleasant, but adjust it to suit your own taste.

Jasmine tea

Serves 4–6

1 sachet ordinary tea*
4–6 cups boiling water
12 dried or 6 fresh jasmine flowers
twist lemon peel

* Experiment here with different brands until you get the flavour you like best

Pour the boiling water over the teabag, the peel and the jasmine flowers. Stand for 5 minutes.

Pour into small cups and serve without milk or sugar as an after-dinner digestive to settle and digest a heavy meal. It can be sweetened with a little honey if desired.

Jasmine bath vinegar

1 bottle white grape vinegar
1 cup jasmine flowers

Make several bottles of the vinegar while the flowers are at their springtime best.

Push the flowers into the vinegar (pour off some vinegar to make place for the flowers). Stand in the sun for 1 week. Strain off the vinegar and add more fresh flowers the following week. Strain again and add a few fresh flowers for identification. Use half a cup in your bath, or in your hair-rinsing water.

Jasmine pot-pourri

10 cups dried jasmine flowers (dry the flowers on newspaper in the shade, turning every day)
2 cups minced dried lemon peel
1 cup lightly crushed cinnamon sticks
jasmine oil

This is one of the headiest pot-pourris and you can combine it with other spring flowers but at its most basic it is really very special.

Mix all the ingredients. Store in a large jar and seal. Shake daily for 3 weeks. Adjust the fragrance if needed by adding more oil, then place in bowls or sachets.

Kattekruie

Bollota africana (formerly *Marrubium africana*)

ENGLISH   Cat herb
AFRIKAANS   Kattekruid

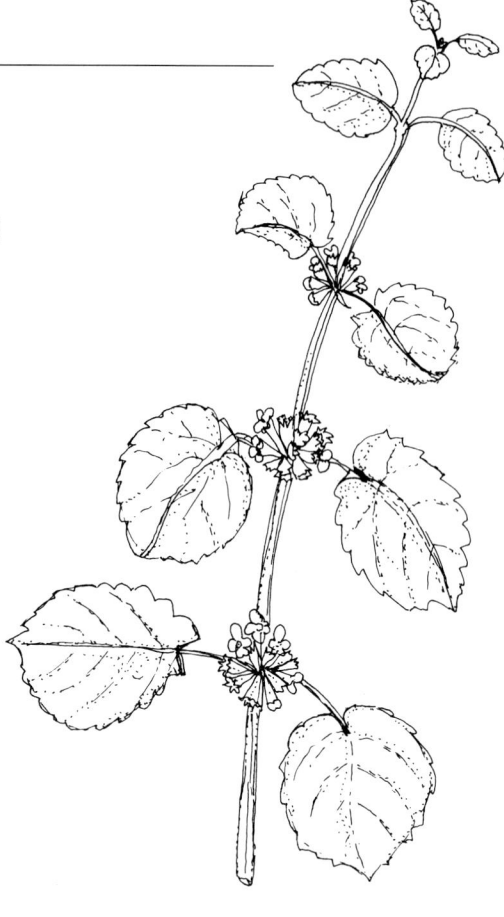

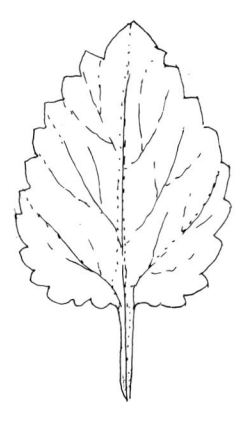

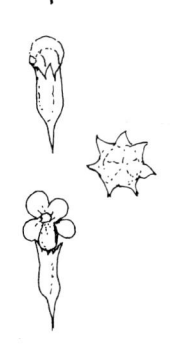

THE pungent kattekruie is similar to the European horehound *(Marrubium vulgare)* and, like its overseas counterpart, has a wide range of uses. It grows wild in the central and western Cape and is cultivated elsewhere. In the hotter parts of the country it seems to need partial shade, and it propagates easily from seed sown in spring. It is of medium height, usually up to 1 metre, and has flower spikes of pale pink, mauve or greyish white arranged in whorls around the stem, alternating with the small leaves which are serrated and hairy.

The whole plant is used in medicine by making an infusion or tea: ¼ cup of fresh leaves and a few flower whorls to 1 cup of boiling water is the usual dosage, steeped for 3 to 5 minutes, then strained and drunk. The dried herb is stronger and 2 to 3 teaspoons in 1 cup of boiling water, steeped for 5 to 6 minutes, then strained, would give the approximate strength to take as a medicine. This tea is an excellent treatment for coughs, colds, sore throats (as a gargle), influenza, asthma, bronchitis, colic, typhoid fever, hysteria and over-excitement. Half a cup of the tea can be given at intervals 3 or 4 times a day to help the above ailments.

The early colonists must have recognised the similarity between kattekruie and horehound, and were taught by the Khoikhoi to use the plant with other 'salies' or sages for a number of ailments. It was an extremely useful herb to have in the medicine chest, and they often gathered it and dried it to take on their treks. Often a cup of leaves and flowers would be steeped in a bottle of brandy, which was then taken a small dose at a time (usually 1 tablespoon in water) as an internal treatment for haemorrhoids, and a lotion made from the leaves was also used as an external application.

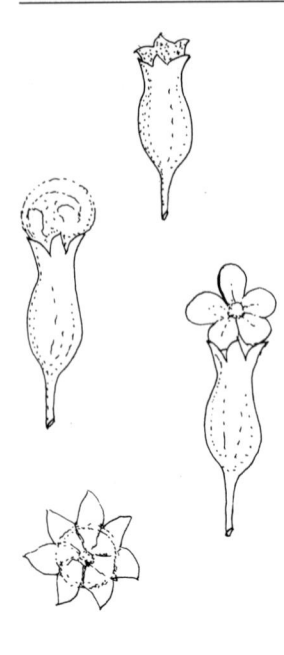

The lotion (made like the tea) is still used today for thrush — rinse the mouth with it at frequent intervals and dab it onto sores and wounds. Certain African tribes use it as a colic remedy and as a treatment for snakebite, as well as using the steamed leaves to pack around the chest for heart lung ailments, pneumonia, and severe colds and coughs. This poultice is still used today, but it does contain volatile oil and is considered toxic if too much is taken.

The kattekruie is still used and much respected in the Western Province, particularly for asthma, hoarseness, throat infections, heart conditions and sleeplessness. It is particularly helpful in the treatment of night coughing and chronic coughs and chest conditions, and if the tea is sweetened with honey and a little fresh lemon juice is added, it is not quite so bitter. I have taken half a cup 3 times a day for a week to clear an obstinate cough — within the first 2 days it had eased and by the end of the week it was completely gone. I'd been coughing for a month prior to that, so I am as much convinced of kattekruie's amazing healing qualities as the colonists were!

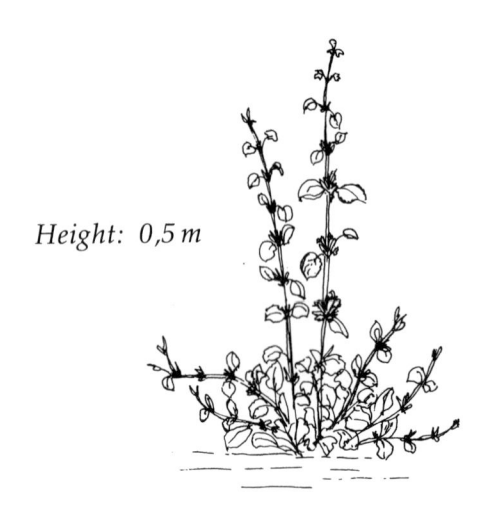

Height: 0,5 m

Kattekruie cough syrup

½ cup kattekruie leaves and a few flowering tips
1 cup brown sugar
6 cloves
½ cup lemon juice
½ cup water

Boil the ingredients together in a closed stainless steel pot for half an hour. Stand to cool. Strain through a fine sieve. Store in a screw-top jar in the fridge. Take 1 teaspoon diluted in a little warm water every 2 to 3 hours.

This will also help to bring down fever and ease the aches and pains of flu, and a little at night will help you sleep, taken in warm water.

Kiepersol
Cussonia spicata

ENGLISH	Common cabbage tree, cabbage palm
AFRIKAANS	Nooiensboom, sambreelboom, gewone kiepersol
LOVEDU	Mosetshe
PEDI	Lerole
SHANGAAN	Musenje, musengele
SHONA	Mufenje, musheme, mushenje
SWAHILI	Mtindi
SWATI	Umsenze
TSWANA	Mosetshe
VENDA	Musenzhe
XHOSA	Intsenge, umsenge
ZULU	Umsenge

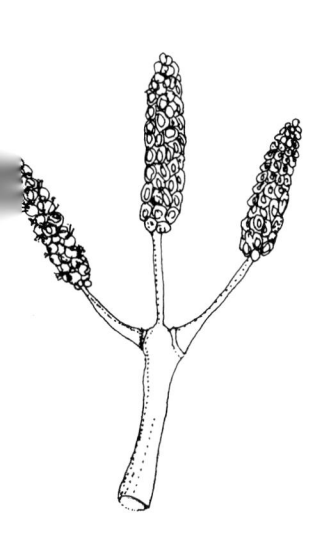

THE interesting shapes of the *Cussonia* species make them much sought after garden specimens. They all have attractive, cork-like trunks and are crowned with a tuft of beautifully shaped leaves, resembling a featherduster. The leaves in all the species are predominantly deeply lobed, palmately compound and are greyish green, and this unusual leaf structure is one of the most fascinating in the plant kingdom.

The flowering candlelike inflorescence is odd-looking as it emerges in thick spikes from the central tuft of leaves and the small greeny yellow flowers are packed along the spike during mid and late summer, turning into purply green fruits of about 5 mm in diameter at the end of the summer.

There are many species of kiepersol in South Africa but the best known and most widely distributed, in a broad band from the Cape through the eastern Cape, Natal, Orange Free State and the Transvaal, is *Cussonia spicata*. There are five forms of this species, which is often confusing, but the tree is so interesting that seeking out the different varieties makes a fascinating hobby.

The kiepersol needs good drainage, it does not like overwatering and for fast and luxuriant growth rich composted soil is necessary. It does need some protection from heavy frost, and in the colder areas of the country it chooses a protected spot to grow — near a rocky outcrop or on the warm side of a hillside or mountain, or in a group of protective trees.

Many African tribes respect and revere the tree for its many medicinal uses, and transplant seedlings with great care to their homes as they believe it will ensure healthy, fat children! The bark is shaved and rasped and used in hot water as a poultice, bound in place (often with an agapanthus or eucomis leaf) over the area to relieve cramp and muscle spasm. Some tribes wash a newborn baby with an infusion of the crushed and pounded root. This is done daily until the infant is taken out of the hut to view the outside world — usually about a week, but in some cases up to 6 weeks. This is believed to make the baby strong and to prevent skin rashes and pimples from forming. The cold tea is also used as a lotion for adolescent pimples and oily skin.

The roots are fleshy, tender and fatty and are eaten as a survival food in times of scarcity, and the Zulu use a drink made by shaving pieces of root into hot water — usually 1 cup of pieces of root to 4 cups of water. This is left to draw, strained when cold and taken in small, frequent doses to treat malaria and to bring down fever. The bark is also used to treat malaria and fevers taken in tea and as a wash.

The flowers, dried and ground, are added to snuff or tobacco to give it more taste. The thickened stem at the base of the flower is considered to be helpful, taken in the form of a tea, as an emetic for nausea and biliousness. The root can also be used in this way, and is considered to be an effective treatment for venereal disease. Pounded pieces of root are worked into a paste with warm water and applied to the area twice daily, and small doses are taken internally daily until the condition clears.

The Xhosa chew underground portions of the root for strength and virility, avoiding the portions above ground as these are bitter. The root has a succulent sweetness and is sucked and chewed like sugar cane.

Some tribes use the leaves in a weak tea as a colic remedy, and some apply pulped leaves to blisters on the feet to ensure quick healing of the raw area. In Zimbabwe the leaves are also used as a fish poison, but are much enjoyed by cattle and goats and in times of drought are a very important stock feed.

The kiepersol is offered for sale in nurseries, and also germinates easily from seed, which is ripe when the whole inflorescence falls from the tree in early winter. In my frost-free herb garden the trees have reached up to 6 m in height and are constantly admired. Ideally they should be grown as specimen trees to get the most from their unusual shape.

The large leaves pressed between layers of newspaper make excellent dried material for flower arrangements and dry to a pale green, remaining so for years. If they are placed in the sun once they are pressed they will turn a pale cinnamon colour and are much sought after by florists.

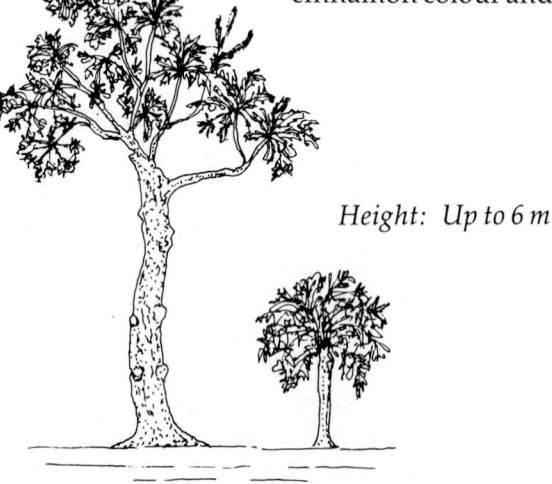

Height: Up to 6 m

Knobwood

Zanthoxylum capense
(formerly *Fagara capensis*)

ENGLISH	Small knobwood, wild cardamom, fever tree
AFRIKAANS	Knopdoring, kardamom, perdeboom, kleinperdepram
NDEBELE	Umnungumabele
SHANGAAN	Khinugumorupa
TSWANA	Monokwane
XHOSA	Umnikandiba, umlungumabele
ZULU	Umnungwana

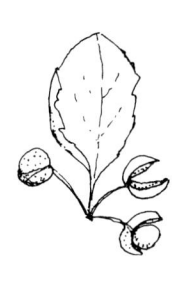

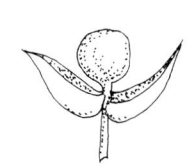

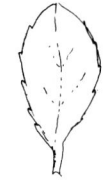

THE small knobwood is a small, attractive, much-branched tree that grows in a variety of habitats along the Cape coast and inland through Natal, into the Transvaal and up along the east coast of Africa into Zimbabwe. It has small, glossy, sweetly smelling compound leaves with the smallest leaflets at the base of the stem and the biggest leaflets at the tip. The grey knobbly bark has thorns or bosses, and the branches have thorns along them too.

In summer small, inconspicuous, pale green flowers in sprays emerge with a sweet scent, and these turn into clusters of small, reddish brown fruits at the end of summer. Each fruit splits open to reveal a black pip which tastes sour, lemony and acrid, leaving a burning sensation that lingers on the tongue for some time.

The chewed fruit is an old folk remedy for treating colic, flatulence, stomach aches, pains and cramps, and ripe berries are often dried and stored for this purpose. The berries were sometimes steeped in brandy and a teaspoonful taken for flatulence and colic and cramp in the legs. The brandy acts as a preservative, so this was kept on the kitchen or bathroom shelf for a quick treatment.

A tea made of the leaves (¼ cup of crushed fresh leaves to 1 cup of boiling water, stand, steep for 5 minutes, then strain) may be taken for stomach upsets, diarrhoea, cramps and intestinal worms. The same brew is also excellent for colds, flu, fever and coughs, and is sometimes also used as a gargle. During the 1918 influenza epidemic a treatment which proved to be very effective was a tea made of equal quantities of knobwood and wilde als (*Artemisia afra*) taken at intervals throughout the day. Pour 6 cups of boiling water over 1 cup of wilde als leaves and 1 cup of knobwood leaves, stand for 5 minutes, then strain. Keep covered in a cool place and drink ¼ cup every hour for the day during the illness, or ½ cup morning and evening as a preventative.

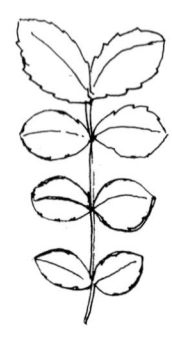

The bark, scraped and pounded and then chewed (this can also be made into a tea) is taken as a tonic for blood conditions, acne and skin eruptions (¼ cup of pounded bark to 2 cups of boiling water, stand until cold, then strain and drink ½ cup twice or 3 times daily). It is also an esteemed and ancient snakebite remedy. A piece of the bark is chewed and swallowed every 15 minutes until the swelling subsides while the victim is kept still and covered. Crushed and pounded bark is also applied to the bite. The Zulu rub the dried powdered bark into the cuts after lancing the bite and give powdered bark in water to the victim to drink.

A treatment for gall sickness in cattle using knobwood has long been used by both black and white farmers — the powdered bark is administered in water, often with ginger bush (page 87). The dosage is usually 1 cup of powdered bark in 8 cups of warm water, 1 cup being given at a time 4 to 8 times during the day. A cup of leaves and young shoots of the ginger bush could also be added to the 8 cups of warm water.

The fresh bark has pain-killing properties and the powdered bark or pieces of bark are used as a dressing for toothache. The hollow tooth is packed with the pieces, which is believed to loosen the tooth. It also forms the basis for medicines for tuberculosis (excellent for a long-standing chronic cough), bronchitis, paralysis and epilepsy. Rural whites have long used a tea made of the bark as a daily medicine for epilepsy.

One of the best known uses of knobwood is to disinfect anthrax-infected meat. Either the meat is cooked with the leaves, or else a tea made from the fresh leaves is taken after the roasted meat has been eaten. The leaves and bark have been used in the treatment of anthrax in a similar way as the gall sickness method.

The Tswana on my farm pound twigs, strip them of their bark and tie them together for use as a toothbrush as they believe it strengthens the teeth and gums. One of the old Tswana grandmothers showed me how she used the bark softened in hot water as a local application for lumbago and sciatica. She made a pad of the bark and bound it over the painful area with a piece of cloth and slept all night with it. In the morning the pain was gone. She also made a strong tea of the leaves and used this as a wash for infected sores, and as a lotion for insect bites, scrapes and grazes.

I have made a soothing massage oil for aching legs and feet by steeping a cup of fresh knobwood leaves in enough sweet oil (available from the chemist) to cover them. I leave this for 4 days, then strain the oil off. I then rebottle the oil and use this as a rub for aches and pains, and around the heels, and find it excellent. My garden staff frequently ask for it, as they too find it very soothing for aching muscles, particularly during spring planting!

A kierie or walking stick made of knobwood is considered a protective charm. The herdboys search out strong branches which they cut into kieries, often doing a brisk trade in them.

I love using the fragrant leaves in wild flower pot-pourris. The seeds crushed or whole also absorb the essential oil, so are excellent in a pot-pourri.

This lovely tree is so worthwhile growing in the garden — do search it out and plant it where you can smell it often. A crushed leaf clears the head!

Height: 4 m

Lemon bush

melkbos
Asclepias fruticosa

melkbos
Asclepias physocarpa

pennywort

Nastergal

Nutgrass

mother-in-law's tongue

Insect-repelling pot-pourri

4 cups dried knobwood leaves
3 cups dried wild lavender leaves
3 cups dried wilde als leaves
1 cup dried buchu leaves (particularly pleasant is the lemon-scented buchu)
few knobwood berries
2 cups dried minced orange and lemon peel
1 cup mixed crushed cloves and nutmeg
essential oil e.g. lavender or lemon oil

Mix all the ingredients together. Add a little oil. Shake well in a sealed crock or tin or jar daily for 2 weeks. Add more oil or spices at the end of the time. Shake well 2 or 3 days more, then place in sachets or bowls around the house or in cupboards. Revive with a little essential oil from time to time.

Lavender tree
Heteropyxis natalensis

ENGLISH	Wild lemon verbena, wild verbena tree
AFRIKAANS	Laventelboom
SHANGAAN	Tatazan
TSWANA	Thathazane
ZULU	Umkhuzwa

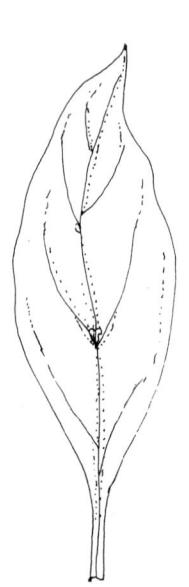

THE wild lavender tree is a delight to have in the garden. The leaves are bright, glossy and exquisitely perfumed when they are crushed. The shape of the tree is beautiful, even when the tree is tiny, as its final shape starts off in miniature. It always looks fragile and dainty, and lovely at all times of the year. It is semi-deciduous, and in autumn the leaves turn all shades of orange, gold and dark red, and in some areas never seem to fall until the new leaves appear in early spring. It is fast growing, with slender branches that are covered in a light bark. Several nurseries offer it for sale and all it needs is full sun and a deep compost-filled hole.

The Zulu have used the lavender tree medicinally and otherwise since time immemorial. A fragrant tea is made by pouring boiling water over the leaves and stems (usually 1 cup of leaves and stems to 6 cups of boiling water), then leaving this to stand until it is luke warm. The tea is then strained and a little drunk to ease heartburn, colic and flatulence. It is also considered to be a tonic tea that is strengthening and reviving, and is considered to be of special importance for old people, to refresh travellers, and to help a new mother regain her strength if her labour has been prolonged. The Venda and Shangaan drink the tea for digestive upsets and colds and use the tea as a wash. The Zulu are fond of making a strong brew by boiling up the leaves and twigs in water and using this as a wash or lotion, particularly under the arms, as the brew lightly perfumes the skin.

Crushed leaves are added to Vaseline or mutton fat and used as a nourishing treatment for cracked heels and dry skin on the legs, feet and hands and the Zulu and Venda rub the crushed leaves onto the feet to help tiredness and to soften callouses. The Tswana on my farm rub the leaves onto their pillows and blankets to keep insects away, and dry and powder the leaves to sprinkle as a talc or dried perfume amongst clothing.

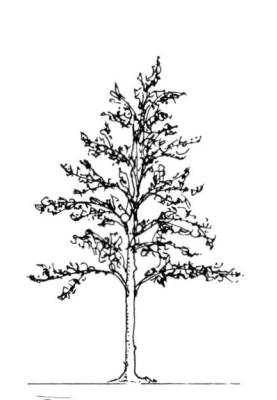

The leaves and twigs boiled in water make an excellent inhalent to clear the nose and chest if a towel or blanket is held tentlike over the head while bending over the steaming pot, and the Venda and Zulu use this method to treat a bleeding nose and gums. They also use the brew as a mouthwash for mouth and gum infections, for toothache and after tooth extraction.

The Zulu also make a strong brew of the dried powdered leaf which is then used as a drench for tick-infested cattle, donkeys and goats, and for treating sores and scratches and infected tickbites on animals.

I use the dried leaves as a base for a wonderful wild flower pot-pourri which I use as a cupboard freshener and find the leaves retain their fragrance when mixed with a little lavender oil.

Height: 4–6 m

Lavender tree pot-pourri

4 cups dried lavender tree leaves
2 cups dried pompon tree flowers and seed pods
1 cup dried wilde als
½ cup cloves and cinnamon sticks, lightly crushed
¼–½ cup dried anise buchu leaves
1 cup minced dried lemon peel
lavender oil

Mix all the ingredients, and keep sealed in a large tin or bottle. Shake daily for a fortnight, adding more oil at the end of the time if necessary.

Tie into small sachets or bags, or place in open bowls throughout the house.

Lavender tree bath freshener

2 cups fresh lavender tree leaves and twigs
piece of old pantihose or square of muslin
1 cup coarse sea salt
1 cup large flake oats

Tie all the ingredients together in a muslin square or knot into a piece of pantihose. Toss this into the bath under the hot tap and use it to wash with. You'll enjoy both the fragrance and the invigorating bath!

Lemon bush
Lippia javanica

ENGLISH	Fever tree, fever tea, wild tea
AFRIKAANS	Koorsbossie, maagbossie, beukesbos
LOVEDU	Mosungwani
MANYIKA	Mosukubyane
MASAI	Ol magirigirieni
NDEBELE	Umsuzwana
SHANGAAN	Musuzwane, m'suzwani, m'suzungwani
SHONA	Mumara, mosukubyane
SWATI	Musutane, mutswane
TSWANA	Bokhukhwane
VENDA	Musudzungwane
XHOSA	Inzinziniba, umzinzinibe
ZULU	Umsuzwane

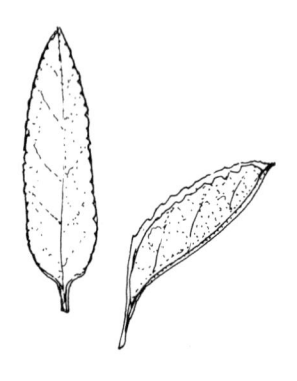

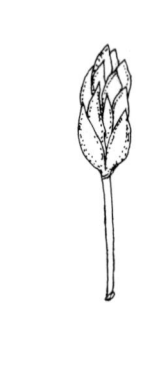

THE lemon bush is one of the most intensely aromatic of the indigenous shrubs. It grows from the Cape up into tropical Africa, in the open veld, in the bush, and at the edges of forests. It is much used medicinally and in some areas is used as a perfume, and like its close cousin the beautiful lemon verbena *(Aloysia triphylla)*, is a prized ingredient in pot-pourris and cupboard fresheners.

Lemon bush tea is a favourite cough and cold remedy (1 tablespoon of fresh leaves in 1 cup of boiling water, steep for 3 to 5 minutes, then strain and drink). This tea sipped a little at a time at frequent intervals is also excellent for bringing down fevers, and with wilde als it is also used for influenza, measles and chest ailments. The cooled tea is used as a lotion to dab frequently onto the forehead, and as a tea sipped at frequent intervals is still believed to be the best fever treatment for malaria.

The tea is also excellent for treating rashes, scratches, stings, bites and heat rash and is used by the Tswana to treat scabies, head lice and scalp infections. For these last three the tea is made very strong and is used as a lotion and dabbed on frequently.

Some tribes inhale the smoke from burning lemon bush leaves and stems for asthma, chronic cough and pleurisy, and the Zulu and Ndebele infuse the leaves in hot water and use this as a wash to ease chest ailments and coughs; sometimes soaking a piece of cloth in the tea and then wringing it out and binding it over the chest.

The crumpled leaf stuffed into the nose to stop bleeding and to ease colds is an old remedy. The leaf is also warmed in fat or Vaseline and this is used as a rub for sore muscles. The easiest way of doing this is to warm ½ cup of Vaseline or aqueous cream in a double boiler, and add ½ to ¾ cup of crushed lemon bush leaves. Simmer for half an hour. Strain and cool. Keep in a screw-top jar, and use this for sprains, bruises and aching muscles.

The Xhosa use the lemon bush to disinfect anthrax-infected meat, and some tribes store meat with lemon bush sprigs between the pieces, as they believe it will keep the meat fresh for longer periods. It will certainly impart a delicious flavour!

My favourite way of using the lemon bush is to tie a small bunch of leaves in a large handkerchief or piece of muslin and drop this into my bath under the hot tap as it runs in. I find it wonderfully soothing and relaxing after a busy day.

Fresh sprigs placed in cupboards and amongst clothes and linen will freshen and perfume them, and the leaves retain their scent for a long time.

The plant grows easily from cuttings in a sunny position, and seems to do well in all soil types. This is one of the most rewarding of all the wild herbs to grow and is one of the best ingredients for pot-pourris as it retains its glorious scent for years.

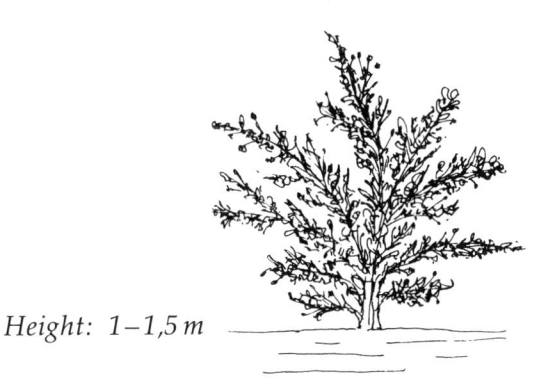

Height: 1–1,5 m

Maidenhair fern
Adiantum capillus-veneris

ENGLISH   Fine-leaf maidenhair
AFRIKAANS   Nooienshaarvaring, swart vrouehaar
SOUTHERN SOTHO   Pata-lewana, pata-mawa

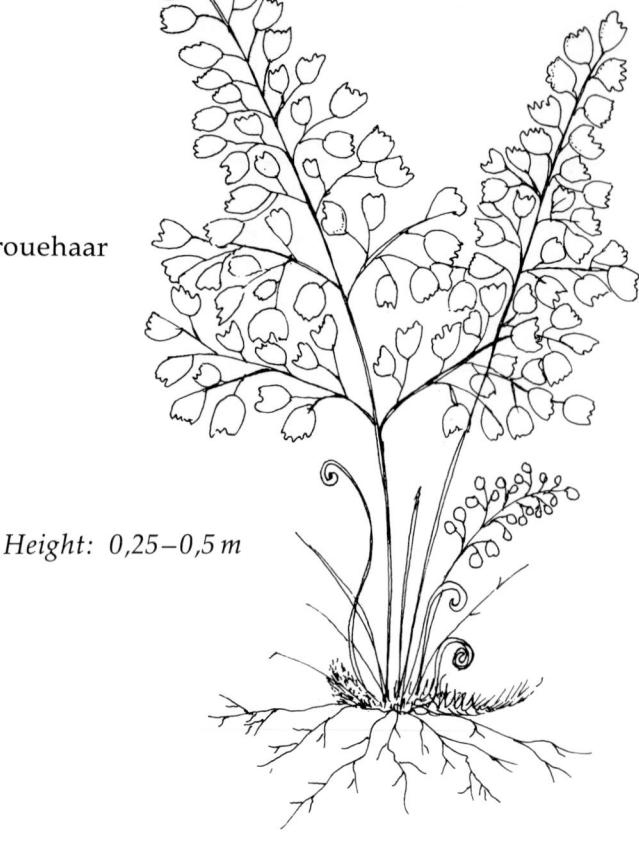

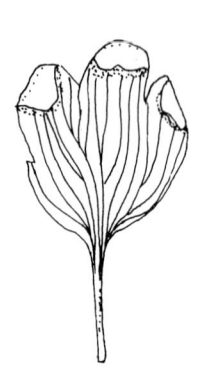

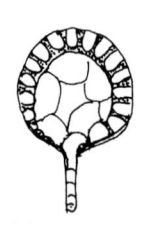

Height: 0,25–0,5 m

THE exquisite maidenhair fern is widely distributed throughout the world and is also indigenous to South Africa, where it is found in all the provinces in damp, shady places. It is a favourite potplant in many homes and in some temperate areas it is grown as a garden plant. Tiny spores resembling green confetti on the underside of the leaves seed themselves in every moist nook and cranny, and the established mother plants continuously send out more rhizomes, making the clump bigger and bigger. These can be divided in spring for although the fern needs to feel its roots against stones or rocks in the garden, or against the sides of a pot, if the roots appear above the surface of the soil this is a sure sign that it needs to be transplanted into a bigger container or to be given more space.

   The ferns need shade and moisture in the garden, and a little protection from wind and intense cold. I have found maidenhair growing along the banks of rivers where they seem to thrive in dappled shade, not minding the cold winds or summer heat, but as a potplant they don't like a draught or too warm a room. One has to experiment with them to find the right position.

   Medicinally, the maidenhair fern was a much used treatment for coughs, colds, pleurisy, bronchitis, catarrh and respiratory ailments. In Europe it was made into a syrup called 'capillaire' which was a favourite drink of the ladies of the court. This was made by boiling the fern fronds in sugar and water until it thickened, and this was then diluted with water. It had a strong, very typical taste, and was considered to be a fashionable social drink.

   I have had much pleasure from making capillaire, which I use for coughs and colds, and have taught many of my students to make it. All of them, without exception, have found it wonderfully beneficial. I grow maidenhair ferns prolifically in a great plant house I have created, and they come to pick fronds for their winter coughs.

The colonists coming to South Africa and finding this familiar fern growing wild would have used it to make syrups and drinks as they had in Europe, while the indigenous people were already making their own medicines with the fern — also using it for chest complaints and colds. One wonders who knew about it first!

The Southern Sotho have smoked the dried leaves for many generations to ease head colds and chest ailments. The Zulu boil up a quantity of leaves in water until it is gummy and thick, and this is taken a little at a time for chest conditions, coughs, bronchitis and a tight chest. The Tswana and Venda sangomas use it boiled in sugar water to give to children with colds and burn the leaves on the fire to clear stuffy sinuses. Fresh fronds tucked into the shirt are also considered to be beneficial for chest ailments.

Young maidenhair fronds are edible in salads, and small sugared fronds make exquisite cake and pudding decorations — simply dip the frond into syrup and then into icing sugar and place it on the cake.

Pressed fronds make beautiful additions to pressed flower pictures, and I also use them in dried arrangements. They start off brilliantly green, changing to beautiful ambers and golds as they mature, and always look exquisite. If you use maidenhair fern in fresh flower bouquets and arrangements do not forget to burn the last 2 or 3 cm of the stems to make it last longer.

My favourite way of using maidenhair fern is as a hot night-time toddy when I feel chilled and overtired or if a cold is coming on.

Capillaire

3 cups white sugar*
3 cups water
2 cups maidenhair fern fronds, well pressed into the cup
juice of 3 lemons and little grated rind (optional)

* I use natural brown treacle sugar for the medicine, but this makes the syrup brown, which some may not like.

Dissolve the sugar in the water and boil for 3 minutes, then add the fern fronds and stir with a wooden spoon until they are all submerged. Boil gently for about half an hour until the syrup thickens slightly.

Add the lemon juice if desired and a little grated rind. Boil for a further 10 minutes, then allow to cool.

When cool strain, bottle and seal well. Use this as a cough syrup, a teaspoon at a time taken frequently, or dilute with water or soda water for a wonderfully refreshing, healthy drink.

Maidenhair fern hot toddy — Serves 1

1 cup boiling water
½ cup maidenhair fern fronds
4 cloves
½–1 teaspoon powdered ginger
2 teaspoons honey
juice of half a lemon

Pour the boiling water on the fern, the cloves and the ginger. Allow to draw for 5 minutes, stirring gently. Remove the fern, add the honey and lemon juice, stir well, and sip slowly. You'll sleep beautifully!

Marula

Sclerocarya birrea subsp. *caffra*

ENGLISH	Cider tree
AFRIKAANS	Maroela
NDEBELE	Umganu, iganu, ikanyi, umkano
PEDI	Morula, merula (plant), lerula, marula (fruit)
SHANGAAN	Inkanyi
SHONA	Marula, mafuna, mufura, muganu, mukwakwa, mupfura
SOTHO	Murula
SWAHILI	Mugongo
SWATI	Umganu
TONGA	Tsua, tsula, umganu
TSWANA	Morula
ZULU	Umganu

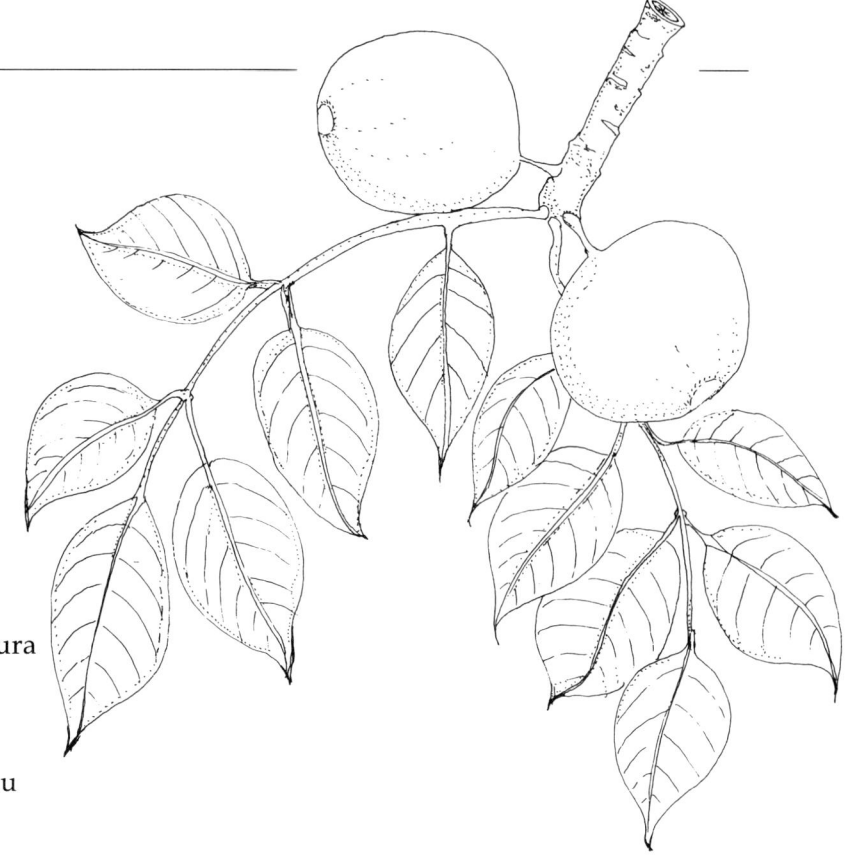

THE marula is a huge and beautiful deciduous tree found in the warmer, frost-free areas of South Africa, often in somewhat poor, shallow soil. Their abundance in some areas is often due to the fact that this tree is so highly esteemed, valued and revered by the African and is therefore seldom cut down. The species extends widely into Central Africa, probably owing to the fact that the fruit with its large seed or pip is much loved by elephants, monkeys and baboons and the seeds germinate well having passed through them as they move from one area to another.

The tree is beautiful in all seasons and at all stages of its growth. It has a rough bark when it is mature but the pallid bark in young trees is smooth and unnotched. The leaves are compound, 5–13 foliate and the small, nondescript flowers are yellowy green and sweetly scented. The wood is light, soft and easily worked but in larger articles warps badly, and is often infested by wood borer. The traditional 'stampblok' is often made of marula wood as it seldom cracks, and is used with much vigour in the hollow receptacle to pound the grain. The wood is also carved into spoons, plates, bowls, drums and headrests. In curio stalls and shops in the eastern Transvaal one often sees beautifully carved chains up to 3 m in length with various sized links — some even 20 cm long — that have been carved from marula wood.

The fruit ripens in March and is much prized not only by humans but by elephants and baboons as well. When it is ripe it is soft, round and butter yellow. The greater part of the fruit is taken up by the pip, which contains an edible 'meat' rich in oil. This valuable and tasty protein can be used in cakes, biscuits, puddings and sweet dishes. The seeds themselves are hard, with three holes at one end which can be prised open and the 'meat' removed, leaving small skull-like seed shells which are used by African children in a game, or strung into a necklace or belt which is believed to protect the wearer.

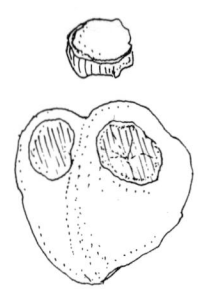

Zulu women boil the pulp from the seed into a mass with water until an oily residue emerges. This precious substance is used to rub onto skins and belts to soften them, and is also used as a beauty treatment for cracked skin on the hands, feet and lips.

The oil from the seed has been commercially exploited and at one time it was sold in Madagascar as 'sokoa oil', which was rich in protein and iodine. The cooked kernel is excellent as a food, tasting a bit like a peanut, and the farming community in the Transvaal and Natal where the trees grow abundantly hold 'marula feasts' where all sorts of delicious jams, jellies, sweetmeats, drinks and pickles made from marulas are for sale.

Medicinally the marula is one of Africa's most important trees. A fragrant tea made from the bark (usually 1 cup of pieces of bark boiled in 3 litres of water for 3 hours, then left to cool, strained and bottled) is used in small doses to treat diarrhoea, dysentery, malaria, gonorrhoea, or is taken as an enema for abdominal upsets.

In the treatment of malaria both prophylactically and curatively the bark is gathered in spring just before the leaf appears on the tree, usually in late September, preserved in brandy, and taken daily in small doses 3 to 6 times a day, or the bark is ground and powdered and taken a teaspoon at a time in water twice a day. Many rural whites and blacks strongly believe this to be a remarkably effective treatment for malaria, although medical tests have not found it to be particularly effective.

The inner bark is astringent, and this is boiled and applied as a poultice to ulcers, skin eruptions and smallpox pustules.

The Zulu and the Tswana regard the fruit as a potent insecticide and use it to make a strong infusion for bathing tick-infested cattle or goats. They boil the fruits for an hour in enough water to cover them and splash the cooled brew on the animals, or they split open the fruit and rub the juicy peel over the ticks and infested skin.

The outer skin of the fruit is pungent-smelling and the fruit has been described as tasting like a cross between an apple, a litchi and a gauva! Its smell is so strong that one or two fruits will scent a room for days. My children used to collect the ripe fruits from under the marula trees growing near their Rustenburg school and lovingly store them in their school cases until they could be eaten. Their books would smell of marulas for weeks afterwards!

The fruits are rich in vitamin C and the fermented fruits are used by blacks to make a highly intoxicating and nourishing beer. The Tswana and Venda have a somewhat disconcerting practice of not working until the beer-making ceremony — known to the locals as the 'marula binge' — is over, and the drinking goes on steadily for several days!

The marula fruit is considered to be intoxicating if consumed in large amounts, and the drunkenness of elephants and baboons after gorging themselves on the fruit has been the cause of great amusement. Interestingly, it has been put forward that the marula may not be intoxicating but may work on the nervous system, causing temporary nerve paralysis. This theory is borne out by an incident that occurred when a large group of children once camped on my farm. One morning several children ate 20 or 30 marulas, and some hours later started to show marked signs of drunkenness — heavy eyes, slurred speech, erratic gait and uncoordinated movements. In great concern the children were rushed to hospital where the diagnosis was not intoxication but partial nerve paralysis. It wore off several hours later.

The marula bark exudes a gum which is rich in tannin, and this can be dissolved in water and mixed with soot to make a useful ink for drawing and writing.

Height: 10 m

Marula jelly

Wash and pick ripe marulas and put them in a saucepan. Cover with water and boil slowly for about 2 hours to extract the juice from the fruit. Strain the juice through a cloth or a sieve lined with a double layer of cloth.

Test the juice for pectin in the following way: add 10 ml of strained marula fruit juice to 10 ml methylated spirits in a small glass. Do not stir, but whirl the glass in order to mix the two liquids. Leave for a few moments, then pour carefully into a saucer.

Note the firmness of the clot of jelly, and add sugar accordingly: 250 ml sugar to 250 ml juice if one solid clot of jelly is formed; 200 ml sugar to 250 ml juice if two clots are formed; and 200 ml sugar to 250 ml juice if numerous small clots are formed, and in this case lemon juice must be added to the jelly just before it is ready to be bottled.

Place over heat and stir to dissolve the sugar before the jelly starts to boil. Boil fairly rapidly until a little of the mixture put in a saucer forms a firm jelly when cool. Leave to stand for a few minutes before bottling, and seal.

Serve with meat, poultry and baked beans, or use as a jam on bread or with rice puddings.

Melianthus

Melianthus major, M. minor

AFRIKAANS	Kruidjie-roer-my-nie
TSWANA	Ibonya
XHOSA	Ubulungubemamba, ubutyayi
ZULU	Ibonya

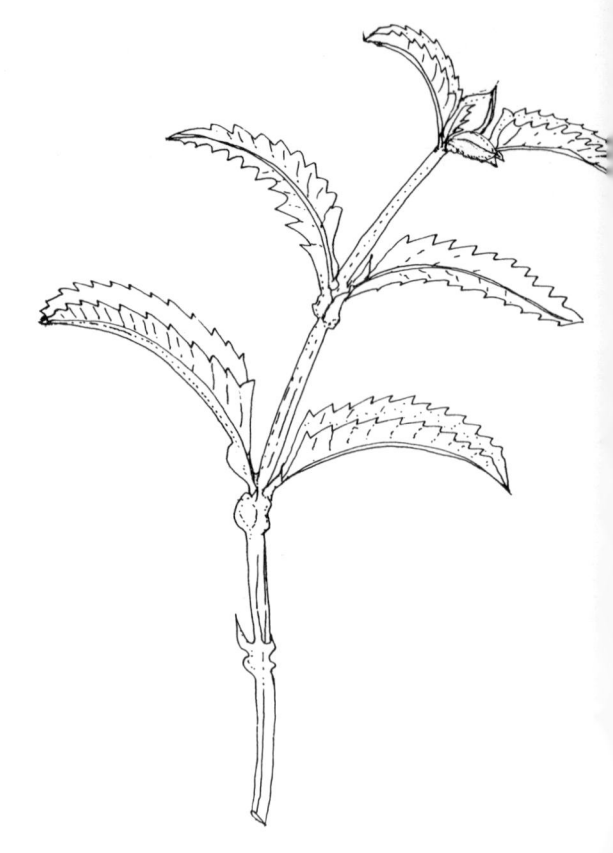

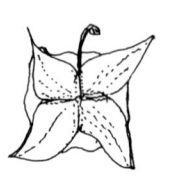

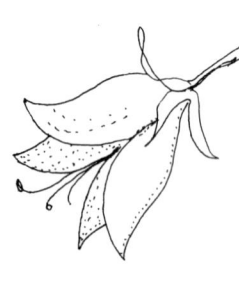

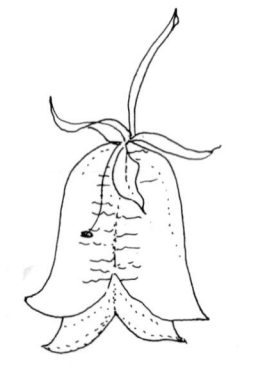

MELIANTHUS is a fairly common garden shrub. *Melianthus major* grows up to 1,5 m high and makes a dramatic display of deep red-brown flower spikes in summer, which contrast beautifully with the serrated grey leaves. *M. minor* is a smaller bush, growing about 0,5 m high. *Melianthus* comes from two Greek words meaning honey and flower, and the copious amounts of rich nectar that the flowers secrete is much loved by bees and wasps and by sunbirds. I had a family of exquisite whitebellied sunbirds nesting near my herb garden who spent all day happily visiting flower after flower of my melianthus.

The Khoikhoi were the first to use melianthus medicinally, and they must have taught the colonists its uses. Until fairly recently the external use of the leaves was recommended by doctors — particularly in the Cape — for aches and pains, backache and rheumatism, as a gargle for sore throats and mouth infections, and amazingly even for snakebite. I have used warmed leaves over an aching lower back, held in place with a crepe bandage, and found this compress to be soothing and comforting. It can also be used for boils and abscesses to bring them to a head.

An excellent wash for wounds, ulcers and sores is made by boiling 6 big leaves in 4 litres of water for 10 minutes, then allowing the mixture to stand and cool until it is comfortable enough to use as a wash. A piece of lint or gauze dipped into this solution can also be used as a compress over sores or ulcers, but needs to be changed frequently.

The leaves are toxic so take care with this plant, and never use it internally. Its common name 'kruidjie-roer-my-nie' — 'little-herb-don't-touch-me' — alludes to its unpleasant smell.

I have found the dried winged seeds very useful in a pot-pourri as they seem to absorb the fragrant oils, and they lose their rather pungent smell once they are dry.

An old coloured woman who lives in the Saldanha Bay area where melianthus grows easily told me she uses the dried flowers and leaves with other wild herbs — like Hottentotskooigoed and buchu — to keep insects out of her cupboards.

Crushed and dried melianthus leaves sprinkled into ants' holes is helpful in getting rid of them and a strong brew of melianthus leaves, wilde als leaves, khakibos and klipdagga leaves (equal quantities of each steeped overnight in enough boiling water to cover them) is an effective spray for aphids and fruit flies. Remember to wash the fruit or vegetables extremely well before eating!

Melianthus grows easily from seed and from suckers, but give it a lot of space as it can be very vigorous. I start new plants every alternate year to replace my old ones as they tend to become rather tatty after a time. Cut back the old stems and let the new suckers coming up under the branches develop.

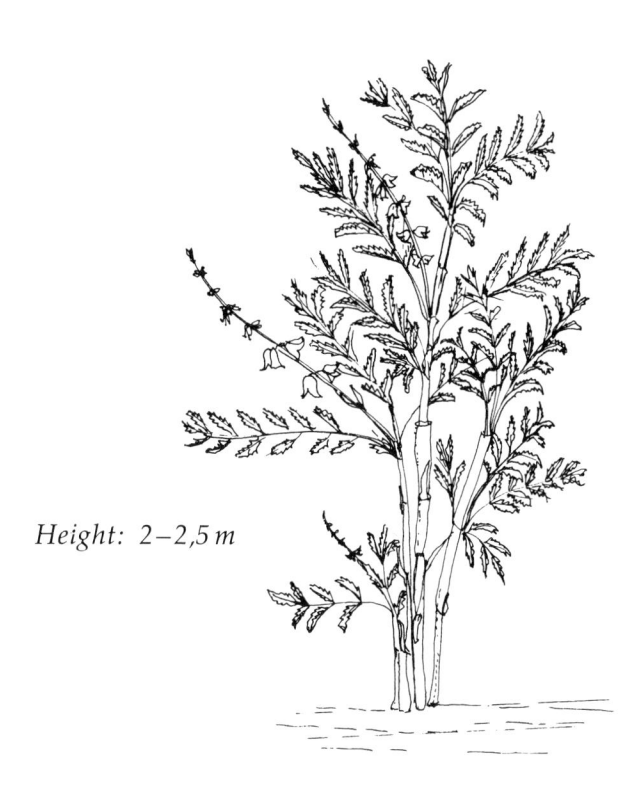

Height: 2–2,5 m

Melkbos
Asclepias fruticosa

ENGLISH	Milkweed, wild cotton, cotine
AFRIKAANS	Tontelbos, gansies
SOTHO	Modimolo
ZULU	Usinga, lwesalukasi

A. fruticosa

A. fruticosa

Height: 1–2 m

THIS common South African weed is found along roadsides throughout the country and is considered to be an exotic garden plant overseas. As it sows readily from seed and seems to adapt quickly to all soil types and climates, it is now in cultivation far and wide.

The white milky latex gives it the name melkbos or milkweed and this latex is considered to be an excellent treatment for warts if it is dabbed frequently onto the area.

The Zulu make a tea of the leaf for children with tummy ache and diarrhoea, and the dried and powdered leaf has been used as a snuff for treating tuberculosis. The Sotho and Tswana use this tea as a purgative. The dosage varies between ¼ to ½ cup of fresh leaves infused in 1 cup of boiling water. This is allowed to steep until it is luke warm and is then drunk.

A strong brew of leaves in water is a respected farm remedy for easing distemper in dogs. One cup of fresh leaves is boiled in 1 litre of water for 5 minutes, then left to cool and strained. The cool water is given to the dog to drink, even if he takes only a little at a time. Keep the bowl available all day.

In the Orange Free State the fresh or dried roots are made into a tea and used as a remedy for diabetes. The approximate dosage is 1 tablespoon of chopped fresh roots to 1 cup of boiling water, stand, steep for 10 minutes, strain, then drink 2 to 3 times daily. Or use 2 teaspoons of chopped dried roots boiled in 1 cup of water for 20 minutes — then strain and drink 3 times per day. I have not used this, so as in all home treatments be sure to consult your doctor before you start. Fascinating though these remedies are, I am always cautious, and urge you to be too.

Although the melkbos is browsed upon by cattle, sheep and goats too much can be poisonous, so farmers keep their pastures as free of the bush as possible.

Asclepias physocarpa is an attractive species of melkbos with large, balloon-like seeds. The milky juice from this species is also used to dab on warts, and plants cut fresh and stuffed into mole holes is said to keep the rodents far away. I have grown a thick hedge of melkbos and have had very little trouble from moles — but that may only be co-incidence! Several farmers in the Transvaal use melkbos in this way. It is worth trying if you are plagued by moles.

The plant is easily propagated from seed, and grows into an attractive shrub. I trim the lower branches so that it grows into a standard. In midsummer the balloon-like seeds are at their best, and these are much sought after by florists. I pick long sprays of the lime green, hairy, ball-like fruits, and use them in attractive green and white arrangements. Remember to plunge 10 cm of the stems into boiling water for 5 minutes. The balloons also look lovely floated in glass bowls and the dried seeds, when they burst open, make wonderfully soft and silky pillow stuffings. Many tribes use all varieties of melkbos seeds to stuff pillows, and this is a much sought after luxury. Rather large quantities of seed down are needed to stuff a small pillow but it is worth the effort — it is a delight to rest on a delicate melkbos seed pillow!

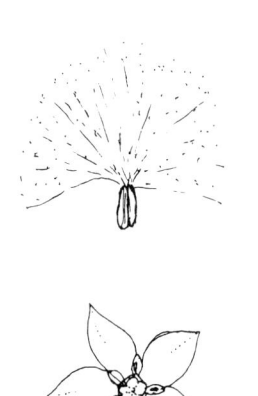

A. physocarpa

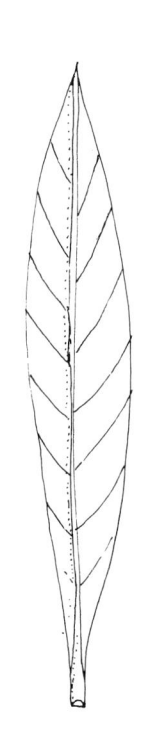

Height: 1–2 m

Mother-in-law's tongue
Sansevieria hyacinthoides (formerly *S. thyrsiflora*)

ENGLISH	Piles root, piles bush
AFRIKAANS	Haasoor, aambeiwortel
NDEBELE	Isikusha
TSWANA	Mogaga
XHOSA	Isikholokhoto
ZULU	Isikholokhoto

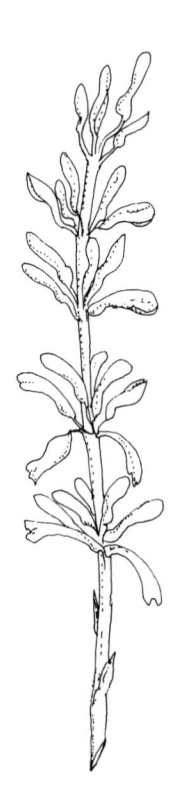

Height: 0,5–0,75 m

THIS remarkable plant is much used and respected medicinally. There are a large number of different plants falling under the heading *Sansevieria hyacinthoides* and it seems that most are used medicinally. All the varieties have thick, leathery, sword-shaped leaves. Some have a narrow yellow margin, and the leaves are dull grey-green with white speckled markings. They grow wild in the forests of the eastern Cape, on hillsides in the Transvaal where they are lower growing with a folded grooved leaf, and under trees and against rocks in Natal.

The best-known African use of mother-in-law's tongue is in the treatment of earache. A cut leaf is heated and the juice dripped into a teaspoon and dropped gently into the ear. A piece of warmed leaf is held behind the ear as well until the pain subsides. Some tribes also use the leaf in this way for toothache.

The juicy root is a well known treatment for haemorrhoids and intestinal worms. The root can be eaten raw, or boiled in water or milk and a cup drunk for the relief of piles. The dosage is 1 tablespoon of chopped raw root, chewed well and the juice swallowed — the rest discarded — each morning until the piles subsides, or until the worms are removed. For the decoction, 2 tablespoons of chopped root boiled in 1 litre of water for 10 minutes, then strained and cooled, is used in the prevention of miscarriage where there has been a history of miscarriages during pregnancy.

The Tswana use the finely sliced root of the Transvaal variety as an external application to haemorrhoids and varicose veins. They also cook the root and eat it with other vegetables for easing the pains of childbirth and soothing contused veins.

Pelargonium radens
Rasp leaved geranium

Pelargonium betulinum
Camphor leaf geranium

Pelargonium peltatum
Ivy leaf geranium

Pelargonium quercifolium
Oak leafed geranium

Pelargonium cucullatum
Hooded geranium

P. triste
Sad geranium.

Pelargonium tomentosum
Peppermint geranium

Pelargonium bowkeri

Pelargonium fragrans
Nutmeg geranium

Pelargonium graveolens
Rose geranium

paintbrush lily

pig's ear cotyledon

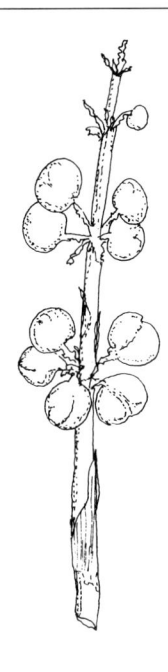

The Zulu and the Tswana consider it to be a protective plant and drink a little of a cold infusion to protect themselves against lightning, as well as if any member of the family is believed to have been bewitched or influenced in any way.

The leaf is pounded, washed and twisted into a strong and durable string or rope, and is used in building, for weaving baskets and mats, for binding fractures in humans and animals and making fishing nets. During circumcision ceremonies certain tribes make a ceremonial garb from the woven fibres, believing the boy wearing it will be endowed with strength, virility and courage in his adult years.

The leaf fibre can also be made into an excellent paper with a fine and smooth texture, and for those who wish to experiment with the ancient art of papermaking, this is an excellent plant to use.

The plants are easy to grow in both sun and partial shade, and propagate by roots and suckers. They have beautiful racemes of whitish flowers in spring and summer and are able to withstand frost and drought.

They have been exported overseas for many years where they are much sought after as hothouse plants and pot plants. A fascinating collection could be made of the various species of *Sansevieria* and as many nurseries offer the plants for sale one could easily take up a new hobby.

Nastergal
Solanum nigrum

ENGLISH	Black nightshade, woody nightshade, garden nightshade, hound's berry, common nightshade, deadly nightshade, nightshade, pretty morel, stubbleberry
AFRIKAANS	Nagskade, nagskaal, galbessie
LOVEDU	Mofye
NDEBELE	Ixabaxaba
PEDI	Lethotho
SHANGAAN	Kophe
SHONA	Musaka
SOTHO	Seshoa-bohloko, sehloabohloko, momoli
SWATI	Msobo, umsobo
VENDA	Muxe
XHOSA	Seshoa-bohloko, umsobo, umsobo-sobo
ZULU	Umsobo, ugwabha, isihlalakuhle, udoye, umaguqa, umgwaba, umqunbane

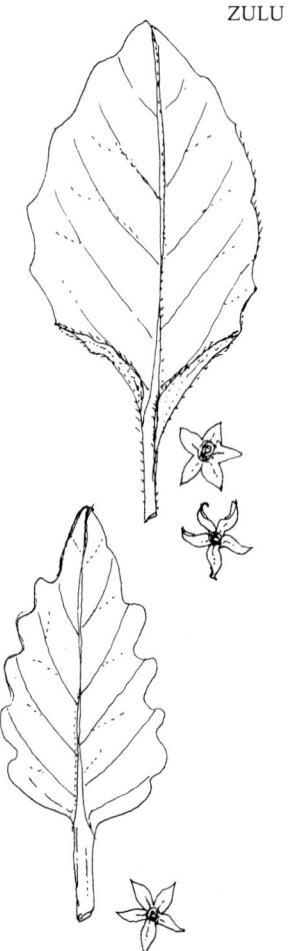

NASTERGAL grows in many places in the world and throughout South Africa, where it is a much respected medicinal plant. It is native to Europe and Africa and may have been introduced to South Africa — probably as far back as 1652 — where it has become naturalised, but there is some doubt as to its real origins. It has been used by the African people here for so many generations that there could well be an indigenous form.

It has many forms, but the one most often found in South Africa has an egg-shaped leaf with a wavy edge. The young leaves are often simple without the wavy margin, which sometimes causes confusion in young seedlings. The plant is spreading, fairly untidy, and usually grows to about half a metre in height and in width, but will grow larger if it has good soil and a lot of water. The small white flowers in clusters are followed by small berries, which are green at first, ripening to black. It grows in any soil, and enjoys full sun and can take some shade in hot areas.

The plant has been used for centuries for a wide variety of ailments. A tea made of the leaves (¼ cup of fresh leaves to 1 cup of boiling water, stand, steep for 5 minutes, then strain and drink) is used for fevers, malaria, convulsions, headaches, dysentery, diarrhoea, as a sedative, a wash for wounds, ulcers, and infected rashes, scratches and bites, and is administered as an enema for abdominal ailments. The leaves are also a soothing dressing for haemorrhoids, varicosities and bruises.

The leaves cooked as a spinach are considered to be blood cleansing, revitalising and energising, and can be cooked on their own or combined with other herbs such as nettle, amaranth species, purslane, sow's thistle and fat hen

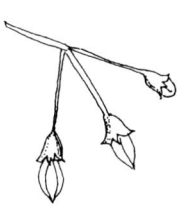

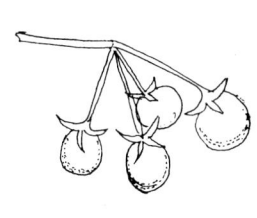

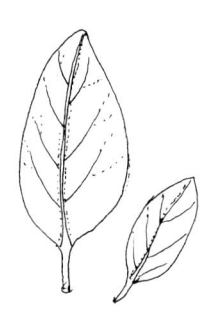

(Chenopodium album), all known to blacks as 'moroggo' or 'morog'. Thunberg recorded in 1772 that an excellent salve could be made by mixing the leaves of nastergal and sow's thistle (sydissel, *Sonchus oleraceus*) with lard or fat, and applying this to wounds and ulcers for cleansing and healing.

The leaf can be used as a pot herb, although I find it a little bitter. But added to spinach dishes, soups and stews it is nourishing and delicious, and the Xhosa and Zulu often dry it for winter use, adding it to gravies and stews after soaking it in water for a few hours.

The colonists made a wonderful gargle for a sore throat by steeping the ripe berries in brandy, then crushing them and mixing this with honey. For ringworm an old remedy was to make a paste of the unripe berry and apply this to the area frequently. The Xhosa also used this as a remedy for anthrax, and for ringworm on dogs and children. A paste made of leaves and green berries is used by the Xhosa and the Zulu as an external application for wounds and ulcers.

The green berries are poisonous, but the ripe fruit is not toxic and is relished by everyone, including the birds! In the Orange Free State the most delicious jam is made of the ripe berries during the summer and is always the very first to be sold at agricultural shows or farm stalls.

The leaves and fruits contain fairly high quantities of vitamin C, and often the ripe fruit crushed and mixed with honey is administered as a cough mixture for lung ailments, tuberculosis, post-nasal drip, and coughs and colds.

The fruit is also used to treat heart conditions, liver ailments, eye diseases and as a tonic. Here it is usually made into a tea by crushing approximately 1 tablespoon of fruit in ½ cup of water, and drinking this 2 to 4 times a day.

Interestingly, the ripe fruit is used by the Tswana and Sotho rain doctors as an important part of their rain ritual, the black fruit symbolising the black storm clouds. It is common knowledge amongst South African rural children that the unripe green berries are poisonous (to humans as well as stock) and that the ripe black berries clear the breath with their delightful liquorice taste. An American friend of mine once described the berry as 'a neat liquorice tomato'!

Nastergal is an easy and rewarding plant to grow in the garden and gives so much for so little. Once you have a plant it will seed itself all over the garden with the help of the birds. It is one of nature's most remarkable medicine chests and we should grow it far more than we do.

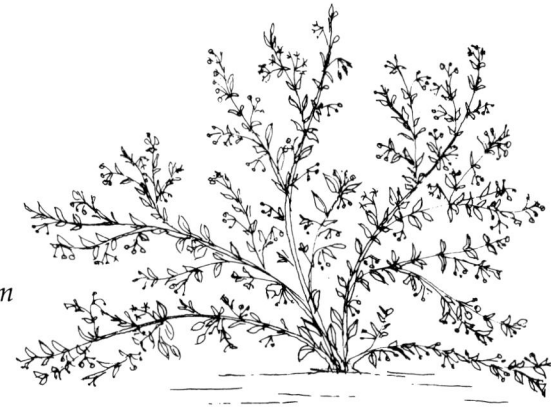

Height: 0,5–0,75 m

Nastergal sore throat gargle

1 cup ripe black nastergal berries
3 cups brandy
½–1 cup honey
½–1 cup lemon juice

Pour the brandy over the berries, cover and leave overnight. Next morning crush the berries in the brandy and strain through a sieve. Mix the honey and lemon juice into the brandy mixture, pour into a bottle, screw on the lid and shake vigorously.

Use a little of the mixture diluted with warm water as a gargle and with hot water (I find 2 tablespoons in 1 cup of hot water best) as a hot toddy for a cold or flu.

The brandy acts as a preservative so you'll find this keeps well in the fridge for a long time.

Nastergal jam

1 kg ripe black nastergal berries, washed and stalks removed
1 kg brown sugar
1 thumblength piece ginger, or 1 cinnamon stick
juice of 3 lemons

Place the berries and sugar in a stainless steel pot with a cup of water. Heat gently, ensuring that the sugar dissolves before bringing to the boil.

Add the ginger or cinnamon and stir well. Then add the lemon juice. Simmer slowly until the jam starts to thicken, stirring with a wooden spoon every now and then to prevent burning.

When the jam is ready (usually after about an hour of cooking) pour into hot bottles, seal with waxed paper dipped into brandy and screw the lids on well. Label and date the bottles.

Nutgrass
Cyperus esculentus

ENGLISH	Earth almond, watergrass
AFRIKAANS	Hoenderuintjie
SOUTHERN SOTHO	Mothoto
TSWANA	Intsikane
ZULU	Indawo

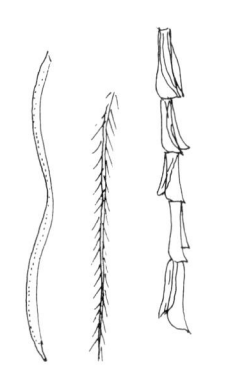

ALTHOUGH this common garden weed from the sedge family looks like a grass, it is in fact a completely different species with solid stems and a bulbous root. It appears in moist, cultivated areas such as ploughed fields and roadsides, and although it is a most irritating weed and difficult to eradicate in the garden, it is an important source of food and medicine for the indigenous people of Africa. Along the Gold Coast and in Central Africa it is sold in markets, cleaned and packaged, as 'tiger nuts'. In West Africa the plant is cultivated, for its sweet tuber is used to make a milk or gruel which is high in nutrients and rich in carbohydrates and minerals.

Some farmers reserve areas of nutgrass for feeding pigs and cattle, and serve it as a delicious steamed vegetable on their own tables. I have tasted the nutty tubers in a tomato bredie and found them absolutely delicious, and a West African recipe for nutgrass curry taught to me by a scientist who lived in the area for some time has become a firm favourite.

Nutgrass or 'tiger's milk' is mixed into mealie meal or oats or wheat porridge and eaten for strength, stamina and endurance. The traditional way of making this milk is to pound a handful of cleaned nutgrass tubers in a little water to form a thick mucilagenous 'milk', which can also be mixed with goat's milk or cow's milk, or with dried milk powder. Sweetened with honey it is delicious and satisfying.

The nutgrass tuber yields a nutritious oil similar to olive oil which contains linoleic, oleic and palmitic acids, amongst others. The roasted tuber can be ground into a powder which is an excellent substitute for coffee or cocoa, and mixed into warm milk is delicious and very digestible.

Medicinally nutgrass is primarily a digestive. Raw tubers chewed will aid the digestion, clear the digestive tract, cleanse the mouth, sweeten the breath and relieve flatulence and colic.

The Zulu are particularly fond of nutgrass, and young girls add the boiled tubers to their porridge to hasten their first menstruation.

Several African tribes believe the plant to be magical, giving health, vitality and protection, and dried tubers are often stored as an insurance against famine and witchcraft.

Nutgrass curry — Serves 6

1 kg good stewing beef (preferably rump)
1 cup flour
little sunflower oil
4 onions, peeled and chopped
6 tomatoes, skinned and chopped
water (about 1 litre)
½–¾ cup honey
½ cup vinegar
1 tablespoon curry powder (I use Masala)*
salt and pepper to taste
4 large potatoes, peeled and diced
4 carrots, peeled and diced
½ cup raisins or sultanas
2 cups nutgrass tubers, cleaned
½ cup dried split peas
2 mashed bananas
½ cup coconut

* Too much curry powder spoils the subtle, nutty flavour of the nutgrass, bananas and coconut, so start with 1 tablespoon and add more if you prefer it stonger.

Cut the meat into cubes, and roll each piece in the flour. Heat the oil in a large, heavy-bottomed pot. Brown the meat and onions in the oil. Add the tomatoes, stir well, and add a little water.

Mix the honey, vinegar, curry powder, the remaining flour and the salt and pepper and add to the stew. Once it is well mixed, add all the other ingredients.

Turn down the heat and simmer, stirring every now and then until the meat and the vegetables are tender. Add more water if necessary to keep the stew moist and succulent.

Paintbrush lily
Haemanthus coccineus

ENGLISH	Sore eye lily
AFRIKAANS	Maartblom, velskoenblaar, seeroogblom, poeierkwasblom
ZULU	Idumi-lika-ntloyile

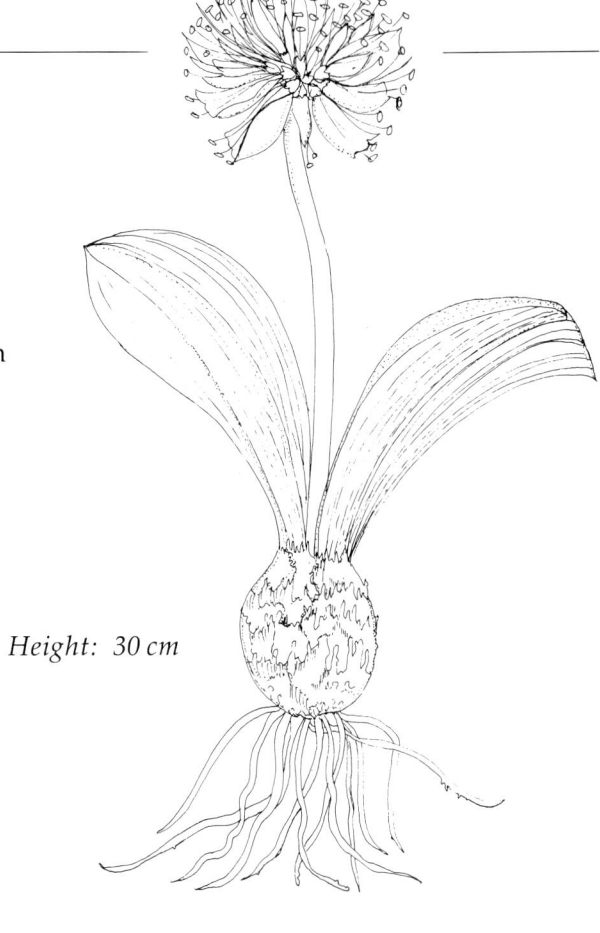

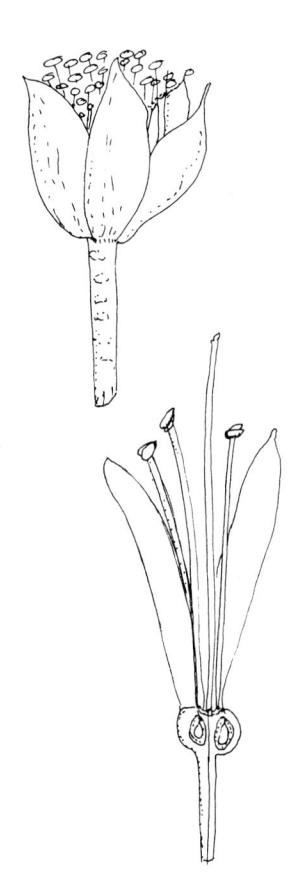

Height: 30 cm

ONE is so often startled by the brilliant bright scarlet of the paintbrush lily flowering against a rock or under a bush in the veld at the end of summer. The flower on its stiff, thick stem seems to come up out of the bare earth, with its mass of needle-like petals liberally dusted with pollen. The flat leaves emerge after the flower has faded, their shape resembling the sole of a shoe — hence the name velskoenblaar.

There are many varieties of *Haemanthus* and most of them are poisonous, so do take care how you handle them. *H. coccineus* was cultivated extensively in Holland from 1611 and the colonists recorded finding the flowers growing abundantly in the white sands of the Cape. They used the bulbs, sliced and steeped in vinegar for 2 to 4 days and then drawn off and diluted, to treat dropsy. It has remarkable diuretic properties.

The Khoikhoi also used the bulb broken open, warmed and then smeared with fat to apply to bruises and sprains, and this remedy was adopted by the colonists. The fresh leaves have always been highly esteemed in the treatment of ulcers, wounds and anthrax in cattle. They are bound over the wound where they act as an antiseptic and drawing agent, and the area is quickly healed. The leaves of other species of *Haemanthus* — *hirsutus* and *rotundifolius* — have also been found to have soothing effects when used as a poultice, and are used for keeping wounds moist.

The early settlers made a medicine of sliced leaves covered in vinegar and honey and left for 2 to 3 days, then boiled down gently for 15 to 20 minutes, to treat asthma.

The word 'seeroogblom' is derived from the fact that the pollen causes intense burning and irritation to the eyes. I always caution visitors not to touch their eyes near the plant or to inhale the pollen, as I found that whenever I brought a flower in from the veld my son would develop violent hayfever and within minutes would have streaming red sore eyes and a runny nose.

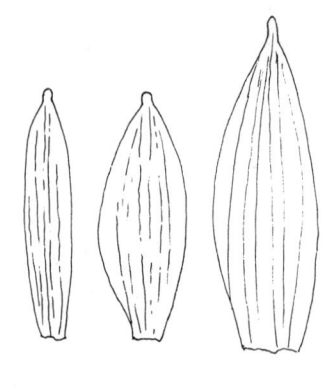

The paintbrush lily grows easily from seed and bulbs, and needs a semi-shaded position. I have a plant house with filtered light in which they thrive in the open ground.

The bulbs are available commercially, and the plants will give you much pleasure in the garden from midsummer until the first frosts.

Papyrus
Cyperus papyrus

ENGLISH	Sedge, watergrass, old man's beard
AFRIKAANS	Papirusriet, biesie

Height: 2 m

THIS remarkable water grass with its long, smooth, pulp-filled stems and its mop of thread-like grassy 'flowers' is a popular garden plant. It grows in a wide band from southern Africa up into Egypt and was the reed that the ancient Egyptians used to make paper. It does well in marshy areas, and equally well in ordinary garden rockeries and beds. I have had so much pleasure from my clumps of papyrus that I have experimented with growing it in all sorts of places, and have found it survives admirably even in the driest places. It seems to do best in full sun, but I also grow it under trees along my water garden and it does well in the partial shade.

Medicinally, the split stem, pounded and made supple, is used by several African tribes as a bandage or poultice for broken bones, as a binding with the stems in a splint, and as a bandage to hold dressings in place over sprains, strains and injuries. The Tswana believe the pulp is healing and sometimes use it directly applied and bound over a wound.

The ash from the dried, burned stems and tops of the papyrus is an excellent fertiliser, and can be added to sprays and foliar feeds to keep plants mildew-free.

I love using the tall mops in great big flower arrangements — my favourite combination is with white agapanthus and white arum lilies, and at Christmas time the arrangement lasts well into the new year.

In the garden propagate papyrus by digging out the spreading rhizomes from the existing clump — lots of nurseries stock the plant to get you started. Do keep the clump tidy, as you will find the older stems turn dry and brown and need to be cut off regularly to keep the clump looking neat. When planting out, give the new rhizomes plenty of space as in a good position with plenty of water the plant towers to a majestic 2 m!

I have always been fascinated as to how the ancient Egyptians made their papyrus scrolls and to my great delight an artist friend of mine, Jeanette Horn, taught me how to make my own.

Papyrus paper

mature papyrus stem
1 cup cake flour
3 cups water

Cut the stem of a mature papyrus into pieces 24 cm long. Split them with a sharp knife or blade down one side. Open them up and pound them with a wooden mallet until they are flat. Lay the flattened pieces on a board side by side. Make a thin paste of the flour and water. Paint or dab this with a sponge over the flat pieces, then lay another layer of pieces crosswise over these. Once the pieces are in position, pound and hammer to flatten them as much as possible. For thick paper add another layer longwise again, with a binding layer of flour and water glue. Then weight the paper by placing another board on top of it and a few stones on top of that. Leave for a week, then gently lift the board. Your paper may need a little more drying, in which case keep it lightly weighted. Once it is completely dry it is ready for writing or painting.

Parsley tree

Heteromorpha arborescens

ENGLISH	Parsnip tree
AFRIKAANS	Wildepieterseliebos, kraaibos
MPONDO	Umbangandlela
SOTHO	Maka-tlala
TSWANA	Mongwane
XHOSA	Umbangandhala
ZULU	Umbangandhala

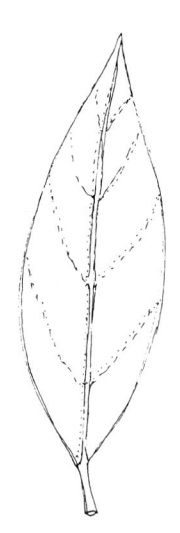

THE wild parsley tree is a straggling, open-branched, rather frail-looking small tree or shrub that occurs from the Cape through all the provinces up to the far northern Transvaal, and then on into Zimbabwe and Zambia. It is found on open hillsides, in rocky wooded ravines and at the edges of forests.

The bark is its most characteristic feature — it is glossy, waxy to the touch, dark purplish brown and peels away in thin papery flakes. The leaves are compound and variable, and in summer are a light grey-green, turning to reds and yellows in autumn, and these have a pleasant carrot fragrance when crushed. The flowers are small, creamy green and smell strongly of carrots and parsley. They form in dense round umbels at the tips of the branches in December and January, and are excellent in pot-pourris. The seeds too are strong smelling and these are also wonderful additions to pot-pourris. They also germinate easily into tough, sturdy seedlings.

The parsley tree is much prized in African medicine, and is important in the treatment of mental disorders. The wide number of medicinal uses that the tree is put to is quite remarkable, and one hopes that in the future controlled medical tests will give scientific credibility to the healing properties of this remarkable tree.

The Zulu and the Tswana make a tea by boiling 1 cup of leaves in 4 cups of boiling water for 10 minutes, straining it and taking a little at a time for stomach upsets, colic, flatulence, and for intestinal worms taken on an empty stomach first thing in the morning. This tea is also used as an enema for constipation and abdominal upsets.

The Sotho make a tea in the same way for nervousness and mental disturbances, and give it to children who have nightmares or who are easily frightened. Several tribes believe that the smoke inhaled from the burning wood will clear headaches and sinus congestion.

A brew made of the root and bark is also used by both blacks and whites for shortness of breath, asthma, bronchitis, coughs, colds, dysentery and for intestinal worms, and this is also used for threadworm in horses and dogs (1 cup of mixed root and bark, peeled and washed and chopped, boiled in 4 to 5 cups of water for 20 minutes then cooled and strained). Some whites in the eastern Transvaal use the bark to treat skin ailments such as eczema, where horny layers of skin on the face and backs of the hands become inflamed from scratching. A brew from the bark (2 cups of bark boiled in 4 cups of water for 20 minutes) is an excellent lotion if frequently applied. This is also used for scaly scalp conditions as a rinse after shampooing, and some of the liquid is combed into the hair daily, and rubbed well into the scalp.

Crushed leaves are used to treat ringworm and itch, by rubbing them onto the area 3 or 4 times every day until the condition clears; and I have found quick relief from mosquito bites by rubbing a crushed leaf over the area.

This is a fascinating and useful tree to grow in the garden and it is small and undemanding. The tree is considered to bring good luck and afford protection, and is therefore much sought after. Several nurseries have established trees for sale, so keep asking.

Height: 3–4 m

Insect-repelling pot-pourri

approx. 2 teaspoons lavender essential oil
1 cup dried minced lemon peel
2 cups dried parsley tree seeds and flowers
1 cup dried blue sage leaves and flowers
1 cup dried lavender tree leaves
1 cup cinnamon, coriander and cloves

Mix the oil into the lemon peel. Add the spices and store in a sealed bottle for 1 week. Shake well. Add all the other ingredients and mix well. Add more oil if liked. Fill sachets and place in cupboards or between books.

Pennywort
Centella asiatica

ENGLISH	Wild violet, waternavel, marsh pepperwort, gotu kola
AFRIKAANS	Varkoortjies, wilde vioolblaar
SOTHO	Bodila-ba-dinku
ZULU	Udingu

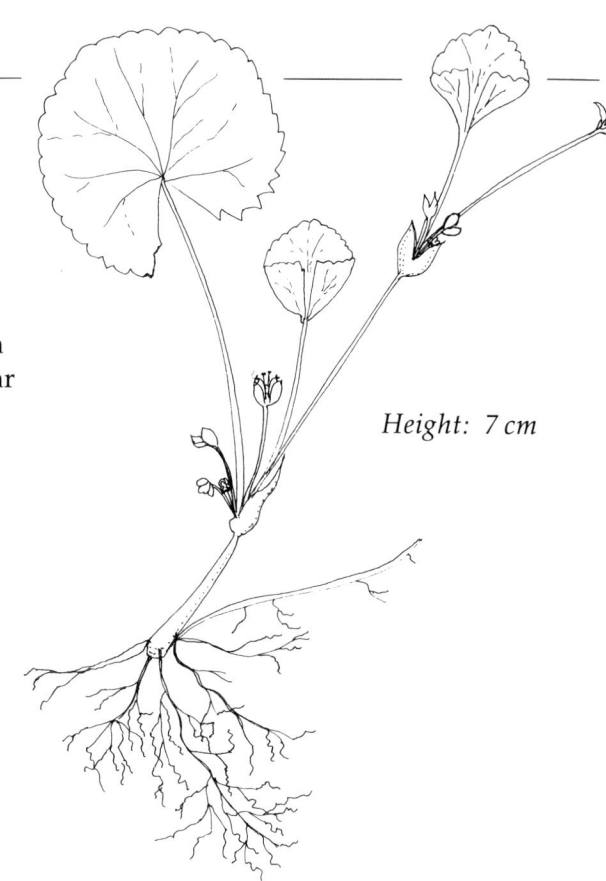

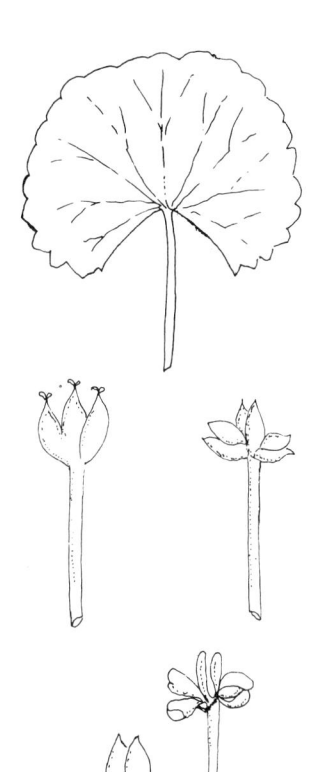

Height: 7 cm

PENNYWORT is a bright green, creeping plant that grows in the shade and loves marshy areas. It has violet-shaped leaves, hence its alternative name wild violet, and runners with tiny nondescript green flowers at the nodes. Once it has established itself in the garden it will creep into all sorts of places, and I find it such an easy and attractive groundcover I leave it to go its own way.

It is indigenous to Africa and India, where it has been used to treat leprous sores, syphilis, leg ulcers, scrofula, slow-healing wounds, venereal sores and tuberculose skin lesions.

The plant is much respected by blacks, who include the leaf in their diet. I was introduced to it by the Zulu caretaker at our seaside cottage, who told me how his grandmother had cured people in his village of bad sores by applying a softened warmed leaf of pennywort directly to the area every day for 6 days, and making a tea of the leaves for them to drink. The treatment cured even the sores the miners brought back — presumably syphilis. He also showed me how to crush the leaf by rubbing it between the hands, and then apply this pulp to a wound, bound in place with a bandage or a castor oil leaf. I proved its remarkable healing powers by clearing up a badly infected scratch in 2 days.

The herb made into a tea (4 to 6 leaves in 1 cup of boiling water, leave to draw for 5 minutes, then strain and sip slowly) is cooling and diuretic and acts as a light purgative. It will also bring down fevers. It can also be used as a wash or in the bath to help a fever, as a lotion and wash to treat sores and wounds and swilled in the mouth for mouth infections and ulcers. Some tribes use the tea as a treatment for tuberculosis, and dry the leaf as a snuff — perhaps it is slightly narcotic, for it is much sought after!

Pennywort was included in the French pharmacopoeia in 1884, and its importance in Hindu medicine is fascinating — one wonders who was the first to experiment with this precious little plant.

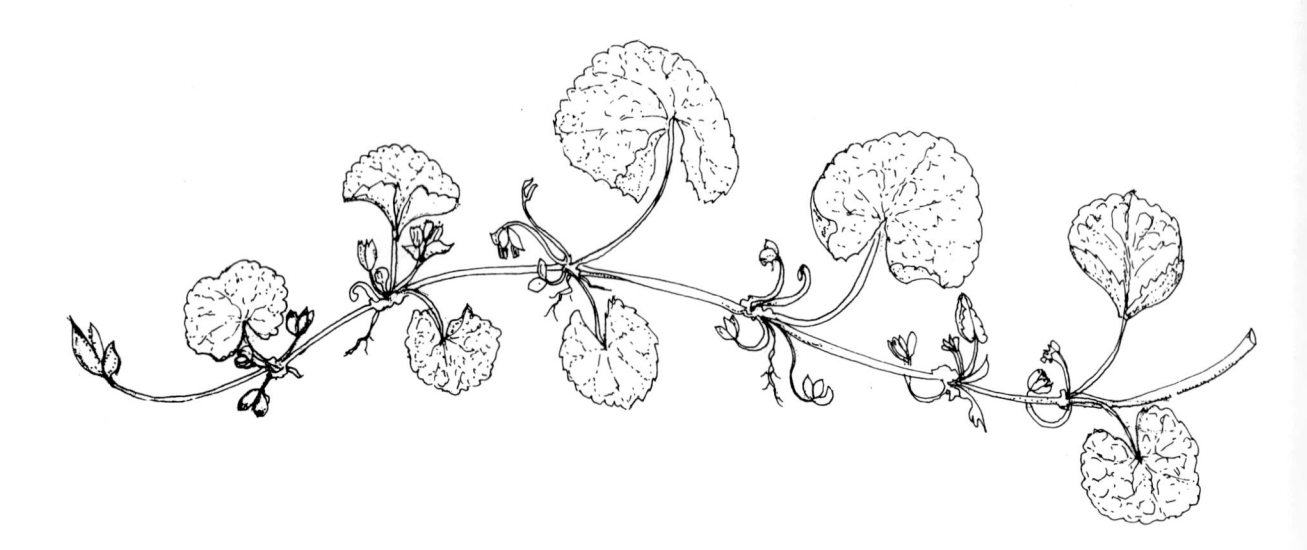

Pig's ear cotyledon
Cotyledon orbiculata

AFRIKAANS	Plakkie, varkoor, varkoorblaar, oorlams plakkie, kouterie
SOTHO	Seredile
TSWANA	Intelesi
XHOSA	Iphewula
ZULU	Intelezi

Height: Leaves 15 cm
Flowers 30 cm

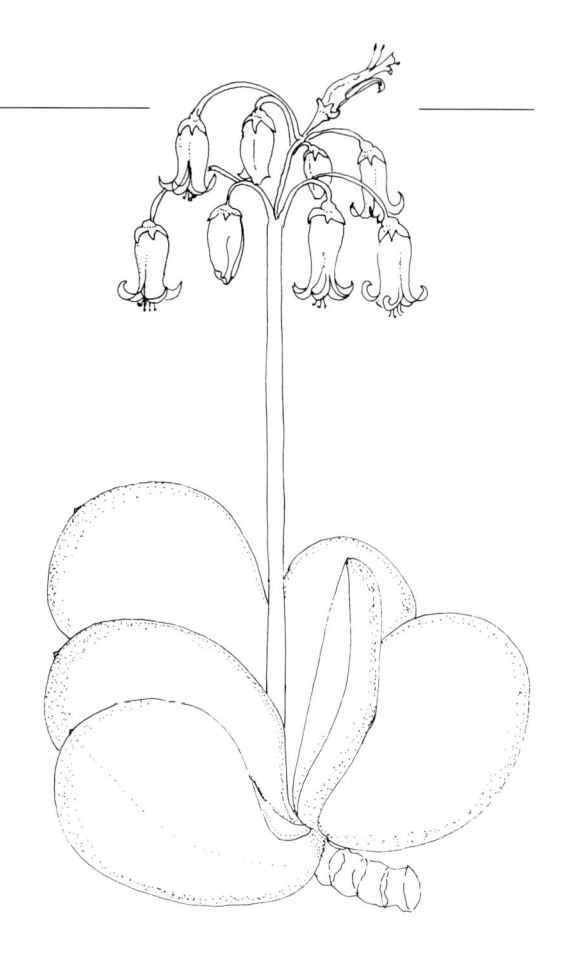

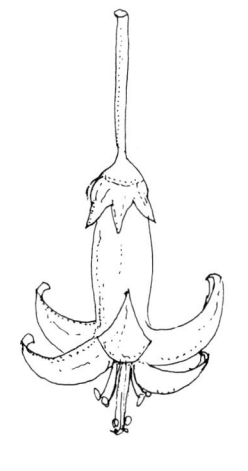

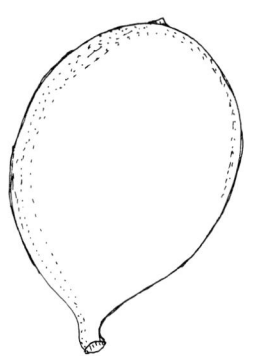

THE Afrikaans word 'plakkie' is aptly used to describe plants that are used as poultices, and one of the best known is the pig's ear cotyledon.

It is found throughout the country, usually growing vigorously in grasslands, rocky koppies and in scrub. There are several varieties of *Cotyledon* with varying leaf colours, but *C. orbiculata* is easily recognisable. It has thick, smooth, succulent, obovate, tongue-shaped leaves, often with a red edge, and a stem topped with pinky-red, drooping, bell-shaped flowers.

C. leucophylla is a poisonous species that grows wild in the Transvaal. The leaves are grey with a defined red edge, and it too makes an excellent drawing poultice over infected wounds, scratches and warts. The Tswana use the leaf warmed in hot water for drawing infection out of wounds and sores. The plant is toxic to grazing animals and at certain times of the year, usually from November to February, is even more toxic than usual.

Most varieties of 'plakkies' are used as a remarkably effective dressing for planter warts and verucas and pig's ear cotyledon is probably best known for this treatment. A piece of scraped, softened leaf — or a piece of leaf with a 'window' the size of the wart — is placed over the wart and held in place with a sticky plaster. A fresh piece is applied every night for 10 to 14 days and the wart is left uncovered during the day. At the end of the treatment the softened wart falls out, and the area is healed. *C. orbiculata* and *C. leucophylla* are the best species to use for this purpose.

The warmed juice of the leaf is used by the Xhosa as soothing drops for earache and toothache and in country districts even today a poultice of warmed leaves is held behind the ear to ease otitis media or placed over a boil to bring it to a head, or used externally for toothache. The leaf can be pulped and hot water poured over it, then drained and used as a poultice, or the leaf can be placed in a folded piece of cloth and placed in a low oven to warm it thoroughly before being used as a poultice.

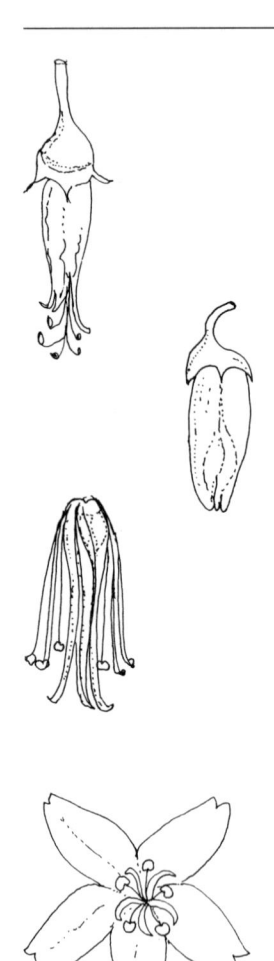

Carl Willem Ludwig Pappe was a physician who came to the Cape in 1831 and his interest in the medicinal Cape flora led him to study the uses of the plants, which he listed in his remarkable little book *Indigenous Plants Used as Remedies by the Colonists of the Cape of Good Hope* in 1847. Among the plants he listed was *C. orbiculata*, which he used in the treatment of epilepsy. The plant is still believed to be beneficial for epilepsy, given as a daily dose.

The Sotho believe the plant to have magical uses, and make a protective charm from a dried leaf, which is tied around the neck or waist of an orphaned child. They also use the plant medicinally for a number of ailments, the most common being as a drawing poultice and wart treatment.

The plant needs full sun, and makes an interesting rockery or pot plant subject on a hot patio. It also grows well in hot rock crevices and in poor soil. To propagate, you merely need to push a leaf into the ground or break off a piece of stem, and it will grow if it is kept moist initially. Once established, the plant needs very little attention or water, and in spring and early summer you will be rewarded by the startlingly beautiful pinky red bell-shaped flowers that suddenly appear amidst the leaves.

I have been thrilled to find how easily and quickly the plant grows in an arid and bleak part of the rock garden, and this is one plant that thrives in the heat and looks good throughout the year. You may even be inspired to start a 'plakkie' collection as I have, which fascinates overseas visitors and needs no attention whatsoever.

Pincushion
Scabiosa
columbaria

plumbago

Pincushion
scabiosa incisa

Plectanthus ecklonii

Sweet thorn

Cross berry

Parsley tree

Raisinbush

Raasblaar

Pincushion
Scabiosa columbaria

ENGLISH	Wild scabious, rice flower, morning bride
AFRIKAANS	Koringblom, bitterbos
SOUTHERN SOTHO	Selomi
TSWANA	Makga
XHOSA	Makgha, isilawu esikhulu

THE wild scabious with its pretty white flowers can be seen all over the veld in midsummer throughout the country. It is a slender perennial plant, and grows up to 30 cm in height in open grassland and on hillsides. The charming pincushion flowers in white and sometimes pale mauve are close relatives of the beautiful pink and mauve hybrids which are cultivated for cut flowers in market gardens all over the world; and some of these originated here in South Africa. All have remarkable medicinal properties.

The flower heads on their long peduncles are reminiscent of an old-fashioned lace-covered pincushion and when the petals fall the seed head, conspicuous and bristly, is much prized as a dried flower.

The plant gets its generic name *Scabiosa* from the word scabies, meaning itch or skin irritation, and this family of plants is used to cure skin irritations.

Medicinally the wild scabious is used by several African tribes, and many whites also respect and use the fresh root as a treatment for colic, dyspepsia and flatulence. The root is scraped, washed and chewed a small piece at a time to relieve heartburn and the Sotho dry and powder the root and take it in water for all these ailments.

An effective ointment for veld sores, venereal sores and skin ulcers is made by the Sotho by charring the root, pounding it in parrafin and applying this paste to the sore. They also use this for sores on cattle and dogs.

The Xhosa make a decoction of the root as an eyewash for sore eyes, but I found this to be rather harsh, so be cautious. The Sotho and Zulu use this same brew for colic, sipping the warm tea slowly, and add other herbs to it, like kiepersol leaves, to aid difficult confinements and painful menstruation.

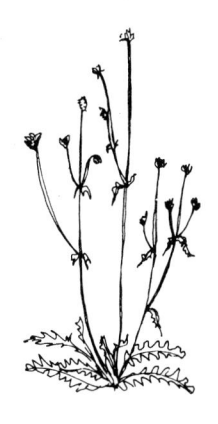

Height: 30 cm

S. incisa

Probably its most famous use is as a dusting powder, as it has a soothing effect on the skin and a pleasant fragrance. The common African name for this powder is 'makga', and this is a favourite for babies after a nappy change to soothe chafed skin and to act as a talcum powder in drying the area. The root is washed, dried in the sun and pounded between stones to make this fine powder.

The colonists found the wild scabious growing abundantly in the veld and used the plants as they had in their homelands — as an antiseptic wound wash and lotion for sores and ulcers. Fresh roots were boiled for 10 minutes in enough water to cover them, and the resultant tea was strained and used as a lotion. The brew can also be mixed with borax and this paste applied to the scalp once a week to remove dandruff effectively.

A favourite tonic which was much prized as a pick-me-up was made by infusing 2 tablespoons of flowers (about 10 flowers) in 1 litre of red wine with 12 cloves, a stick of cinnamon and 2 tablespoons of honey. This was shaken daily and stored in a cool place for about 1 week, then strained, and 2 tablespoons taken daily to restore strength and vitality.

Other varieties of wild scabious include *S. transvaalensis,* which is used by the Tswana as a decoction for sore eyes, *S. africana,* a beautiful mauve-flowered species which is used by the Zulu and Sotho for an antiseptic wound wash, and the beautiful, rare *S. incisa,* whose large purple flowers are prized long-lasting cut flowers and are also used to make a dusting powder and wound lotion.

The plants grow easily from seed and in the garden need full sun and well-drained soil. Once the clumps are established they can be separated out to make more plants, and they make a wonderful midsummer show.

S. incisa
Height: 30–40 cm

Plectranthus

Plectranthus species

ENGLISH	Spurflower
AFRIKAANS	Vlieëbos
SOUTHERN SOTHO	Lephelephele
ZULU	Umadolwana

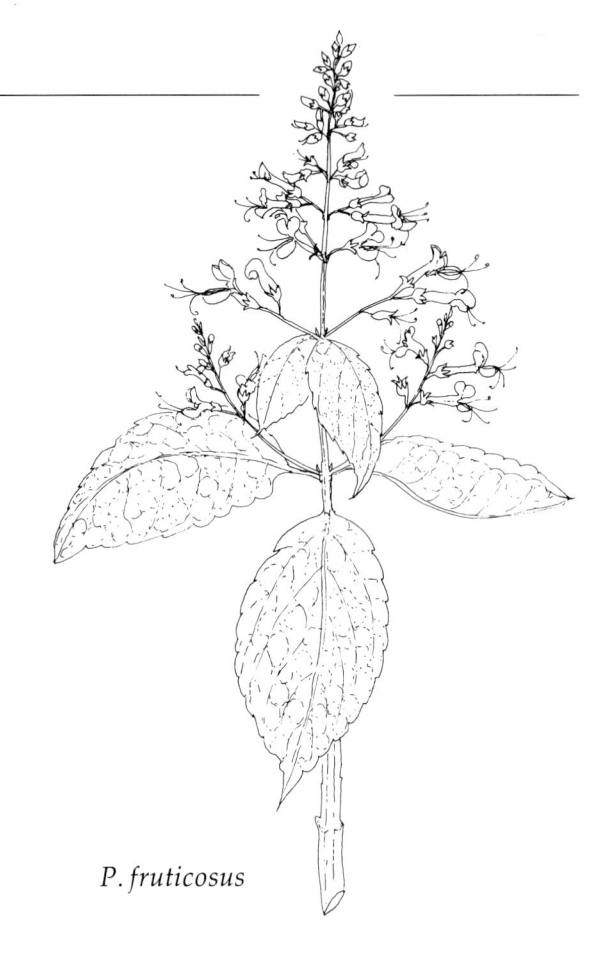

P. fruticosus

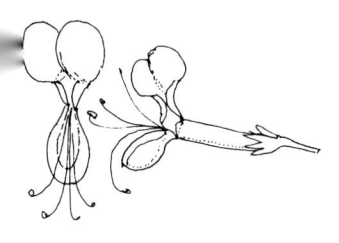

P. fruticosus

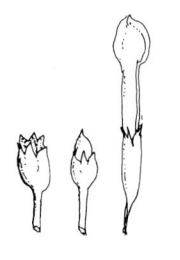

THERE are some 30 or 40 species of plectranthus indigenous to South Africa. All have fragrant leaves, and their exquisite flowers range from mauve, pink and white to purple, and all are much enjoyed as garden plants and pot plants both here and abroad.

Several species are used as a cough and cold remedy and the Zulu make a decoction of the root and the leaf — usually of *P. hirtus* — to treat chest complaints and asthma.

The Xhosa use the crushed leaf for veld sores and scabies by rubbing it into the skin. I have used this to good effect on the dogs to heal a slow-healing sore.

Several species of plectranthus can be used as a soothing, fragrant wash or lotion (pour 6 litres of boiling water over 4 cups of leaves, stand, steep and cool) to bring down fevers, to aid sleep, to soften the skin, as well as to wash sheep skin and cotton garments.

The citronella plectranthus, or lemonspur flower as it is commonly called (*P. laxiflorus*), smells fresh and lemony and is wonderful in the bath. A brew made of the leaves (4 cups of leaves in 6 litres of boiling water) is used by several African tribes as an enema, to bring down fevers, and as a drink for stomach upsets. The same brew is used as a gargle for mouth infections, sore teeth and bleeding gums.

The Zulu use plectranthus abundantly (usually *P. natalensis*) for washing themselves, their clothing and animal skins.

The thick underground tubers of *P. esculentus* (*P. floribundus*) and *P. urticoides* are delicious roasted and eaten like sweet potatoes. A root is pushed into the embers of the fire and allowed to roast slowly until it is cooked

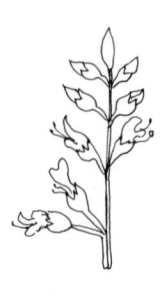

P. fruticosus

through, then peeled, split open and eaten with butter and salt — it is not only an unusual delicacy, but apparently much prized as an energy booster!

Many of the plectranthus species are used as insecticides and flies particularly are repelled by their leaves, which contain volatile oils. Two varieties, *P. fruticosus* and *P. thunbergii*, are particularly effective as fly repellents, hence the name vlieëbossie, and pieces of the plant are often brought into the house and rubbed onto window sills and table tops to keep flies away.

Many nurseries offer the plants for sale and you could start a fascinating collection of plectranthus species. They all root with such ease that it does not need to become an expensive hobby. The tips of the non-flowering branch — about 10 cm long — root best. I press them into wet soil and keep them shaded and moist for a month.

Add dried plectranthus leaves and flowers to pot-pourris (see wild flower pot-pourri, page 268) and steep the flowers in vinegar and use this in the bath as a skin softener. I also add this vinegar to my hair rinsing water and I find it has a wonderfully softening effect.

Height: 0,25 – 1,5 m

Plectranthus insect-repelling pot-pourri

3 cups dried plectranthus leaves
6 cups dried plectranthus flowers (any variety)
½ cup cloves
½ cup broken cinnamon pieces
2 cups dried, minced lemon and orange peel and pips
3 cups sweet basil or wild basil leaves, seeds and flowers
1 cup khakibos
citronella oil

Mix all the ingredients. Add citronella oil (usually about 2 teaspoons). Store in a sealed container and shake daily for 2 weeks. Then add a little more oil, place in sachets or bowls in the kitchen cupboards and drawers, or amongst books and records. (You can also add wilde als and melianthus leaves for added strength.)

Plumbago
Plumbago auriculata

ENGLISH Cape leadwort
XHOSA Umabophe, umashintshine
ZULU Umasheleshele, umshwilishwili

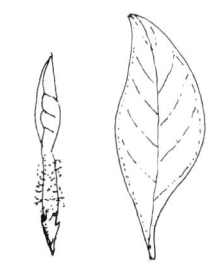

IF you drive along the roads in the eastern Cape, from the Humansdorp area on to East London and even into Natal, you will notice great banks of pretty blue plumbago scrambling over the rocks and hillsides. It flowers at its best over the Christmas period and the flowers with their sticky calyxes make charming flower arrangements which last fairly well indoors. My favourite way of arranging plumbago is to pick fairly short stems and cluster them together in a tight posy. They stick to each other and an exquisite blue 'ball' can be easily achieved.

Best of all, use the flowers in a wild flower pot-pourri (see page 268) or combine plumbago with other garden flowers like delphiniums, blue larkspur, cornflowers or blue hydrangeas to make an unusual and beautiful blue pot-pourri, as these all keep their colour well.

You can eat the flowers too, and I find they are particularly pretty decorating a fruit salad or floating in a cool drink.

The shrub grows easily from rooted runners or from cuttings, and has become a great favourite for the garden both in South Africa and overseas. The white form is more unusual and is an attractive back plant in the border.

My mother trains plumbago into a trellis as a glorious pale blue hedge which she prunes frequently, and cultivated in this way it seems to be a mass of flowers all the time. I have also clipped plumbago into a standard, and support the single stem with a strong pole or fencing strut, clipping away all side shoots as they appear. Keep the top lightly pruned through the first two summers, and in its third year you'll have a most unusual and exquisite feature in your garden that will give you endless pleasure.

Plumbago is used medicinally by many indigenous people. The Xhosa and the Zulu believe it to have magical properties and use it as a charm for all sorts of purposes, including the 'confounding of the enemy'! The Shangaans on the farm tie bundles of plumbago twigs and push them into the eaves of their houses or tie them in the roof to protect against lightning and to ward off evil.

The root dried and then ground into a powder is used by several races as a snuff to clear the head and ease a headache. If the powder is mixed with water and applied to a wart frequently for some days, the wart is said to suddenly disappear. Some Zulu say the powdered root rubbed over the area of a fracture will help knit the bone.

The settlers in the interior used an infusion of the root to treat black water fever in cattle, and as the leaves are relished by sheep, cattle and chickens you'll find plumbago in many old farmyards and gardens.

Height: 2–3 m

Pompon tree
Dais cotinifolia

AFRIKAANS	Kannabas, basboom
MPONDO	Ilozane, intozane
ZULU	Intozane

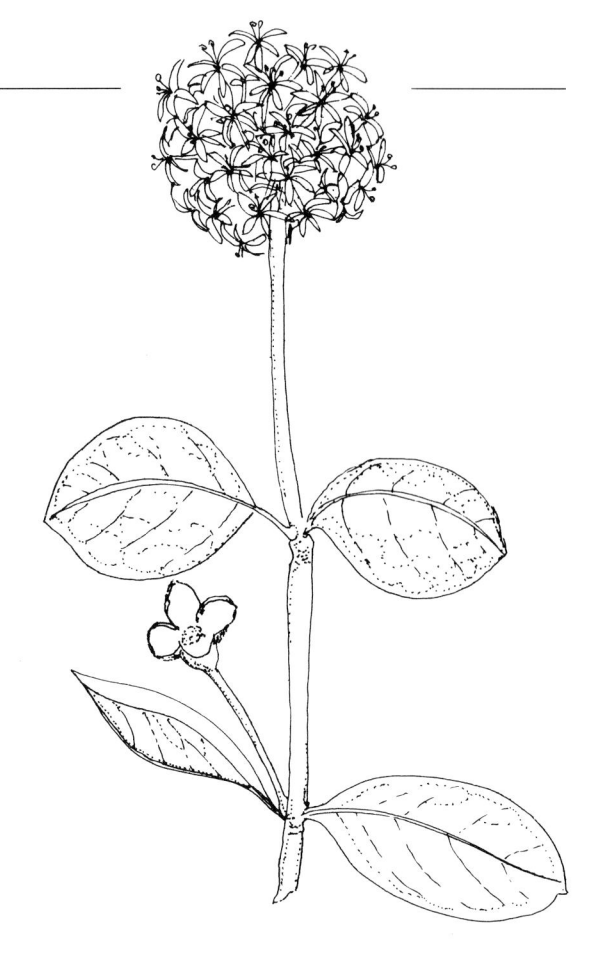

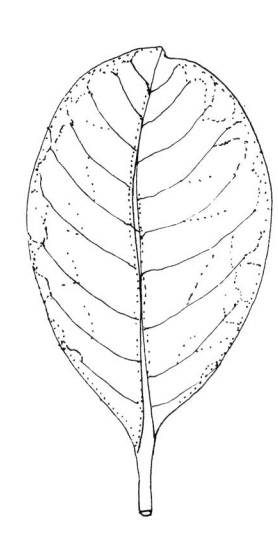

THE beautiful pompon tree, one of the most successful indigenous garden trees, is found wild all through the eastern Cape, Transkei, Natal and into the eastern Transvaal. In early summer its masses of round lacy pink pompons are exquisite and eyecatching, and a sheer delight to come upon as one drives through the open countryside.

A small, attractively shaped tree 5 to 6 m in height at the most, the pompon tree is one of the fastest growers, and seems to thrive in any soil. Where there is no frost, the tree is evergreen, and is particularly suitable for a small garden, as it gives dense shade, has a round crown and looks good all year through. I have seen a driveway lined with pompon trees — planted 4 m apart and well established within 3 years as they grow so quickly — so breathtaking in their early summer flowers that people come from far and wide to see them and photograph them.

The bark can easily be torn off the branches, and is strong, flexible and tough. It is used by the Zulu and the Mpondo as a rope or string. They tear the strips into thin pieces and roll these between the hands, adding new pieces as they roll, twisting the strands together to make a long, very strong rope. The bark also contains tannin, and was used by the Voortrekkers as a useful ingredient in the tanning of hides.

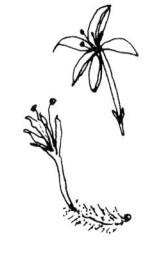

These strong strips of bark are also a most useful and important bandage used by the Zulu, the Mpondo and the Venda to bind fractures and splints, and to hold dressings in place over wounds or sprains. They also believe that there are healing properties within the bark which assist the healing of broken bones.

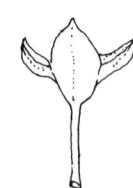

After the summer glory of pompon flowers, the tree is covered with small dried calyx-like seed pods. These are a most useful addition to pot-pourris, as they absorb the essential oil so well that they act as a fixative. The pink pompons also dry well, keeping their colour, and these too make a useful pot-pourri ingredient.

Propagation of the pompon tree is easy either by seed or cuttings — particularly side shoots, and if these are kept in moist sand for 1 month during the summer months, you'll find sturdy roots quickly appear.

To start your tree off dig a large hole and fill with compost mixed with good garden soil. Water deeply every week for quick growth. You can prune the tree into a desirable shape, and it will give you endless pleasure as a garden subject.

It is a much sought after plant overseas, where a large specimen in a tub fetches an enormous price!

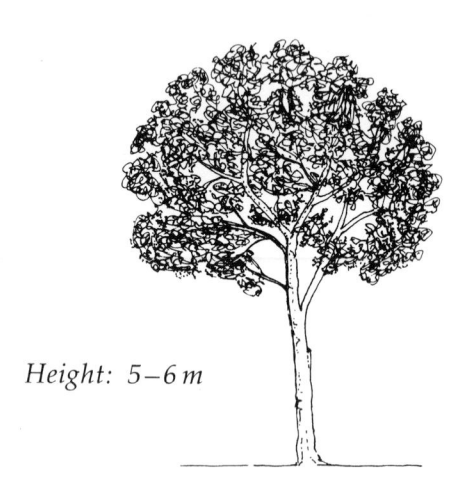

Height: 5–6 m

Pompon tree fixative

5 cups dried seed pods (wait until they are brown and brittle)
2 cups dried minced lemon peel
1 cup broken cinnamon pieces
½ cup cloves
½ cup roughly broken nutmeg
1 tablespoon essential oil (I like rose oil, as I add this to rose petals, or lavender oil if I'm using lavender)

Empty the seed pods into an airtight container and add the other ingredients. Mix together, seal the tin and shake daily.

Store for at least 1 month before adding to your pot-pourri mixtures as a fixative.

Raasblaar
Combretum zeyheri

ENGLISH	Large-fruited bushwillow, large-fruited combretum
AFRIKAANS	Nicholas klapper
NDEBELE	Umbonda
SHANGAAN	Mphuba
SWAHILI	Msana
TSWANA	Lesapo, modubana
VENDA	Modubana

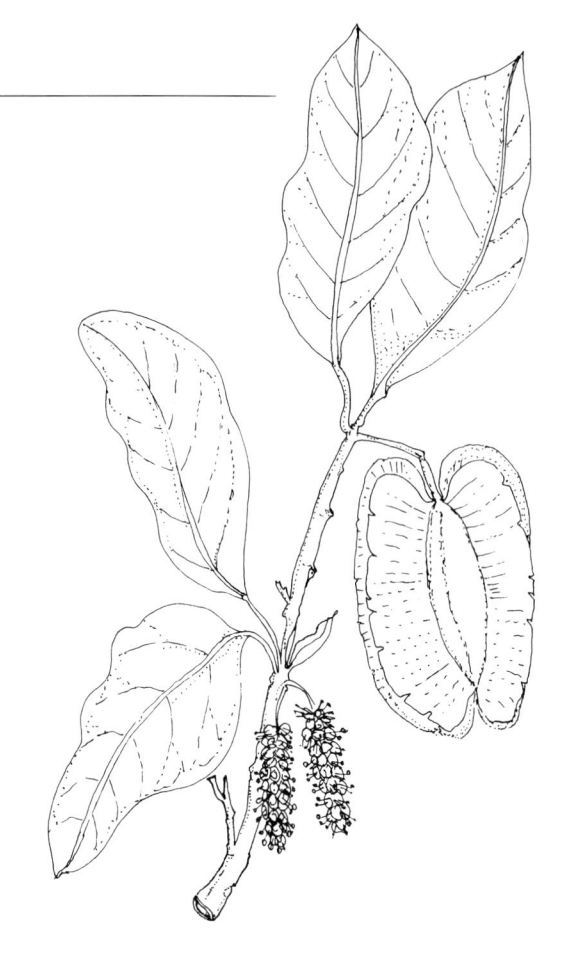

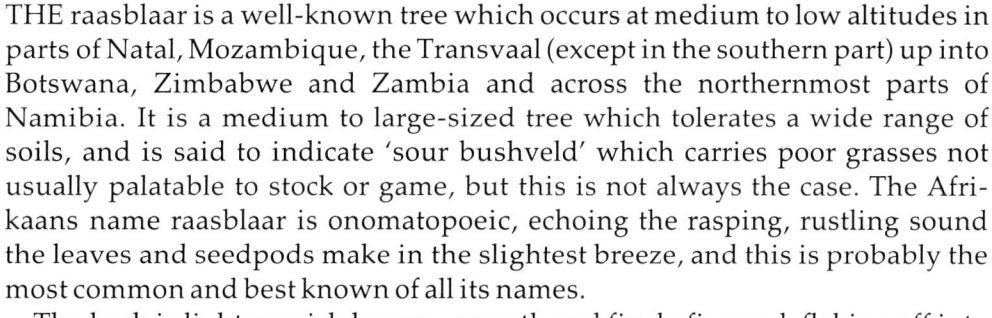

THE raasblaar is a well-known tree which occurs at medium to low altitudes in parts of Natal, Mozambique, the Transvaal (except in the southern part) up into Botswana, Zimbabwe and Zambia and across the northernmost parts of Namibia. It is a medium to large-sized tree which tolerates a wide range of soils, and is said to indicate 'sour bushveld' which carries poor grasses not usually palatable to stock or game, but this is not always the case. The Afrikaans name raasblaar is onomatopoeic, echoing the rasping, rustling sound the leaves and seedpods make in the slightest breeze, and this is probably the most common and best known of all its names.

The bark is light greyish brown, smooth and finely fissured, flaking off into small pieces, giving a mottled appearance. The branches are twisted, forming fascinating shapes, and the slender pliable branchlets seem to droop under the weight of the large, oval leaves. The tree is deciduous, and the bare twisted branches are beautiful in their winter starkness. In early October masses of sweetly scented greeny yellow small flowers with fine orange anthers appear, sometimes with the first tender leaves, and this gives the tree the daintiest, freshest look. By the end of November the tree offers deep shade.

The fruit is four-winged, large (about 6 cm in width and length) and pale green when young, turning to pale golden brown when mature and is much sought after by florists and dried flower arrangers, as it can be threaded to form mobiles or wired for the vase, and always looks attractive.

The roots are tough, fibrous and strong and are used as ropes, to weave baskets, and in Mozambique and Zimbabwe and further north, to make fishing baskets, nets and traps.

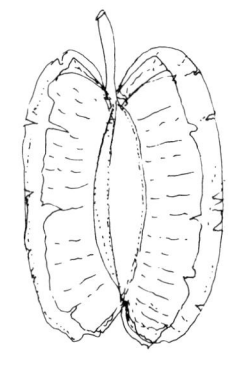

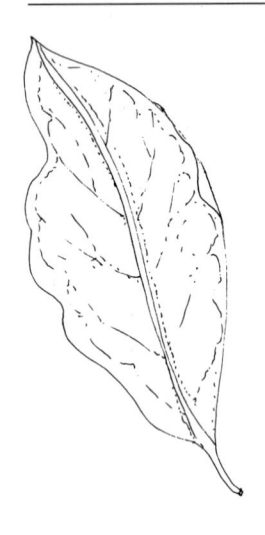

The raasblaar is a much respected medicinal plant. The leaves crushed and macerated into oil — usually sunflower seed oil — or mutton fat are pounded into a paste, then warmed and applied as an embrocation onto the back to relieve backache. This is also used over sprains and bruises or painful contusions.

Some tribes pour boiling water over the leaves and use this resulting brew as an eye lotion for inflamed eyes or styes, or opthalmia in both humans and cattle. Sometimes the leaves are pounded into mutton fat and used as a paste externally around the eye to soothe the condition.

A decoction of the bark is used as a purgative, and by some tribes as a treatment for leprosy, as it is believed to purify the blood. The leaves are also pounded and mixed with water in which pieces of bark have been boiled, and fat or lard, and used as a soothing treatment applied nightly for haemorrhoids. This is an old recipe used by several African tribes and the early settlers, and in some outlying districts is still used today. Many years ago I was shown by a farmer's wife how she applied the pounded leaves made into an ointment with ordinary Vaseline to her varicose veins, and she said how much it soothed and helped the discomfort, particularly if she had been standing all day. She minced about half a cup of fresh leaves first, and then mixed these into 1 tablespoon of Vaseline, pounding and working it together for some minutes. This was then spread on a piece of lint, and held in place with a loose crepe bandage. (Some skins are sensitive so it is always wise to test a little first.) Each morning she washed off the sticky area with warm soapy water to which a little vinegar had been added.

A leaf or two packed into your shoe around the heels will ease aching feet on a long walk. The Tswana make a nest around the heel in the shoe for tired feet. A few leaves placed in the hat will also ease a headache and prevent heatstroke, which is good for hikers to know.

If you plant a raasblaar in your garden give it space, as it becomes large and sprawling, and be prepared for much raking up of leaves in winter and spring. It is one of the most interesting shaped trees and is always attractive. I have a huge one on my lawn and it gives me endless pleasure in all seasons.

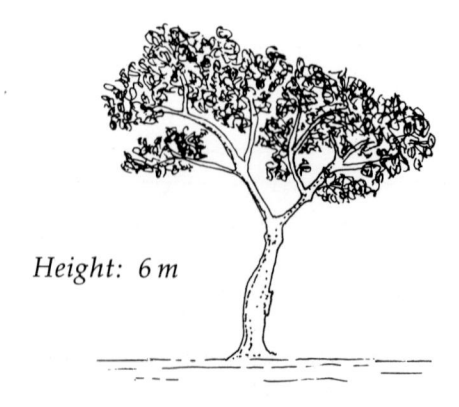

Height: 6 m

Raisinbush
Grewia flava

ENGLISH	Raisin tree, wild raisin, brandy bush, wild plum
AFRIKAANS	Brandewynbos, wilderosyntjie, rosyntjiebos, kafferbessie, sandbessies
BACWANA	Moretellwa
KADE BUSHMAN	Kxom
!KUNG BUSHMAN	/nun
KWANGALI	Mpundu, ngogo
LOZI	Muwane
NARON BUSHMAN	K'um, ini
PEDI	Moretlwa, meretlua
SOTHO	Morethlwa
SWATI	Liklolo
TSWANA	Moreeko, morethlwa, moretlwa
ZULU	Ulusizimezane, umhlalophansi

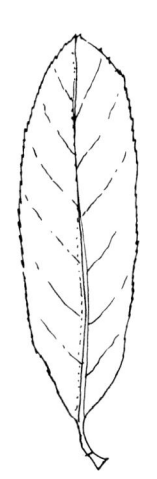

THE raisinbush is a typical bushveld shrub found throughout the Transvaal into the Orange Free State and through the western parts of the country into Namibia.

The freely branching, twiggy bush grows to about 1,5 m in width and height. The stems are tough, pliable and covered with a brown bark that is used to bind over wounds and sprains. The small leaves are greyish and felt-like and are shed in winter.

The presence of the plant in the open veld and in the bush is said to indicate good grazing, and the raisinbush is relished by sheep, goats and cattle when the spring grass has not yet come into growth.

In early summer the starry, bright yellow flowers appear, followed by shiny, round, pale brown fruits which are sometimes two-lobed, but usually pea-shaped. The fruits comprise a central stone surrounded by a thin pulp that is sweetly astringent and much sought after by rural children, birds and monkeys. The fruit is considered to be an energiser and thirst quencher for the weary traveller, so keep a look out for it when you are out walking in the veld.

The fruit is often dried and then pounded and added to porridge. A Tswana delicacy is the stamped dried fruit mixed with dried locusts and fat, and this strange mixture is fried in a pot over the coals. The Sotho and the Tswana make a beer from the fresh fruit, and the Khoikhoi used to distil a spirit from it.

The plant is believed to have magical uses too and several African tribes believe it to be protective. Pegs are made of the thicker stems and stamped into the ground around their huts as a protection against lightning, and the plant is used in certain death rite ceremonies.

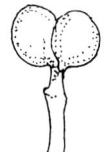

A tea made of the stems and a few leaves (usually 1 cup of chopped twigs to 3 or 4 cups of boiling water) is used as a treatment for kidney ailments — small amounts of the tepid tea are drunk frequently to ease the condition. The tea is also used as a wash for scratches and rashes.

The raisinbush is related to the cross-berry, *Grewia occidentalis*, page 74.

Height: 1–1,5 m

Renosterbos

Elytropappus rhinocerotis

ENGLISH   Rhenoster bush
AFRIKAANS   Renostertoppe, anosterbos, vlieëbossie

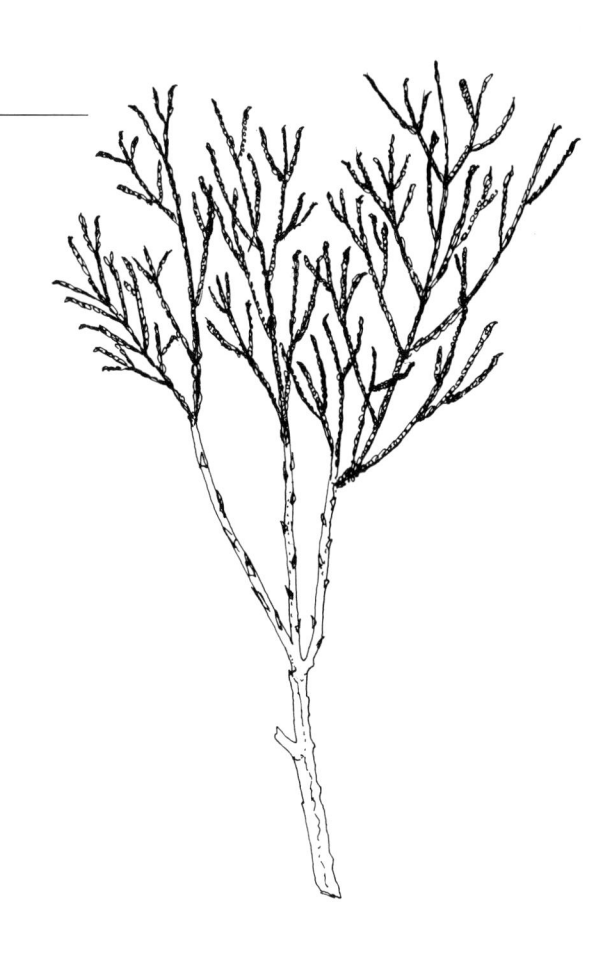

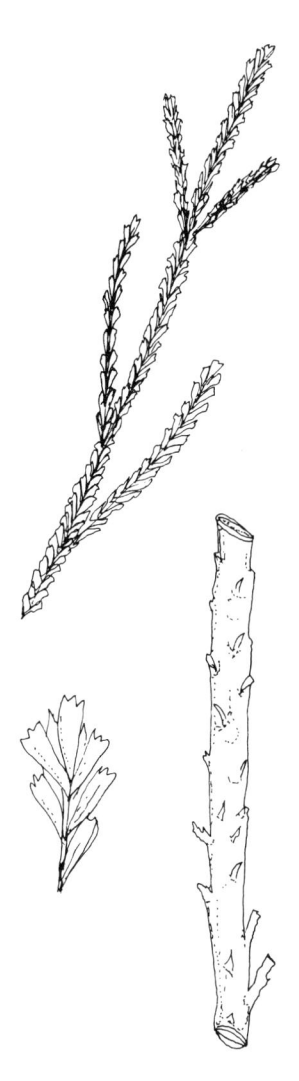

RENOSTERBOS is a resinous, dull olive green bushy shrub that grows up to about a metre in height usually in clayey soil, and is a common sight along the roads and in the fields in the Cape. It characterises large tracts of veld known as renosterveld, and encroaches on cleared lands. It is grazed by sheep and cattle in spite of its resinous content, and is a valued fuel as it burns easily even when wet.

It is probably one of the oldest known Cape plants, recorded by Simon van der Stel in his journal in 1685.

Medicinally renosterbos has long been used for a variety of ailments, from flu and lack of appetite to diarrhoea and convulsions. The usual way of treating these ailments is to pour 2 cups of boiling water over 1 tablespoon of the fresh, stick-like leaves, leave the brew to draw for about 7 minutes, then strain and take 1 tablespoon every 2 to 3 hours.

A twig or two of renosterbos (2 pieces of approximately 10 cm in length) can be pushed into a bottle of brandy and left to draw, and a dessertspoon taken for dyspepsia, flatulent colic and heartburn.

A few twigs steeped in warm water also help to bring down a high temperature in children if the child is washed in the water. A cooled tea of renosterbos is still used in the western Cape today to soothe heatstroke, flu and convulsions and many coloured people still keep a twig in a wine bottle for heartburn!

Renosterbos came into its own during the influenza epidemic in 1918 and old timers still swear by it for coughs and colds.

The tip of the twig dried, powdered and taken with a little warm water a teaspoon at a time is still used as an excellent diarrhoea remedy for children.

The rather interesting smell of renosterbos makes it an excellent pot-pourri ingredient, and I find it holds the essential oil very well in a wild flower pot-pourri. Some farmers' wives in the Cape use renosterbos with lavender in moth-repellent sachets in their cupboards.

Height: 1–1,5 m

Resurrection plant
Myrothamnus flabellifolius

NDEBELE	Ufazimuke
SHANGAAN	Ikalimela
TSWANA	Monnakgang
VENDA	Ufazham-ka

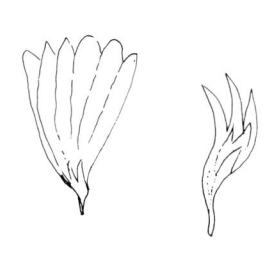

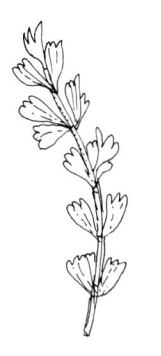

THIS is a most fascinating plant and one which has intrigued everyone who has brought back a dry, brown twig from a walk in the mountains and watched it become fresh and green within an hour or two after placing it in water. During the rainy season the bush is green, but in periods of drought or during the winter months it stands brown and brittle until the rains come again.

I have tried to coax it into growing in my herb garden for I know how it would draw the attention of visitors, but I have had no success no matter where I have placed the rooted cuttings. It seems to favour the granitic quartz-type rock outcrops, and grows strongly between the cracks, with a long fibrous root system. I have found it along the northern slopes of the Magaliesberg, and it grows as far north as Zambia and Zimbabwe, and east towards Swaziland and west along mountain ridges towards Angola.

The plant has important medicinal uses and was much used by the Voortrekkers and the early Rhodesian settlers. An infusion of 1 thumblength sprig of green leaves to 1 cup of boiling water, allowed to stand for 5 minutes and then strained, can be drunk to ease cold and flu symptoms, kidney ailments, haemorrhoids and aches and pains.

A strong brew (1 cup of green leafy twigs to 3 cups of boiling water — stand and draw for 20 minutes) makes an excellent wash for wounds and grazes and a cloth soaked in this hot brew can be applied to ease backache and haemorrhoids; hold in place over the area and replace at intervals.

Some African tribes burn the stems and roots and inhale the fragrant smoke to ease chest pains and congestion in the lungs; and some mix the dried leaves with tobacco and smoke it for relief.

If you are on a long hike over the mountains, the fresh green leaves (soak twigs in water first if you only find the dried plant) mixed into a handcream or massage cream, or even into butter, will do much to ease aching muscles if it is rubbed into the area well.

A leaf chewed will cleanse the mouth and the breath and many have found the plant makes a remarkable tonic tea, energising, uplifting and revitalising. Pour 1 cup of boiling water over 1 thumblength sprig. Stand for 5 minutes, then strain and sweeten with honey if desired.

A sangoma in the Pietersburg area once told me that the Shangaan use this plant to assist milk flow in nursing mothers, and for diseases of the breast. They take a tea internally and apply warmed poultices to the breast externally.

Height: 20–40 cm

Pompon tree

Sand olive

Renosterbos

Resurrection plant

Groot Salie
Salvia rugosa

Blue Sage
Salvia africana coerulea

Afrikaanse Salie
Salvia paniculata

Creeping Sage
Salvia repens

Vrystaat Salie
Salvia verbenaca

Brown Sage
Salvia africana-lutea

Narrow leaf Sage
Salvia stenophylla

Sand olive
Dodonaea viscosa

ENGLISH	Common sand olive
AFRIKAANS	Kaapse sandolien, gansiebos, bosysterhout
TSWANA	Tsekatseki
VENDA	Mutata-vhana
ZULU	Mutzuwe

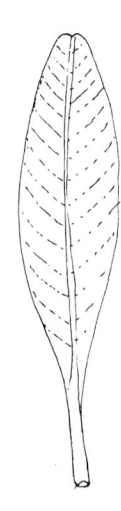

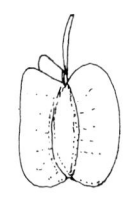

THE sand olive is an evergreen, often multistemmed shrub that is widespread in the Cape, Transvaal and the Orange Free State. It has bright, glossy, sticky leaves and a multitude of small, winged fruits that are pale green and greenish purple when they are mature. It is often planted as a hedge or windbreak and it is excellent in areas with difficult climates — hot summers, bitterly cold winters or dry, arid conditions — and it grows amazingly quickly. Pruned and trained it makes an attractive specimen shrub — I have seen one as a topiary tree with a sturdy straight stem and a perfect ball of leaves and fruits that is kept clipped into its round shape every month. It is quite charming and unusual, and in this garden, which is subject to strong winds and heavy frost, it is the perfect shrub to grow.

The seed is reasonably viable and planted in trays will usually give 100 per cent germination. If you have an established tree in the garden you will find lots of little sand olives coming up all over the garden waiting to be transplanted.

The sand olive has been used medicinally for many years for a great number of ailments and diseases. Probably its best known use is for stomach disorders and fever. A tea made by pouring 1 cup of boiling water over ¼ cup of leaves and a few fruits and allowed to draw for 5 minutes, is sipped slowly to ease colic, gripes, diarrhoea and nausea.

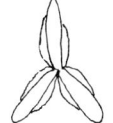

The same brew will help to bring down a fever, and a large quantity can be made and used as a wash when cooled to bring down fever and ease the discomfort of heat rash, heat stroke and inflammations. I have used it in the bath to ease heat rash and sunburn and have found it to be wonderfully soothing, and if a little is kept and applied frequently as a lotion it quickly eases and heals a rash or sunburned area.

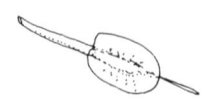

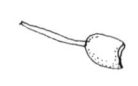

The same tea is excellent as a gargle for a sore throat, and can be sipped to ease chest conditions. A tea made of the root will ease colds, bronchitis and coughs. Boil up 1 cup of well-washed roots in 1 litre of water for 10 minutes, then stand and cool, strain the roots out, and drink ½ cup every hour or so for half a day. Thereafter ¼ cup taken at intervals 4 or 5 times during the day will ease the condition, although so often only 1 or 2 doses are needed before the condition eases.

The Tswana apply the leaf as a dressing for haemorrhoids, first crushing it and warming it in hot water. The leaf is considered to have analgesic properties and some tribes use it as a compress over sprained muscles. A visiting witch-doctor from Pietersburg showed me how he crushed leaves and applied them to help bad chest conditions, and also bound the leaves over the lower back to ease back strain and pain.

The farm children enjoy nibbling the fruits, which are surprisingly used as a fish poison although they are quite harmless to eat!

I use the leaves and the fruits in pot-pourris, and find that the aromatic oils in the pot-pourri are absorbed into those little winged dried fruits wonderfully, and they retain the fragrance for a long time.

Height: 2 m

Sausage tree
Kigelia africana

ENGLISH	Porkwood
AFRIKAANS	Worsboom
NDEBELE	Umvebe
SWAHILI	Mwegea
ZULU	Umzingulu

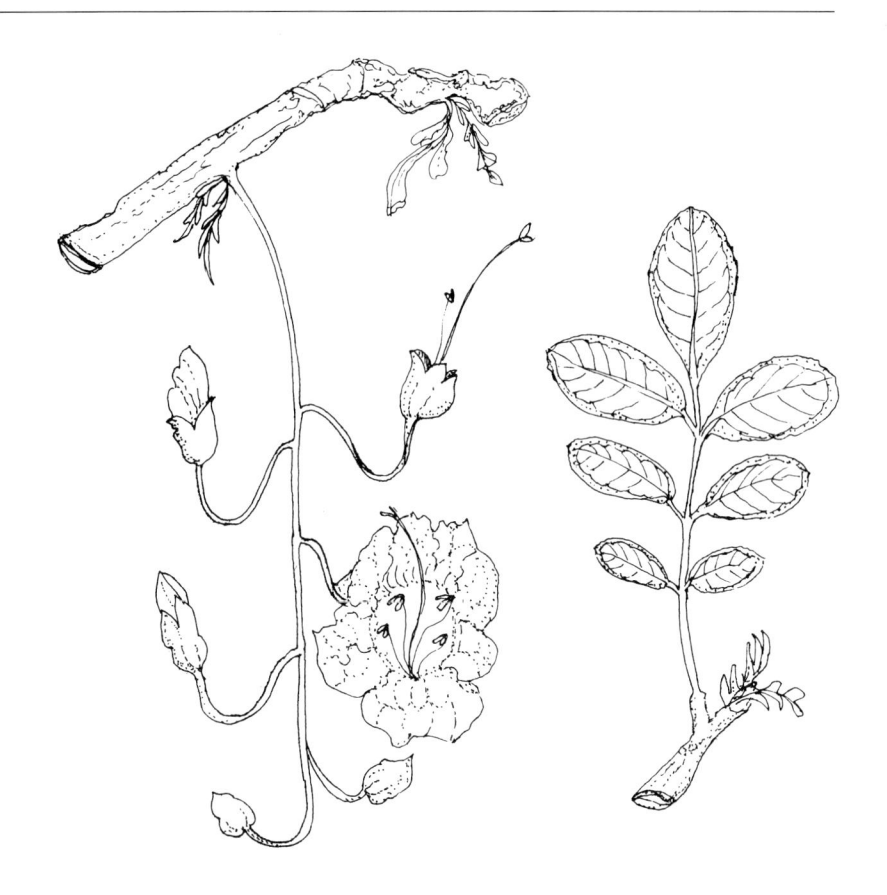

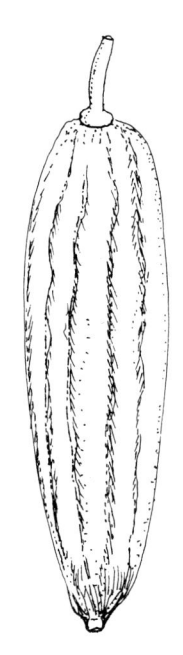

THE whole genus *Kigelia* is indigenous to Africa, but the only species to occur naturally in South Africa is *K. africana*. This is a beautiful medium to large tree which every visitor to the Kruger National Park cannot fail to notice. It does particularly well in the warmer frost-free regions of the Transvaal and Swaziland, but it does well in other areas too.

The tree has compound pale green leaves in summer and exquisite, dark red, cup-shaped flowers that hang in heavy sprays in late spring. The flowers have a lighter reddish green back and long yellow stamens and smell strange, but are laden with nectar and thus frequented by bees and other nectar-loving insects.

The fruit that follows the flowers is large, cylindrical and brown in colour, resembling a large sausage, and can weigh up to 4 kg! It does not burst open but dries within its thick pericarp and the inner pulp is thick with seeds.

The dried pulp of the fruit is powdered and used by the Zulu as a dressing for ulcers, and other tribes use the fruit also dried and finely powdered as a disinfectant, dusting it over slow-healing sores in cattle and in humans. The seeds can be eaten if they are roasted, but are not very palatable and are only used as a famine food.

Slices of unripe fruit are used as a poultice over sores, infected bites, syphilis and sores on the genitalia, and bound over painful rheumatic joints. Some tribes believe that slices of fresh fruit or the juice from the flowers and buds rubbed onto the breasts will help a new mother produce more milk for her baby. The sliced fruit rubbed over a baby is also believed to make it fat and healthy, but it is not applied to the head as this is believed to result in hydrocephalus.

The fruit should never be used as an internal treatment — it is a strong purgative and sometimes causes blisters in the mouth and on the skin.

The tree is regarded as holy, and the area beneath it as a religious place. Many blacks will hold a church service under this natural cathedral, and many believe that if the tree grows near your home you are protected.

The flowers are relished by buck and cattle, and long-tailed hares and porcupines will completely polish off all the fallen flowers during the night before they turn black and dry. In my garden I often keep watch to see who it is that eats the flowers so ravenously. The grass is dotted with fallen flowers at sundown and by morning every single one has disappeared! I have two trees, one flowering earlier than its companion, so the feast is spread over 4 or 5 weeks each spring.

The wood is soft, yellow and easily carved, and the people in Swaziland and in the eastern Transvaal hollow out the large tree trunks for canoes. The crushed twigs and dried fruits make an excellent fixative for essential oils in pot-pourri.

The sausage tree is a fascinating tree to grow and several nurseries offer young trees for sale. Plant the tree in a large, well-composted deep hole into which lots of moisture-holding fibrous plant material has been added. The tree will only do well in full sun and needs a deep heavy mulch of leaves and grass during a dry summer and the winter. After a very cold winter or a very dry summer the fruit does not form, but nevertheless the tree is so interesting and so quick to grow that it does not matter. It needs much water initially to get it started, but once it is established it does well on a deep fortnightly watering and the usual rainfall.

Height: 12–18 m

Scented geraniums
Pelargonium species

AFRIKAANS Wilde malva

THE scented geraniums or pelargoniums are a huge family, and are among South Africa's best loved plants.

A typical pelargonium flower has five petals. Two are broader than the other three, giving it an irregular shape, and it is this arrangement of petals that distinguishes the pelargoniums from the geraniums.

John Tradescant, the seventeenth century gardener and plant collector, was so enthralled with the abundant wild pelargoniums he found at the Cape that he immediately exported them. By around 1688 several species were well established overseas and the first hybrids which now run into thousands were bred from those first few.

The scented leaves were used by the Khoikhoi for poultices, teas and medicines and the early settlers soon learnt to make use of their soothing qualities. I have become so fascinated with their various fragrances that I have started a small collection of pelargoniums, and love using the leaves to flavour dishes and to scent pot-pourris. Some make exquisite pillows to calm one after a hard day, some are wonderful in jam and fruit puddings, and others make fragrant teas to soothe nausea, dysentery, diarrhoea, chest complaints, muscular tension and nervousness.

Cuttings taken from the mother plant are quick and easy to root, so keep a sharp look out for the different varieties at nurseries and in friends' gardens.

Below I give only a short list of the scented pelargoniums that I use and grow, but with an active pelargonium society in South Africa (see page 270) and all the exquisite books on pelargoniums now available, a fascinating hobby could be begun on the study of this beautiful group of plants.

Camphor leaf geranium (Pelargonium betulinum)

AFRIKAANS Kamferblaar

THIS pelargonium grows upright and has a small round leaf and a beautiful magenta flower. The leaves and twigs may be boiled in water and the steam inhaled to clear the head and nose in stuffy colds and sinus attacks. It is also used to clear coughs and chest ailments. The camphor-like fumes greatly alleviate these conditions.

A tea made from ¼ cup of fresh leaves to 1 cup of boiling water, allowed to stand and steep for 5 minutes and drunk a little at a time, is a treatment for flatulent colic, cramps and an over-full feeling.

Height: 30 cm – 1 m

Dysentery geranium (Pelargonium antidystericam)

AFRIKAANS Rooistorm wortel

THIS pelargonium is used in the form of a tea to treat dysentery, diarrhoea, nausea and digestive complaints. (Steep ¼ cup of fresh leaves in 1 cup of boiling water for 3–5 minutes, then strain. Add juice of half a lemon if desired, and sip slowly).

Height: 1–1,5 m

Hooded geranium (Pelargonium cuculatum)

Height: 1,5 m

SHOWY, beautiful and regal, this large pelargonium was one of the first to be cultivated as a hedge in the Cape in 1838. It was taken to Europe in about 1690, and is alleged to be another of the parent plants from which the garden hybrids throughout the world are derived.

The deep pink and magenta flowers can often be seen on mountainsides and along the roads in the south-western Cape where, particularly in spring, they make a spectacular sight. I grow the plant in all situations in the herb garden and find it does well in any soil, and there are usually flowers for salads all through the year. Like most of the pelargoniums, it does need some protection from frost.

The medicinal uses of this pelargonium are many. An excellent tea for colic, kidney ailments, suppression of urine, diarrhoea and aching joints can be made by steeping ¼ cup of roughly chopped fresh leaves in 1 cup of boiling water for 5 minutes, then cooling and straining. It is also used to bring down fevers and a strong brew added to the bath soothes aches and pains in joints and muscles of the legs and back.

The leaf can be warmed in water and crushed a little, and applied to wounds and bruises, stings and bites, cracked heels and abscesses — used as a poultice and bound in place with a crepe bandage or plaster.

The colonists were taught by the Khoikhoi to use the root as an astringent for diarrhoea — probably made into a tea, or even used as an enema.

This is a large, vigorous plant so give it plenty of space. I find it benefits from clipping, so I often use long branches of it in flower arrangements and it lasts for many weeks in water. When the arrangement is over I plant the branches, keeping them damp for a month or so, and so I continue to make hedges just as the colonists did all those years ago.

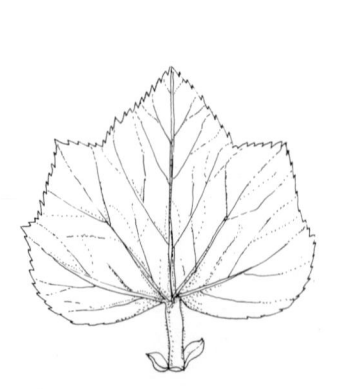

Ivy-leaved geranium (Pelargonium peltatum)

AFRIKAANS Kolsuring

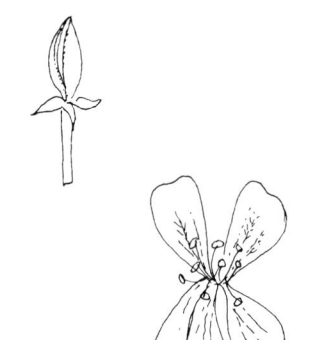

THE ivy-leaved geranium is a climbing variety that can often be seen trailing through bushes and hanging over rocks, particularly between East London and Port Elizabeth. Its pinkish mauve flowers make a splendid sight in spring, and this variety was introduced by Adriaan van der Stel to Holland, from where it became the original mother plant of the many beautiful hybrid ivy-leaved geraniums that are enjoyed all over the world today.

The leaves are somewhat succulent and the acid, sour-tasting sap is used to soothe sore throats. The buds and young leaves are an excellent thirst quencher. The pounded leaves make a good antiseptic for scratches, wounds, grazes and minor burns.

The petals also are astringent, and can be used in hot water to wash greasy skin: steep 1 tablespoon of petals in 1 cup of boiling water until pleasantly warm, then dab this brew soaked into pads of cotton wool over the face and shoulders.

The petals also produce a beautiful grey-blue dye, useful to weavers who use indigenous dye plants. The plants grow easily from slips.

Height: 3 m

Nutmeg geranium (Pelargonium fragrans)

Height: 15–25 cm

THIS is a deeply fragrant and very pretty cushion-shaped plant with tiny white flowers which grows beautifully in pots, forming a lovely grey-green round shape. The flower petals are delicious in fruit salads, and a leaf in a cup of coffee imparts a delightful spicy taste.

Rub a handful of leaves onto aching heels and calf muscles, and you'll be surprised how quickly it eases the ache.

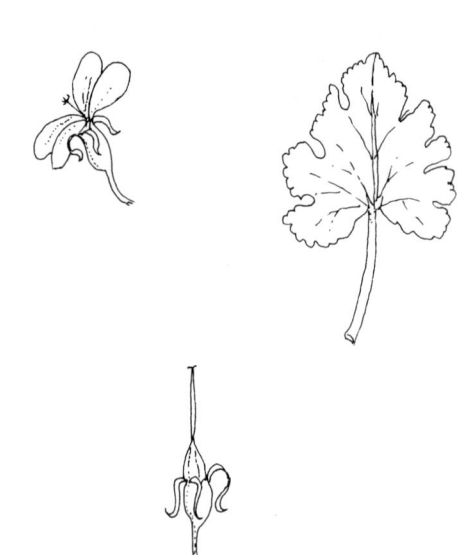

Oak-leaved geranium (Pelargonium quercifolium)

AFRIKAANS   Muishondbos

THIS strong-smelling pelargonium is common in the southern Cape and is often sold in nurseries. Medicinally it is used to treat rheumatism, high blood pressure and heart disease — ¼ cup of fresh leaves in 1 cup of boiling water stand then strain, is the usual dose taken daily.

The attractive oak-shaped leaf has a dark marking along the midrib and has a pungent smell, hence the name muishondbos.

I have found it to be a most useful ingredient in pot-pourris as its strong fragrance blends well with other scented leaves, and it seems to hold the essential oil exceptionally well.

I fill a jar or tin with dried oak-leaved geranium leaves, and a generous dash of scented geranium oil, which I seal and shake well. After 2 weeks of shaking up daily I have a strong, fresh and vigorous ingredient to add to other mixtures, and the oil glands in the leaves hold the essential oil for a long time. I find it particularly effective mixed into a moth-repelling sachet.

Height: 0,5–1,5 m

Peppermint geranium (Pelargonium tomentosum)

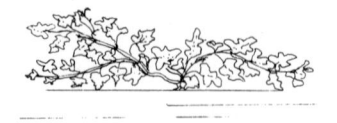

Height: 20–30 cm

OFTEN found in the south-western Cape, the peppermint geranium has become a treasured garden plant which is excellent in pots and hanging baskets. Its big velvety peppermint-scented leaves are exquisite in pot-pourris, rice pudding and in pillows. The leaf also makes a soothing poultice over a bruise or sprain.

I use it as a ground cover and find it needs clipping back every now and then. It grows well in sun and in partial shade, but to keep it at its best I find it necessary to start new cuttings every season as it can become straggly.

Peppermint geranium cool drink   Serves 6–8

10 peppermint geranium leaves
1 litre boiling water
1 litre apple juice
little honey to sweeten

Pour the boiling water over the leaves. Stand and steep until cool, then strain. Add apple juice and honey. Serve with ice.

This is wonderfully refreshing on a hot summer afternoon and this same tea can be made into a cooling jelly by adding 4 tablespoons of gelatine (mix the gelatine with a little hot water to dissolve it first). Set in the fridge.

Raspleaved geranium (Pelargonium radens)

THIS attractive plant with its strong almost chemical-like scent and clusters of pink flowers is found in many places, including the mountains of the south, the eastern Cape and in some areas of Natal.

The flowers are delicious in fruit salads, and as a garden subject it arouses much interest as it is attractive all the year round. It is able to withstand frost, and cuttings root easily.

The leaves are thick with oil and feel almost sticky. They are an excellent fixative in pot-pourris and insect-repelling mixtures as they absorb the essential oils and 'fix' them, making the pot-pourri long lasting.

I also use the leaves in pressed flower pictures as their shape is exquisite and the deep green colour remains unfaded for many years.

Medicinally *P. radens* is used as a rub for aching legs.

Height: 1,5 m

Rose-scented geranium (Pelargonium graveolens)

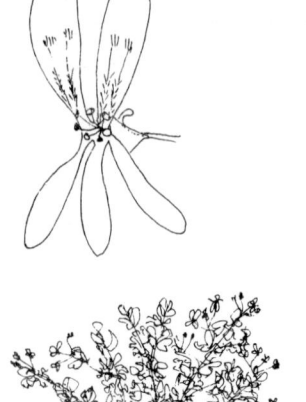

THE fragrant rose-scented leaves of this geranium can be made into a calming, soothing, sleep inducing tea by pouring 1 cup of boiling water over ¼ cup of fresh leaves, leaving it to stand and draw for 3 minutes, then straining it and drinking it sweetened with a little honey. The tea was one of the original remedies in the Cape for diarrhoea, dysentery, nausea, vomiting and colic, taken without the honey.

The leaves are a most beautiful flavouring for cakes, puddings, cooldrinks and scones, and I use them extensively at the cookery school. We also serve the most wonderful scented geranium scones to students at the Herbal Centre, which have become a great favourite.

A bowlful of freshly picked leaves next to the bed will do much to give you a good night's sleep. Crush them a little as you turn out the light. After a hectic day when I am overtired, I pick a few handfuls of fresh rose-scented geranium leaves and fill a baby's pillow case with them, which I tuck behind my pillow. I wake up the next morning clear headed and full of vitality after a peaceful night.

Used in pot-pourris the leaves give a beautiful fragrance and a pure rose-scented geranium pot-pourri is always a very special success.

Height: 1–1,5 m

Rose-scented geranium scones

Makes 1 dozen

3 cups cake flour
1 cup Nuttiwheat flour
250 g butter
2 tablespoons chopped scented geranium leaves
4 tablespoons sugar (optional)
1 teaspoon salt
8 teaspoons baking powder
1½ cups milk

Rub butter into dry ingredients. Add chopped scented geranium. Mix in milk to make a fairly soft dough. Turn out dough onto a floured surface, pat out to about 2 cm in height. Cut out shapes. Bake at 200 °C for about 15 minutes until light brown.

Rose-scented geranium pot-pourri

4 cups dried rose-scented geranium leaves
1 cup minced dried lemon peel
½ cup broken cinnamon pieces
½ cup lightly crushed cloves
scented geranium oil

Mix all the ingredients and store in a sealed jar. Shake daily for 2 weeks, then add a little more oil if necessary, shake well, and place in bowls or sachets about the house. I love it best of all in my night dress case.

Wild rose geranium (Pelargonium capitatum)

Height: 25–80 cm

THIS is a velvety-leaved, low, spreading plant with soft wood that is commonly found along roadsides, on rocky outcrops, in flat sandy stretches all over the southern and eastern Cape and into Natal. It is a remarkable skin softener, and the sweetly scented leaves can be rubbed into the hands to soothe callouses and scratches, into the heels to soften horny, cracked skin and can be tied into a piece of muslin and used in the bath as a wash and skin treatment, which also soothes rashes.

A tea made of the leaves is an old remedy used by the people of the Cape for treating kidney and bladder ailments, stomach cramps, nausea, vomiting, diarrhoea and flatulence. A quarter of a cup of fresh leaves in 1 cup of boiling water is the standard brew — stand, steep for 5 minutes, then strain and drink. This is one of the species cultivated in France for oil of geranium.

Sickle bush

Sorrel

Silver leafed vernonia *Star flower*

Spekboom

Toothache root

Stork's bill geranium

Sour fig

Edible pelargoniums

Several species of pelargonium are edible. Either their 'tubers' or thickened roots are eaten for their high starch content, or their leaves for their flavour. Amongst them are:

P. acetosum

ENGLISH   Sorrel leaf

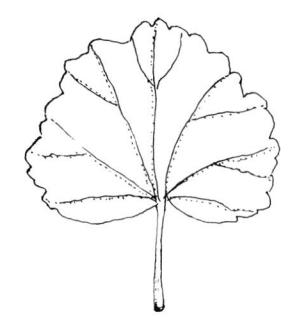

THE young leaves and buds are eaten in salads, and added to soups and stews. The acidic lemony taste is pleasing, and the coloured people and the Xhosa use this pelargonium in their cooking.

Height: 60 cm

P. bowkeri

SOTHO   Bolila khomo, khvaara, kxoara

THE leaf is eaten as a vegetable, often cooked with other vegetables, and the root or 'tuber' can be made into a porridge.

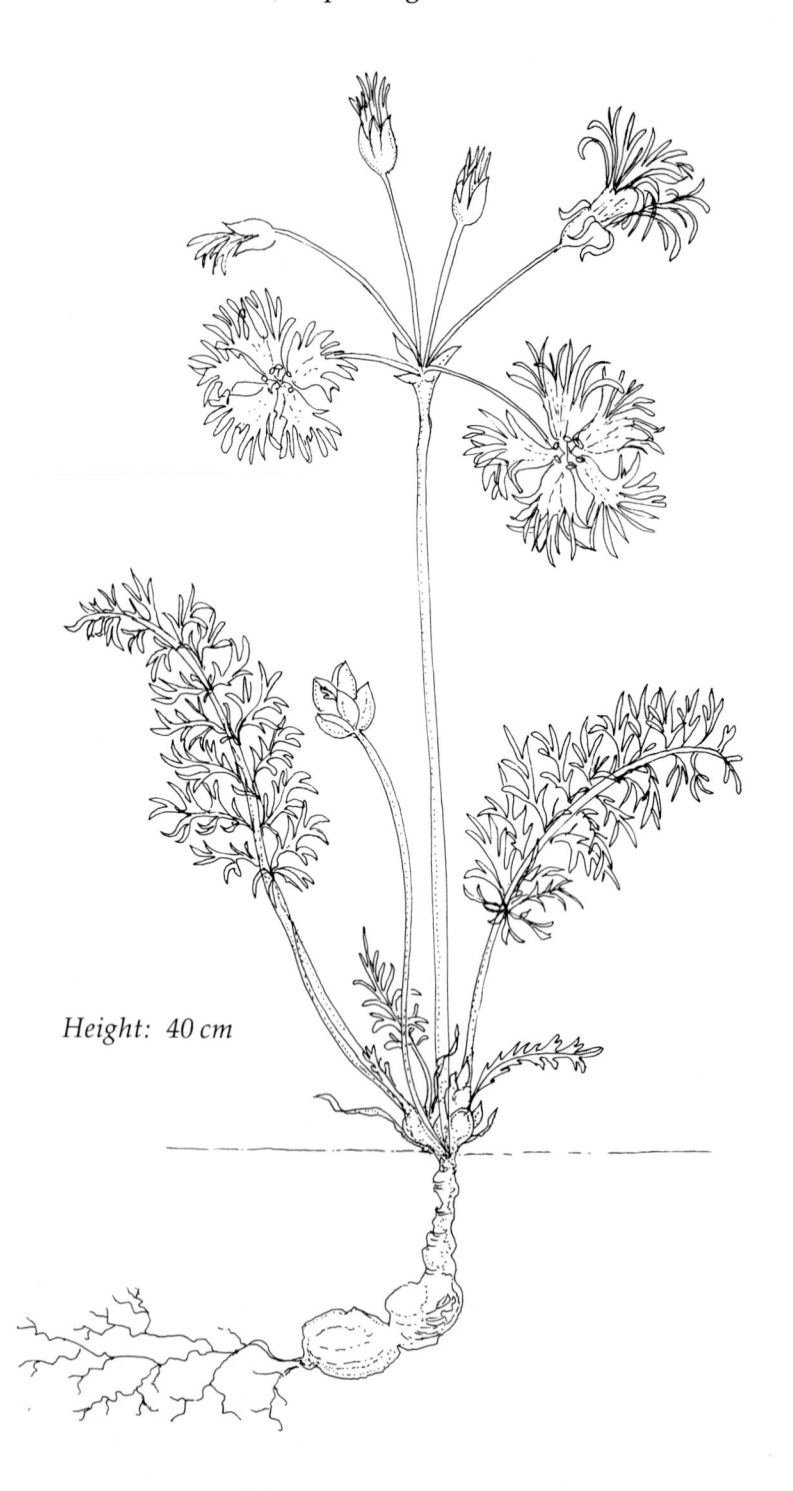

Height: 40 cm

P. lobatum

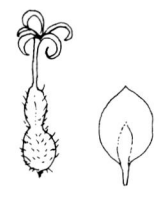

THIS is used in cooking in the same way as *P. triste*.

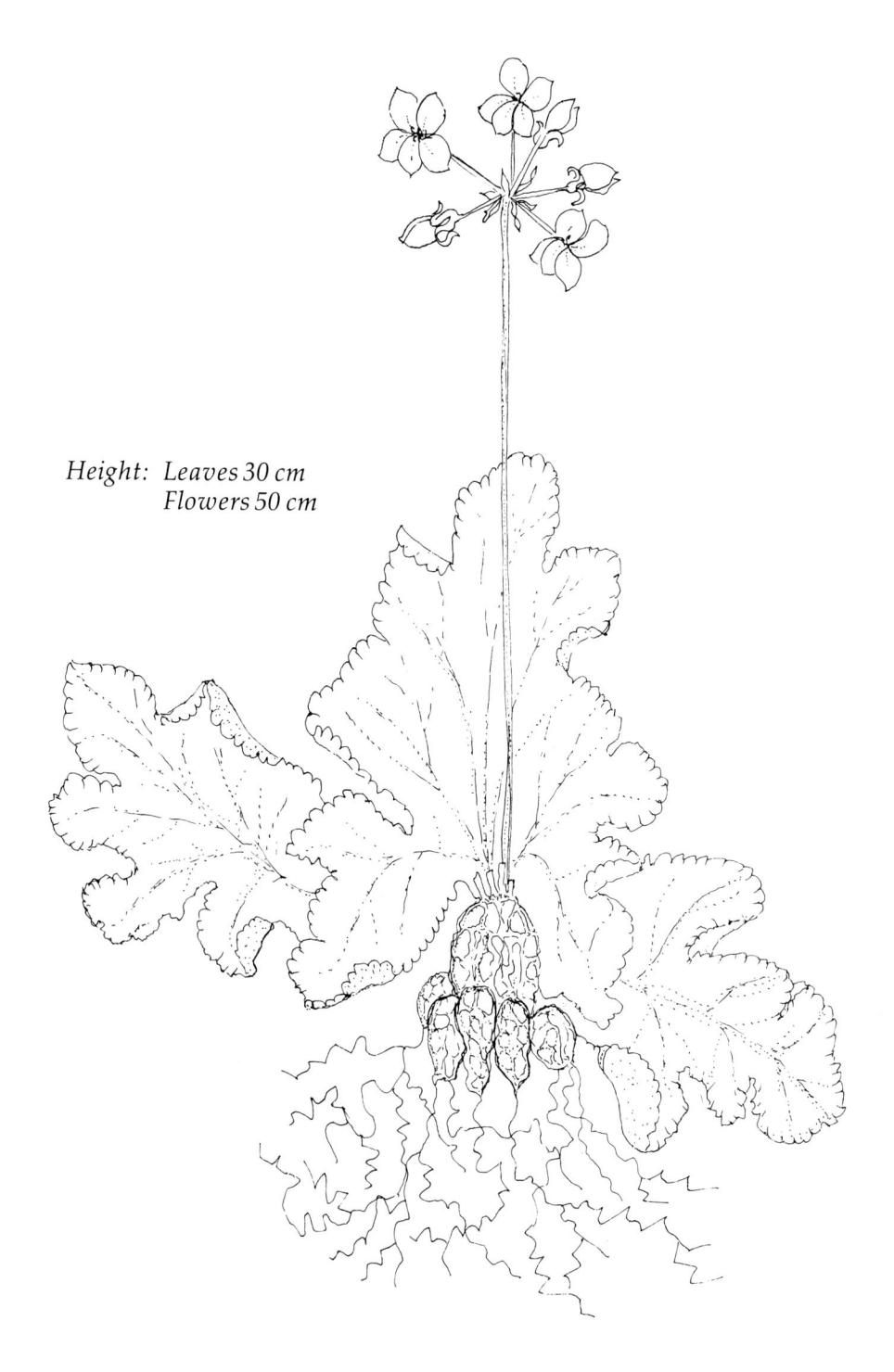

*Height: Leaves 30 cm
Flowers 50 cm*

P. rapaceum

AFRIKAANS   Bergpatat, bergaartappel, norra, norretjie

THE people of the Bokkeveld and Namaqualand eat the roots of this geophytic species, roasting it in the ashes like a sweet potato. The skin is then split open to reveal the fragrant flesh inside.

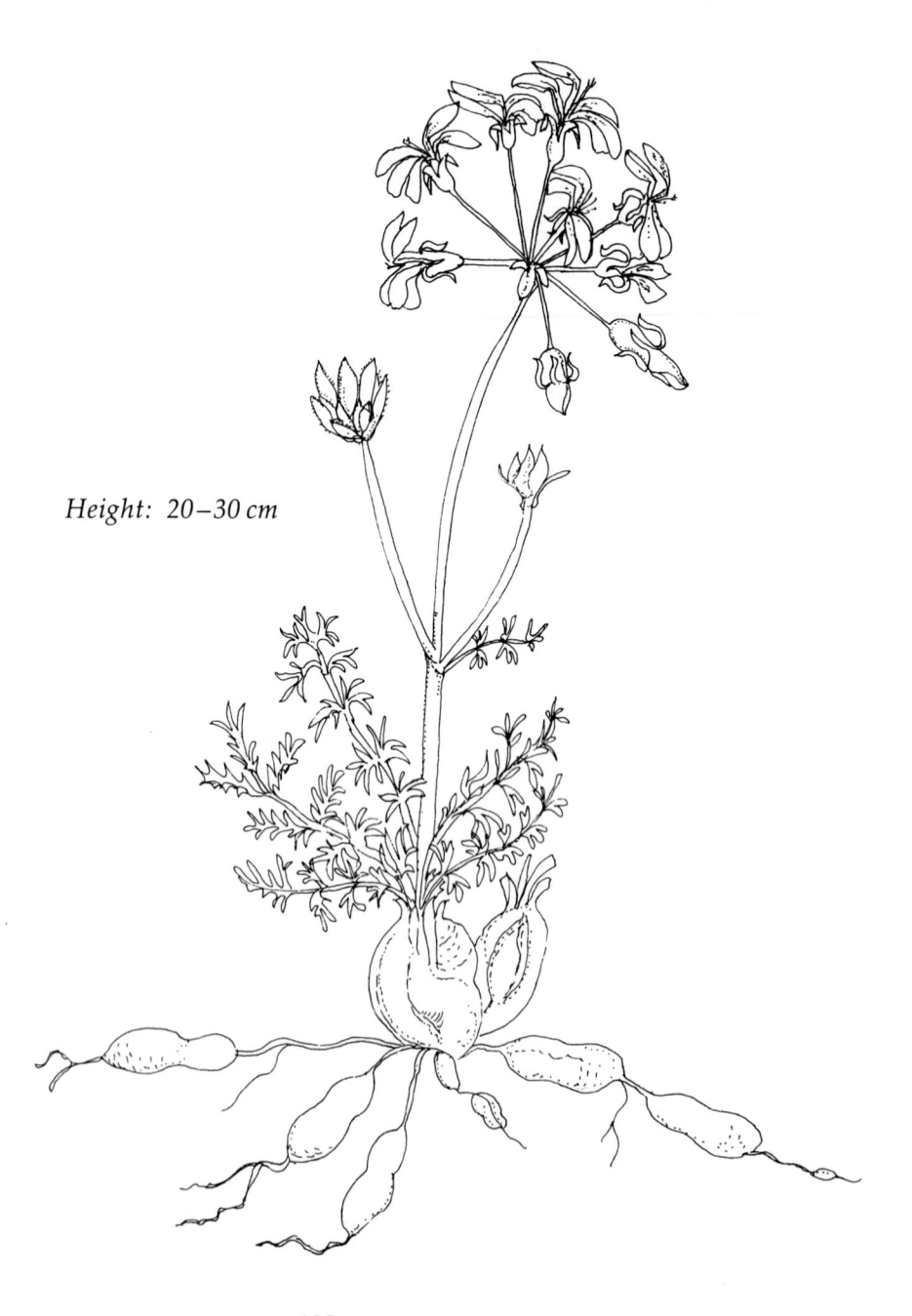

Height: 20–30 cm

P. triste

ENGLISH Sad geranium
AFRIKAANS Kaneelbol

THE young roots are used chopped in stews and soups and mixed with other vegetables like potatoes and onions. This species is used in the form of a tea as a treatment for dysentery, diarrhoea, nausea and digestive complaints.

Height: 30 cm

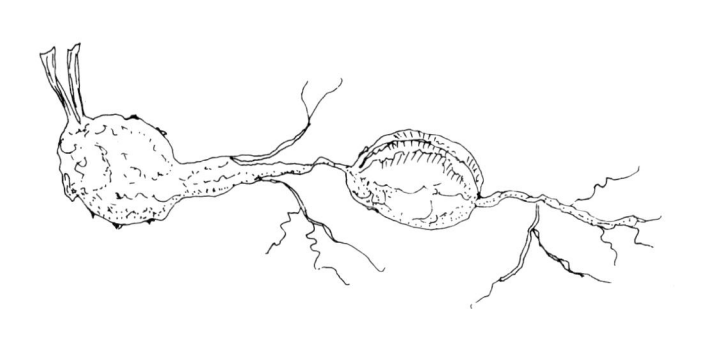

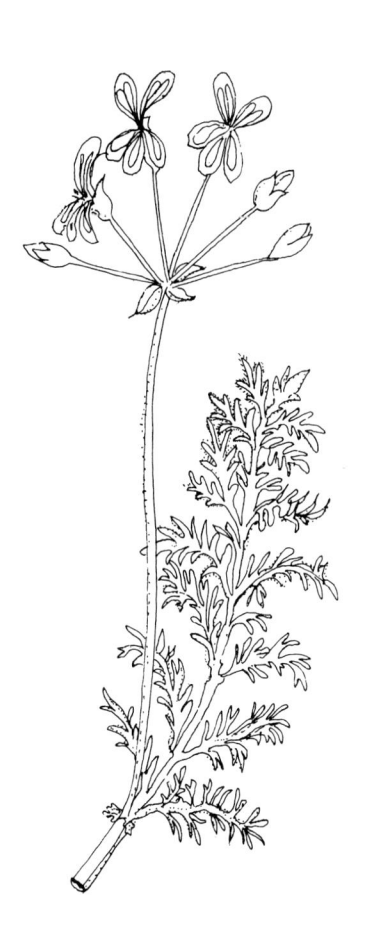

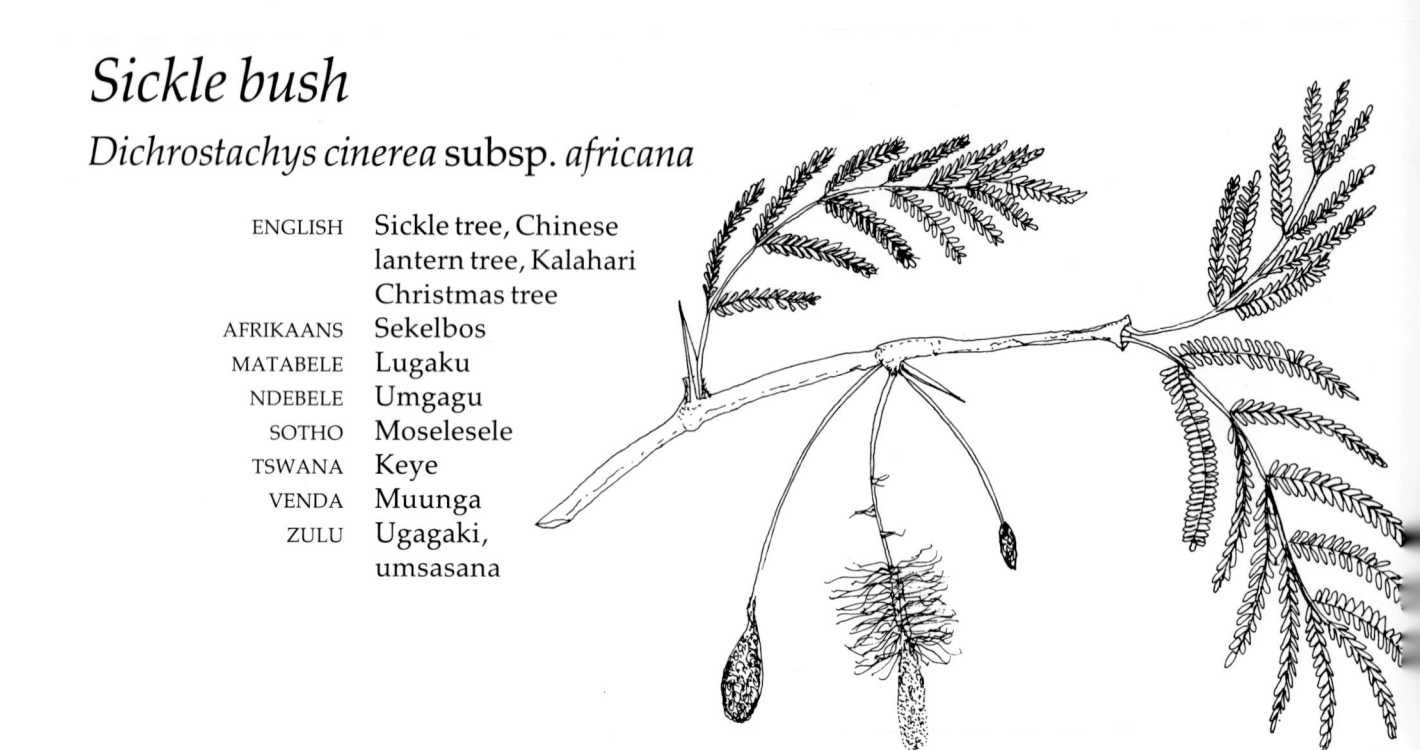

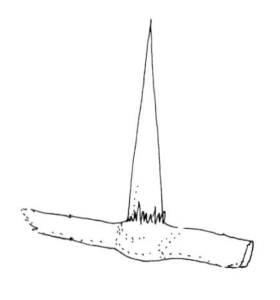

Sickle bush
Dichrostachys cinerea subsp. *africana*

ENGLISH	Sickle tree, Chinese lantern tree, Kalahari Christmas tree
AFRIKAANS	Sekelbos
MATABELE	Lugaku
NDEBELE	Umgagu
SOTHO	Moselesele
TSWANA	Keye
VENDA	Muunga
ZULU	Ugagaki, umsasana

A SPINY, deciduous shrub that grows up to 3,5 m in height, the sickle bush is characterised by its pink and yellow catkins — the upper third is mauvy pink and the lower two-thirds yellow. The branches are spreading, with acacia-like leaves, and the drooping pendulous flowers appear in spring and summer, followed by oddly contorted sickle-shaped pods borne in clusters, giving rise to its common name. The wood is hard, strong, termite and borer proof and is used in the Transvaal as excellent fencing posts, as it is also able to withstand veld fires. The bark is tough and pliable and makes strong ropes and string.

Its medicinal uses are many. A lotion made of the leaves and bark is a wound cleanser and healer and dried powdered bark is sprinkled onto the wound to promote healing. The same lotion or a cold water infusion is used to rinse out the mouth and to soak a cloth which is bound round the head as a soothing headache remedy. In Zimbabwe the dried leaf and root are smoked for pulmonary tuberculosis and chest ailments, and to clear the head during a cold.

The leaf can be chewed to ease colic and heartburn, and made into a tea for stomach ailments and diarrhoea. (Pour 1 cup of boiling water over 4 leaves, stand and steep for 5 minutes, then strain.) Dried powdered bark is used by many blacks as a direct application to skin eruptions, sores, blisters and abscesses for both man and animal.

A decoction of the plant root is often administered to a woman after giving birth to relieve pain, as the plant is considered to be a natural painkiller. Snakebites, scorpion stings and insect stings are also treated with the leaf and bark — the leaf is chewed well and then applied and bound over the area. The inner bark was once also used as a cord tourniquet in the case of snakebite.

The sickle bush occurs in the Transvaal bushveld, Natal, on into East Africa and westwards into Zimbabwe, Botswana, Namibia, and northwards into tropical Africa. In each country the people have evolved their own remedies, and it is fascinating to discover how these differ. For example in French Guinea it is used to heal urinary infections and as a purgative, in Sudan for urethral ailments, in Liberia for sore throat and venereal diseases, in Central Africa for leprosy and syphilis, and in Tanzania it is eaten with mealies as an aphrodisiac!

The dried seedpods are said to make a soothing eyewash for red, tired eyes, by steeping 2 or 3 pods in a cup of warm water for an hour. The Pedi and the Lobedu use it as an anti-witchcraft charm and in tribal ceremonies to ward off evil. This plant is respected and revered by primitive blacks, who believe that if it grows in a garden or near a house no evil will befall the owner.

Among the long list of ailments the sickle bush is used to treat are catarrh, elephantitis (taken internally as a tea, and externally as an application), circumcision wounds, bronchitis, pneumonia, epilepsy, internal abscesses, dysentery and kidney ailments. I often look at the shrub and marvel at those who knew how to use the plant in the beginning. Which race started it all, and what made them experiment with this vast number and variety of ailments? Who could first have discovered that it is a remarkable pain killer?

Perhaps if more farmers knew of its marvellous medicinal values they would not curse it the way they do, as it often forms tough, thorny, impenetrable thickets which are difficult to eradicate.

It needs a warm, frost-free climate, friable soil and tends to spread rapidly in the wild, but in the garden it can be pruned into an attractive small spreading tree that makes a most interesting garden subject. It can withstand long periods of drought, which makes it valuable in water restriction areas. It grows easily from seed, which falls in dark brown clusters beneath the shrub in winter. So do consider one of nature's wonderful medicinal trees for your garden, if only for a talking point!

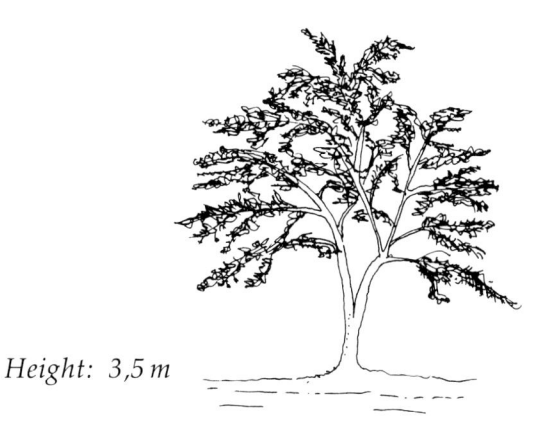

Height: 3,5 m

Silverleafed vernonia
Vernonia oligocephala

AFRIKAANS	Maagtee, amarabossie
SOUTHERN SOTHO	Mofefabana
TSWANA	Sefafatse
ZULU	Inyathelo

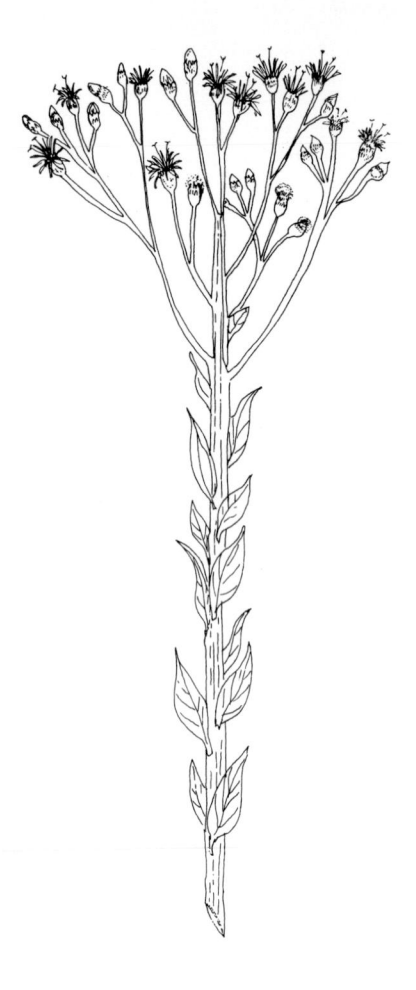

THE attractive silver leaves and purple flowers of the silverleafed vernonia are conspicuous in the early summer all along the roadsides and in the veld in the Transvaal and the Free State, Natal, the northern and eastern Cape and Namibia. They are at their best after the spring rains in October, when their long straight stems with evenly spaced silverbacked pointed leaves and tufts of small maroon thistle-like flowers draw the eye. In some places they grow in great sweeping drifts of silver dusted with purple.

There are many species of vernonia indigenous to South Africa, but the silverleafed variety is the most common, and much used medicinally. Of all its uses the one I find most appealing is the African belief that if you tie a bunch of twigs to a stick and wave it in the direction of an approaching storm it will divert the hail. Every time I see a silverleafed vernonia on the farm I make a mental note of where to find it when I need it, as where I live the hail is often devastating, but after its flowering period the plant is difficult to relocate! The Tswana and Sotho also make a fire of it, the smoke of which is supposed to divert a hailstorm.

Medicinally the silverleafed vernonia is important in the treatment of stomach ache, dysentery, diarrhoea, rheumatism, diabetes and to bring down fevers. The plant is made into a tea by pouring 1 cup of boiling water over a thumblength sprig of leaves (less than ¼ cup of fresh herb), steeping for 5 minutes and this is then drunk, usually sweetened with honey or sugar as it is so bitter. Drink it without sugar for diarrhoea, dysentery and diabetes.

The leaves can be boiled as a vegetable, and the Tswana on the farm make this as a spring vegetable to cleanse and strengthen the blood and to clear the system of winter coughs and colds. To lessen the bitterness of the leaves and stems boil them up once and strain the water off immediately, then boil them again in fresh water until they are tender (about 8 minutes). The leaves are very bitter when eaten raw, but the boiling seems to take much of the bitterness away.

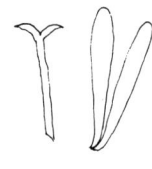

The Tswana also use the plant in the form of a tea, taken as a purgative. A cup of tea is drunk first thing in the morning on an empty stomach and repeated after lunch if necessary.

The Tswana and the Venda living near the farm where I live believe that the leaves are disinfectant and cleansing and use them in a wad to wipe a dirty child. A sangoma told me that this is believed to strengthen the child too.

Rural whites set much store by the silverleafed vernonia in the treatment of chronic constipation and more serious bowel complaints like ulcerative colitis. I find the old country remedies fascinating for so often the simple treatments handed down from generation to generation have their foundation in old-fashioned trial and error, and the treatments are often remarkably effective.

The silverleafed vernonia is so worthwhile to grow in the garden. I pick the mature flowers for dried flower arrangements and pot-pourris. Watch for ripening seed, as this is the only way of propagating the plant. It needs full sun and to be left undisturbed, and seems to do well in all soil types.

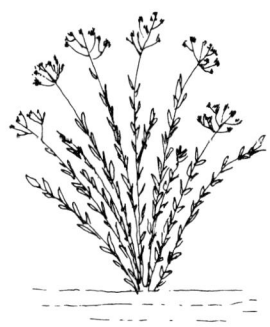

Height: 30–40 cm

Sorrel

Oxalis pes-caprae

ENGLISH	Wood sorrel
AFRIKAANS	Suring, wilde suring, geelsuring, klawersuring
ZULU	Isithathe, isitate

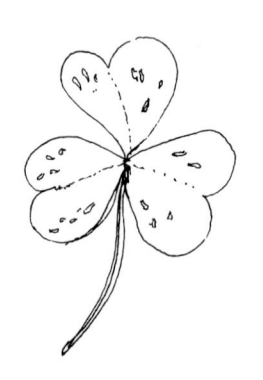

SORREL is a general name given to several indigenous and some introduced species that have an acidulous sap. These are usually the *Oxalis* and *Rumex* species, and this often gives rise to confusion. But here I am writing of the indigenous yellow sorrel that is traditionally used in South African cooking, and which makes a most attractive garden plant.

The triangular, lobed leaves have a sour taste and are used as a salt substitute by both the Xhosa and the Zulu. The leaves can also be eaten fresh in salads, and the finely chopped corm makes a delicious garnish for grilled fish.

The leaves make soothing dressings on burns and scrapes, and leaves warmed in hot water can be applied as a poultice on boils, abscesses and suppurating sores, held in place by a crepe bandage or sticky plaster.

Like all the sorrels with their high oxalic content, they should be used sparingly in the diet, for although they are nourishing, too much may cause digestive disturbances and heartburn.

With its high oxalic content the plant was much prized in the treatment of scurvy, and the corms or 'bulbs' were taken on board the ships calling at the Cape to keep the sailors healthy. If they were dried they could be soaked in water and still used effectively.

The corms are said to be a remarkable vermifuge, but I have not tried this. A dessertspoon on an empty stomach first thing in the morning seems to have been the usual dose used by the early Cape farmers for the farm dogs every spring, and they swore by it.

I am always delighted to see the first yellow flowers in early spring, for the sorrel makes a beautiful edging plant in the garden, and I grow it successfully in pots too. The flowers are not only delicious but enchanting sprinkled onto fish dishes and salads and I always have a few right up until May.

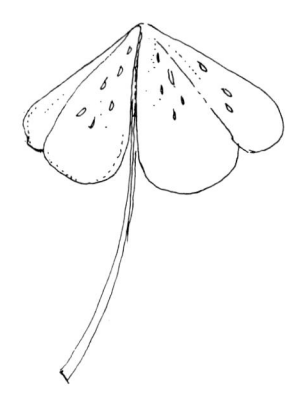

The leaves can be added to soups, stews and sauces for piquancy, and in traditional Cape cookery 'suring' has a special place. I love it best in waterblommetjie bredie (page 209).

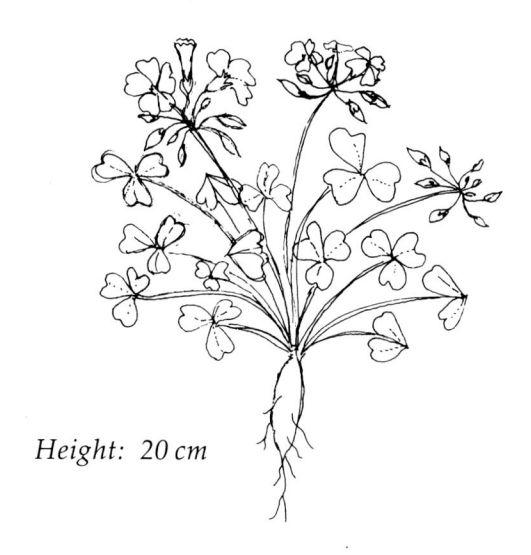

Height: 20 cm

Sweet and sour yellow sorrel mayonnaise

²/₃ cup condensed milk
2 teaspoons mustard powder
1 egg yolk
¼ cup oil
¼ cup vinegar
½ cup chopped sorrel leaves and a few flowers
dash cayenne pepper

This makes a delicious sauce or dressing for chicken, fish and salads. My favourite way of serving it is over steamed cauliflower — the hot vegetable absorbs the delicate taste of the sorrel, making it a dish fit for a king!

Whisk the condensed milk, mustard, egg yolk, oil and vinegar together until they start to thicken. Fold in sorrel and cayenne. Store any left over in a screw top jar in the fridge, where it keeps well.

Sour fig
Carpobrotus edulis

ENGLISH	Hottentot's fig
AFRIKAANS	Hotnotsvy, ghaukum, gouna, perdevy, suurvy, vyerank, vygie
SOTHO	Moriana-wa-ditsebe
ZULU	Ikambilamabulawo, ikhambi lamalawu

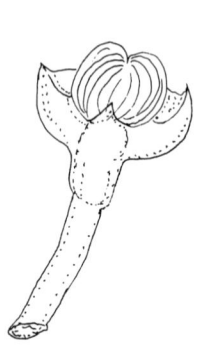

THE sour fig is a useful plant to have in the garden, not only for its remarkable medicinal values but for its ability to survive almost anywhere with the minimum of attention. It is a neat, undemanding groundcover that covers bleak ungardenable ground, steep banks and rocky outcrops with its attractive succulent leaves. It needs no attention, very little water and the rewards are numerous, making it a must for every garden.

The Khoikhoi or Hottentots were the first to make use of the plant, hence its name, and generations of black people have made use of its remarkable medicinal qualities. The three-angled succulent leaves contain an astringent juice which is antiseptic and can be mixed with water and taken internally for treating diarrhoea, dysentery, as a gargle for treating sore throats and mouth infections (just the leaf tip chewed will quickly ease a sore throat), as a lotion for bruises, scrapes, cuts, grazes and sunburn. It is also effective as a daily application for ringworm and in the treatment of infantile eczema. Probably its best known use is as a soothing application for blue bottle stings, and you will find it grows prolifically and conveniently along the sand dunes. The crushed leaf needs to be constantly applied until the sting eases in intensity. It also makes a soothing application for sunburn — just squeeze the juice from the crushed leaves over the sore area. Surprisingly, it also acts as a purgative if brackish water is drunk after eating a few of the 'figs'.

An old remedy for tuberculosis which was much respected and highly nutritive consisted of equal qauntities of honey, sour fig leaf juice and olive oil, mixed well, diluted in water and taken 3 times a day approximately 2 tablespoons at a time.

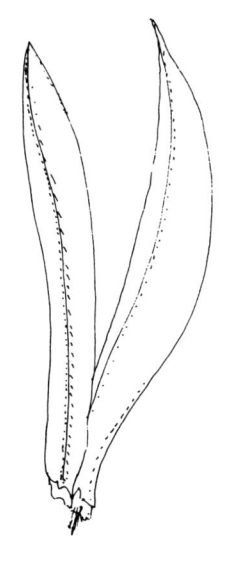

The Khoikhoi used an infusion of the leaf during pregnancy to ensure a strong, healthy baby and this was also an effective diuretic.

I have used the juice of the leaf to relieve the itch of mosquito bites, spider bites and tick bites and I have found that the juice applied at least 3 times a day to infected tick and flea bites on the dogs is the best way to stop them scratching!

There are several varieties of sour fig, some with a yellow flower and some with a brilliant magenta flower. Some produce a more succulent 'fig', and where I live in the Transvaal, the magenta-flowering one produces the best fruits. Often you will find baskets of dried sour figs in the Cape and Natal markets and these made into jam and syrups are a traditional sweetmeat much prized overseas.

The plant was taken to England and Europe in about 1700 and cultivated in greenhouses there, and it now grows prolifically on the shores of the Mediterranean and along the south coast of England — I often wonder if those people know of the sour fig's wonderful medicinal qualities! The flowers open in the sunlight and are industriously worked by the bees, and the spring honey from the sour figs has a wonderfully unique taste. A little stirred into hot water and sipped is used to soothe sore throats. In spring the flowers are at their brilliant best and it is at this time too that the honey is at its most glorious.

Some years ago a friend living in Durban told me of an effective treatment for vaginal thrush taught to her by her grandmother. In hot, humid climates this very irritating and uncomfortable condition does not seem to clear up easily and many women living in the heat of the Natal coastal region verify this simple yet remarkably effective treatment.

Height: 10–12 cm

Sour fig douche

¼–½ cup fresh finely chopped sour fig leaves
¼–½ cup apple cider vinegar (or plain dark brown vinegar)
2 litres warm water

Steep the sour fig leaves in the vinegar for an hour or two (for a strong brew use ½ cup leaves soaked in ½ cup vinegar for 2 hours), strain, pour into the warm water, and use this as a douche.

Use a fresh mixture every evening for 5 to 7 days, then leave off the treatment for a week. Should the condition not yet be cleared repeat for another week. Remember, however, as with all home treatments, to consult your doctor first.

Sour fig jam

ripe, dried figs
500 g sugar for every 500 g fruit
1 litre water to every 500 g sugar
3 tablespoons lemon juice for every 1kg fruit
cinnamon stick
1 thumblength piece ginger root for every 500 g fruit

Soak the ripe dried figs in boiling water and leave overnight. Next morning pare off the hard bottom section and boil the fruits in enough water to cover them until they are tender. Drain.

Boil the sugar and water, then add the fruit. Add the cinnamon and ginger root. Boil until the syrup becomes thick and dark. Discard the ginger and cinnamon and pour into hot, sterilised bottles. Seal.

Eat on hot buttered toast, or add a spoon to mealie meal porridge at breakfast, or stir into rice pudding.

Spekboom
Portulacaria afra

ENGLISH	Porkbush, elephant's food
AFRIKAANS	Olifantboom
XHOSA	Qyanese, igwanitsha
ZULU	Intelesi, isicococo, isidibiti esikhulu

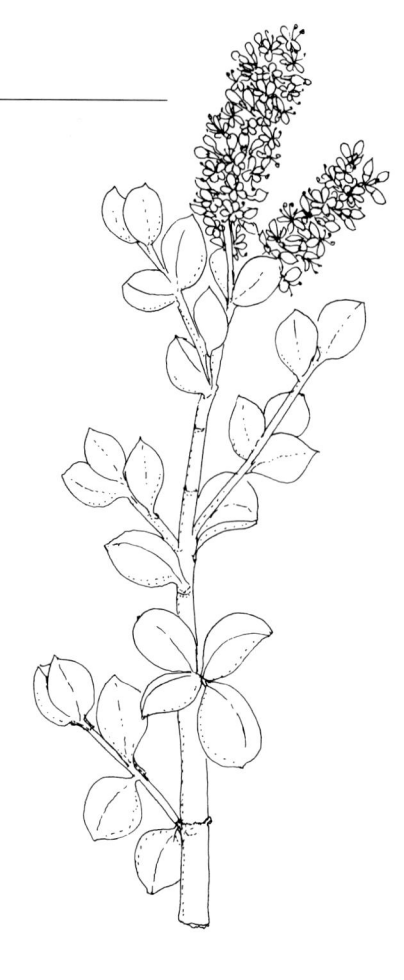

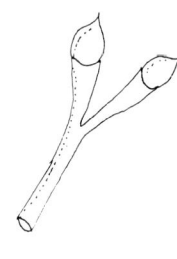

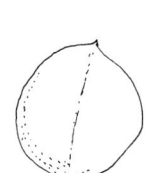

THE spekboom is an attractive shrub or small tree that grows up to 3 or 4 m in height, occurring on rocky hillsides and often in succulent scrub. It is a valuable stock food, also much loved by elephant and buffalo.

I have found spekboom to be an attractive and rewarding garden subject as it roots easily from cuttings and is drought, heat and frost resistant. It can be clipped into a neat hedge and its froth of pink flowers in midsummer are an absolute joy. It is much loved by bees and the honey from the spekboom — if you are lucky enough to find a farmer or beekeeper willing to sell it to you — is unsurpassable in flavour and texture. The farmers in the Karoo have found that there are in fact two varieties of this honey, which seem to be distinguishable in taste only — one is sweeter than the other.

The spekboom leaves with their astringent, lemony taste are thirst-quenching and if a leaf is held in the mouth and sucked it will help over-exhaustion, heatstroke and dehydration. This is a helpful trick for hikers and mountain climbers, and a sprig should be tucked into a pocket before setting off. The juicy leaves rubbed over blisters and corns on the feet will quickly soothe and heal.

Medicinally the leaf is chewed and sucked as a treatment for sore throat and mouth infections, and the astringent juice is soothing and antiseptic for skin spots, pimples, rashes and insect stings. The Xhosa in the eastern Cape use the leaves as a poultice for sores and infected bites, and also squeeze the juice of the leaves onto the sore. The juice is also an effective sunburn treatment, squeezed onto the area.

A spekboom sprig laid over a gently simmering tomato bredie a few minutes before it is served, imparts a delicious flavour. The sprig is 'steamed' above the bredie and discarded before serving. I first tasted the recipe below sitting around a camp fire after a long day's hike in the Barberton area, with the sun setting pink and gold behind the mountains, the firelight glowing and the glorious smell of cooking filling the air. The elusive and tangy taste of the spekboom twig made that fireside meal unforgettable.

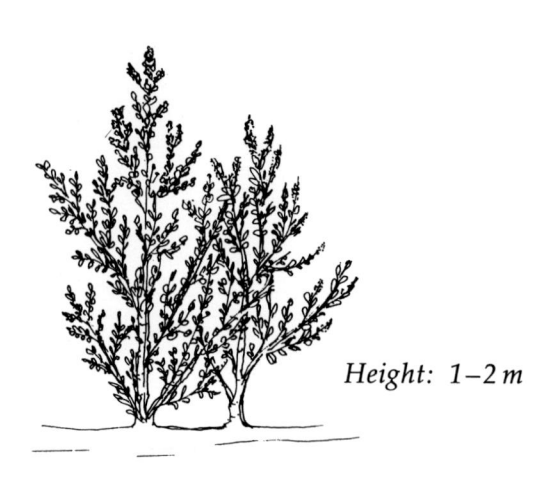

Height: 1–2 m

Tomato bredie with spekboom — Serves 8

8 large tomatoes, skins removed
6 carrots, diced
6 large potatoes, diced
6 large onions, chopped
3 celery stalks, thinly sliced
sunflower cooking oil
2 kg thinly diced stewing steak
little brown bread flour
juice of 2 lemons
3–5 tablespoons honey

Peel and chop all the vegetables. In a large deep pot, heat enough oil to cover the bottom of it. Roll the meat in the flour and brown in the hot oil — turn and fry the other side. Then add the onions and brown these.

Add all the other ingredients, stirring with a wooden spoon. Cover the saucepan with its lid, and turn the heat to low. Check every now and then and add a little water if you see it is becoming too dry.

Simmer for ¾ hour. If the tomatoes are very juicy, thicken the gravy just before serving with 2 tablespoons of maizena mixed into ½ cup of water.

Just before the cooking time is up, place a sprig of spekboom about the length of your hand on top of the stew. Cover and simmer for 5 more minutes. Discard the twig before serving. Serve with mealiemeal ('stywe pap'). *Bon appetit!*

Traveller's joy

Tumbleweed

Water blommetjie

Bulrush

Water lily

Star flower

Hypoxis species

ENGLISH	Yellow star
AFRIKAANS	Geel sterretjie, sterblom
SOUTHERN SOTHO	Lotsane
TSWANA	Tshuka
XHOSA	Inongwe
ZULU	Inkomfe enkulu

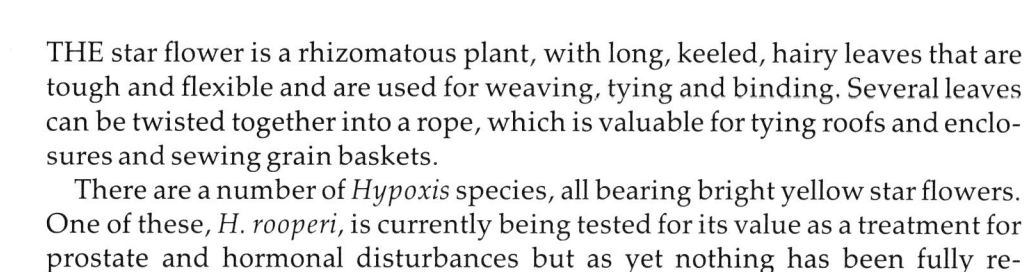

THE star flower is a rhizomatous plant, with long, keeled, hairy leaves that are tough and flexible and are used for weaving, tying and binding. Several leaves can be twisted together into a rope, which is valuable for tying roofs and enclosures and sewing grain baskets.

There are a number of *Hypoxis* species, all bearing bright yellow star flowers. One of these, *H. rooperi*, is currently being tested for its value as a treatment for prostate and hormonal disturbances but as yet nothing has been fully recorded. This plant has been used by blacks for many generations as a tonic in the treatment of physical weakness, to encourage strength in frail children and old people and as a purgative.

A strange ancient Basuto cure for a headache was to hollow out the rhizome or rootstock into a small receptacle, into which a few drops of blood taken from the head of the sufferer would be dripped and the rootstock buried — and with it the headache!

The sliced rootstock contains a soothing sap, which is used by several African tribes as an application to burns. This is useful knowledge for campers and hikers, and I have found a thin sliver of the rhizome applied to a blister on the heel to be very comforting on a long hike. The juice rubbed into the blister area is also soothing.

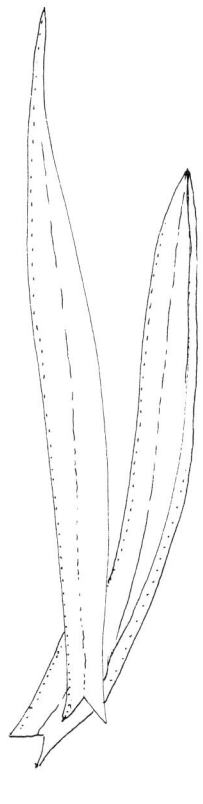

H. villosa (also known as inkbol) is particularly respected by the Southern Sotho as a charm against thunder, and as a medicine. *H. rigidula* is excellent for rope making and is also used medicinally. *H. nyasitica* is used as a cough remedy, and *H. oliqua* is used by the Xhosa as a wound lotion. *H. argentea* has a small yellow star flower that is used by the Xhosa for chafes on horses, as a stomach medicine, and in times of food scarcity the dried and powdered rootstock is used as a food.

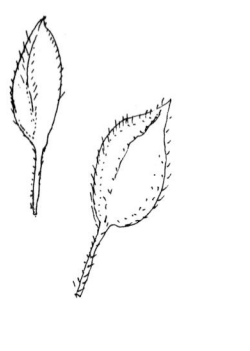

The *Hypoxis* species form a most charming group of plants, and with their bright yellow star flowers are so worth growing in the garden. They are easy to grow from seed, which is obtainable from the Botanical Society of South Africa (see page 270). Sow it in trays in early spring and when the plants are big enough to transplant, select a well-drained site such as a rockery or a slope with sandy soil, in full sun. Plant the rhizomes about 30 cm apart and wait for the bright and beautiful midsummer display of yellow stars!

They don't need much water and can be left for many years to flower summer after summer. Once you are familiar with *Hypoxis* you cannot fail to notice their bright faces in the veld on rocky hillsides and along the roads throughout South Africa.

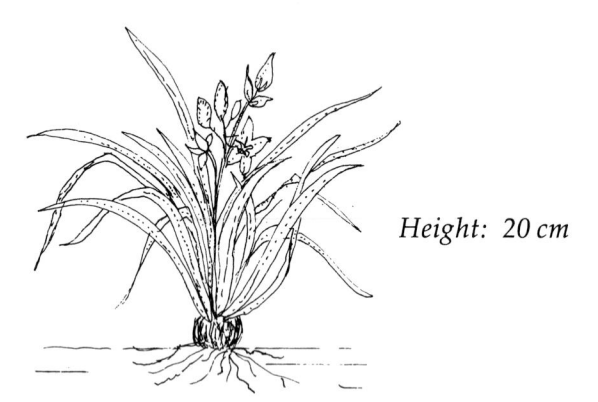

Height: 20 cm

Stork's bill geranium
Pelargonium luridum

ENGLISH   Salmon geranium, dull geranium
TSWANA   Nyamarora
ZULU   Inyonkulu

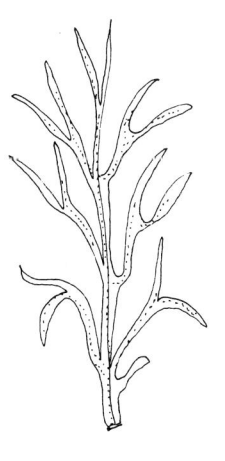

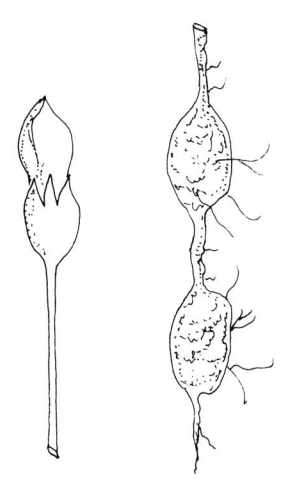

IN midsummer in the open grasslands of the eastern Cape, the Transvaal and sometimes in the Orange Free State and Natal, the pretty salmon pink clusters of flowers of the stork's bill geranium wave above the grasses on long slender stalks. In early summer the flowers are a creamy pale green when they first open, but then turn salmon pink as they mature. What is most fascinating about this wild geranium is that it has a variety of leaf shapes on one single plant.

The first leaves produced each season from the perennial root stock are short, broad and slightly lobed. The successive leaves become longer and more dissected until thin, narrowly divided fern-like leaves appear, up to 30 cm in length. The leaves and the stem are hairy, and the soft, drooping flowers are also slightly hairy and have a faint sweet scent.

The Zulu prize this plant for the treatment of diarrhoea and dysentery. A cup of boiling water is poured over a tablespoon of scraped, crushed rootstock, and left to draw until lukewarm. This is then drunk a little at a time to ease the condition. The woody tuberous rootstock is also dried, pounded and mixed into porridge for treating dysentery, and a tea made of the leaves, stalk and bits of root is also much respected as a treatment for colic, stomach disorders like nausea and vomiting, and to bring down a fever.

The Tswana use a tea made of the leaves as a wash for sores, to cool an inflamed area and to bring down a temperature. Several tribes make a strong lotion of the whole plant and use this to treat skin eruptions in cattle and infected tickbites.

The plant germinates easily from seed, but it does need full sun and well-drained soil. It is unusual and pretty in the midsummer garden and flowers from October to March.

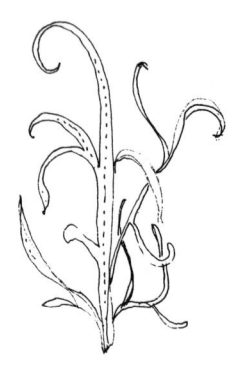

I use the flowers in a wild flower pot-pourri and find they retain their unusual and exquisite salmon colour well; and the leaves and flowers press beautifully for pressed flower pictures.

Height: Leaves 15 cm
Flowers 30–40 cm

Sweet thorn
Acacia karroo

ENGLISH	Gum arabic tree, mimosa, mimosa thorn, Cape gum, acacia, karroo thorn, sour thorn, thorn tree, white thorn
AFRIKAANS	Doringboom, soetdoring, kareedoring, karoodoring, witdoring, suurdoring
HERERO	Orusu
NDEBELE	Isinga, isingawa, isisani, mdoka, mdokana, nganaghu
SHANGAAN	Munga
SHONA	Munenje, muwunga
SOTHO	Mmoka, moko, moka, mookha, leokha, mookhana
SWATI	Isinga, kalimela, umdongolo
TSWANA	Kalagadi, mokana, mokha, mooka
XHOSA	Umga, umnga
ZULU	Umunga

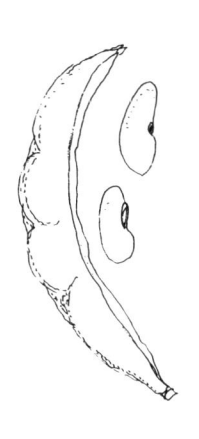

THE sweet thorn grows extensively all over South Africa and is familiar to all South Africans for its many medicinal and other uses. It has a round crown, black branches and trunk, sprays of yellow pompom flowers and characteristic long, white thorns. It will grow in any soil, and is able to withstand frost, veldfires, drought and heavy rains. It varies in size from place to place from a small shrub to a huge tree with flat, spreading branches casting deep shade. Its summer dress of golden yellow pompoms buzz with bees, and sweet thorn honey is probably the most delicious of all South African honeys.

The wood, which is pale, tough and hard, is used for poles, yokes and roof struts. The inner bark, pliable when wet, is an excellent rope and is still used by the rural people for tying roof frames. The outer bark has been used in tanning as it contains a number of valuable tannins, and was a valuable ingredient used by the early colonists in their tanning processes.

The seeds and leaves are a nourishing food for sheep, goats and cattle. The leaves dried, crushed and roasted have been widely used as a coffee substitute.

The Cape colonists used the bark in the form of a tea for diarrhoea and dysentery. Pour 1 litre of boiling water over 1 cup of pieces of bark and leave to cool. Strain and drink half a cup every hour until the symptoms are alleviated.

A Sotho remedy for colic in babies is to crush a little of the fresh root to a fine texture and add it to their food. Some tribes make a tea of the root for colic and flatulence, and the Tswana on the farm boil up pieces of root in water with a little of the bark for about an hour. This brew is then strained off and taken as a tea for heartburn, flatulence and colic, and given to children for coughs and colds and to old people for indigestion.

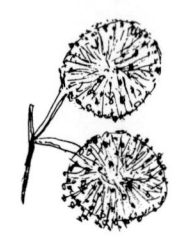

When the bark is damaged or cut in any way, it exudes a beautiful sweet resinous gum of an arabinose-galactose type, which is used as a sweetener in cooking. It is brittle and crumbly, but becomes soluble and mucilagenous when mixed with hot water. The gum is often combined with a tea made of the leaf and bark for coughs and colds, diarrhoea and opthalmia, and is also used to treat cattle and dogs for diarrhoea, coughs and opthalmia. It also clears thrush and mouth infections.

The sweet thorn, also known as Cape gum, was once an important ingredient in confectionery, and was exported from South Africa and Namibia as a gum arabic for the confectionery trade. It is still used today as a sweetener and coagulant in confectionery, and of course is a favourite 'sweet' among the farm children.

An old Transvaal remedy for osteomylitis was to mix Cape gum, chopped capsicum flesh and seeds and vinegar, and to apply this as a plaster or poultice to the area to ease the pain and help the healing process.

The sweet-scented flowers and buds pounded to a pulp in hot water are used by the Tswana, Venda and Zulu as a poultice to draw abscesses and boils and to soothe sprains, replaced twice a day with fresh flowers.

I use the flowers and buds in pot-pourris and find that the dried soft fibres absorb the fragrant essential oil exceptionally well. I keep a quantity in a large glass bottle to which I add essential oil (I find honeysuckle oil the best). I keep this sealed, shaking it daily, and add it to pot-pourris — the scent lasts extremely well.

The tree is fast growing and needs no care, surviving only on rain water once it is established in the garden. It will provide you with shade, beauty, medicine, food and pleasure, and will attract birds to your garden.

Several nurseries offer the trees for sale, but you can propagate your own by soaking the seeds in warm water (place them in a thermos flask) for 3 days and nights; then plant the seeds in bags filled with a mixture of compost and sand. When they are sturdy, plant them in the garden in a large compost-filled hole, and water well until they are established. Thereafter water once a week until the tree is 2 m in height, when it will no longer need watering but will survive on rain water.

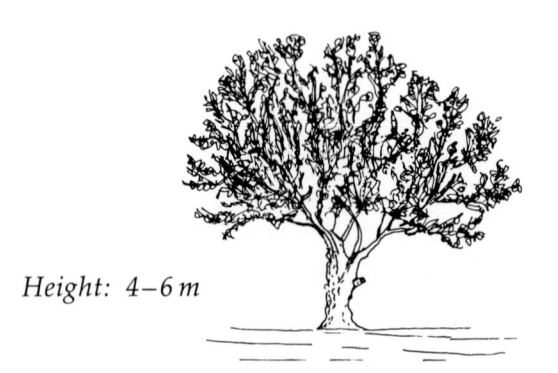

Height: 4–6 m

Toothache root
Berula thunbergii

ENGLISH	Water parsnip
AFRIKAANS	Tandpynwortel
SOTHO	Qaqawe
TSWANA	Lehlatso
ZULU	Lehlatso

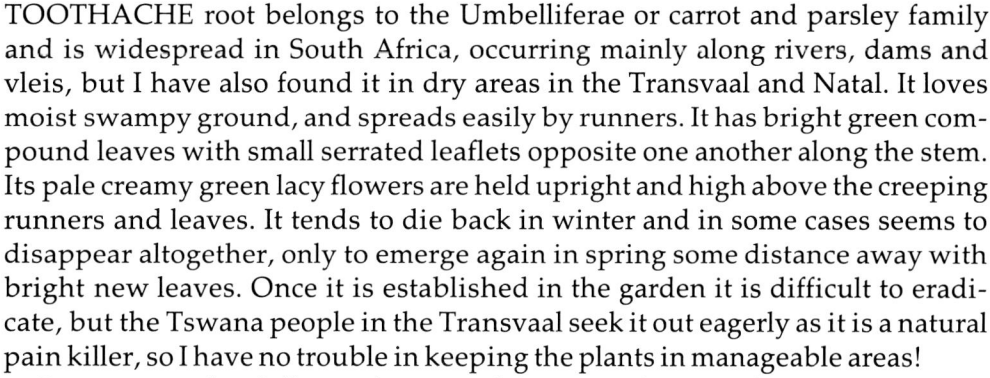

TOOTHACHE root belongs to the Umbelliferae or carrot and parsley family and is widespread in South Africa, occurring mainly along rivers, dams and vleis, but I have also found it in dry areas in the Transvaal and Natal. It loves moist swampy ground, and spreads easily by runners. It has bright green compound leaves with small serrated leaflets opposite one another along the stem. Its pale creamy green lacy flowers are held upright and high above the creeping runners and leaves. It tends to die back in winter and in some cases seems to disappear altogether, only to emerge again in spring some distance away with bright new leaves. Once it is established in the garden it is difficult to eradicate, but the Tswana people in the Transvaal seek it out eagerly as it is a natural pain killer, so I have no trouble in keeping the plants in manageable areas!

Country people of all races have used this spreading weed as a pain killer for many years. The root was renowned amongst the Cape colonists as a toothache remedy, chewed or just held in the mouth to relieve the pain. The Tswana and the Sotho make a brew of the leaves and runners by boiling 3 cups of fresh plant in 10 cups of water for 10 minutes, then allowing the strong brew to cool. When comfortably warm, this is used as a wash for an aching back and legs, or a cloth wrung out in it and bound around the head is used as a comforting headache remedy.

The Zulu and Sotho wash the body in a diluted brew to relieve headaches, muscular aches and pains and rheumatism.

The only internal use of the plant is as a toothache remedy, and a hollow aching tooth can be packed with the crushed leaf and stem to relieve the pain.

The plant is suspected of being poisonous to cattle, possibly producing a condition known as vlei poisoning which affects the nervous system, so care should be taken where it is planted and farmers should be aware of its toxic effects. The plant is not always toxic, however, and is apparently at its most poisonous in spring and in certain soil types. I have grown it successfully in a large pot with generous watering and find it makes an attractive container plant which is also an interesting talking point. It would be interesting to find out the components in scientific research of this amazing little plant.

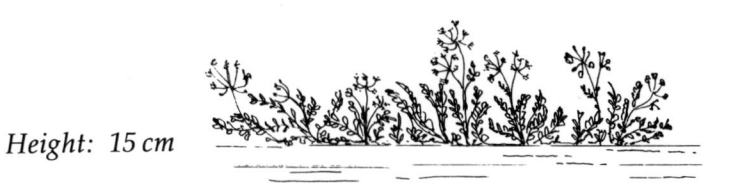

Height: 15 cm

Traveller's joy
Clematis brachiata

ENGLISH	Old man's beard, wild clematis
AFRIKAANS	Klimop
SOTHO	Morara
TSWANA	Mogau
XHOSA	iTyolo
ZULU	Umdlonzo

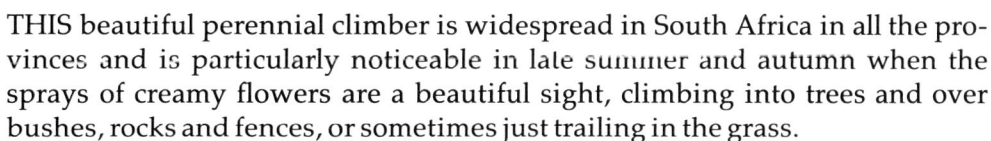

THIS beautiful perennial climber is widespread in South Africa in all the provinces and is particularly noticeable in late summer and autumn when the sprays of creamy flowers are a beautiful sight, climbing into trees and over bushes, rocks and fences, or sometimes just trailing in the grass.

The sweet-smelling flowers soon give way to feathery grey fruits or seeds, which have the appearance of soft wool, frothy and light, cloaking shrubs and trees and drawing the eye with their unusual beauty.

An abundance of flowers in late summer is believed to indicate a good season ahead, and this is probably how the name traveller's joy came into being — apart from the fact that the plant also has wonderful medicinal uses that were helpful to the traveller in days gone by. Leaves packed into the shoes were used to ease blisters and aches and pains, and packed under the saddle prevented saddle sores on horses. Fresh leaves pushed into the crown of a hat in the summer heat kept the wearer cool and energetic, preventing heatstroke and sunstroke.

One hot summer day I walked along an old Boer War road in the Rustenburg district, where the heat, the rough terrain and the dryness of the countryside made me so aware of the difficulties the Boers and the fair-skinned British soldiers must have experienced, until I saw great sprays of traveller's joy and felt much comforted, for I was sure that those weary men would have made use of the plant to relieve their exhaustion! When the colonists first came to South Africa they recognised the plant as being a close cousin of the European *Clematis vitalba* and used our indigenous species, *C. brachiata*, in the same way.

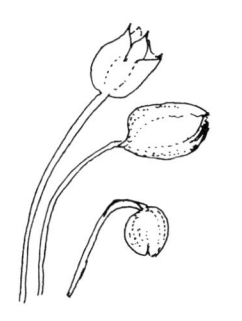

A tea made of the leaves (¼ cup of fresh leaves to 1 cup of boiling water, stand to draw for 5 minutes, then strain and drink, sweetened with honey if desired), is not only refreshing, but is used by the Xhosa, Zulu, Sotho and Tswana to ease headaches, coughs and colds, chest ailments and abdominal upsets. I have found the tea to be refreshing and reviving, and have also used it as a wash for weary feet after a long hike, and felt the difference.

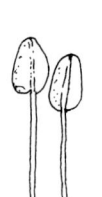

A decoction of the root and stem is used by the Sotho as a remedy for syphilis, and this dried and powdered with the leaf and flower is used as a wash or lotion.

The Xhosa and the Zulu bruise the stem and then inhale the rather pungent smell to clear the head during a cold, ease painful sinuses and induce sneezing.

A hot decoction is made by pouring boiling water over a baisinful of root, stems and leaves and the steam inhaled for easing colds, malaria, sinus infections and asthma. The plant has been used by some tribes for snakebite, as well as for bots in horses, venereal diseases and thrush. Both whites and blacks make a snuff from the fresh leaf to relieve headaches and as a weak tea for stomach upsets and ailments, and this is also used as an enema.

A Zulu gardener once showed me how to dip a towel into the tea and use this, wrung out and bound round the head, to ease a headache. I have found it so refreshing and reviving that I have told many hikers and weekend campers about it and they too have found this to be excellent. Nicest of all, to ease aching muscles make a strong brew of leaves, stems, flowers and even the seeds and add this to the bath water and soak in it.

Traveller's joy grows easily in the garden, but needs a pergola or fence to climb over. I find if I trim it back at the end of winter, the untidy end-of-season growth makes way for fresh shoots, which quickly scramble up into the supports. The beautiful flower sprays and the fascinating seeds make it so worthwhile to grow, and it is undemanding, unfussy and will grow in most soils. It needs full sun, and not much water once it is established.

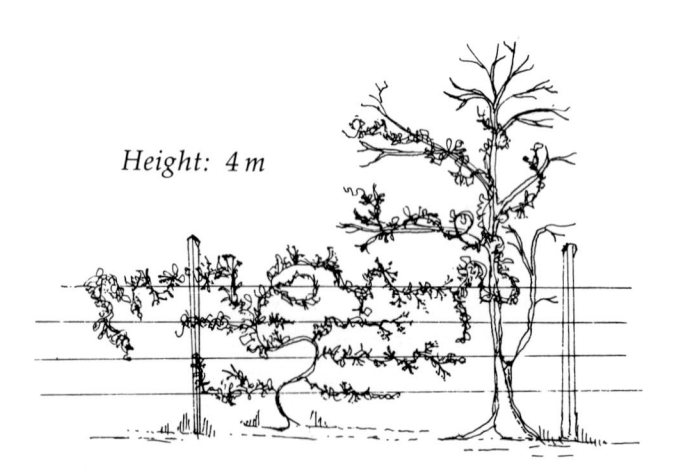

Height: 4 m

Tumbleweed

Boophane disticha

ENGLISH	Candelabra flower
AFRIKAANS	Seeroogblom, perdespook, gifbol
NDEBELE	Incoto
SOUTHERN SOTHO	Motlatsisa
TSWANA	Leshoma
XHOSA	Iswadi, incwadi
ZULU	Incotha, bate

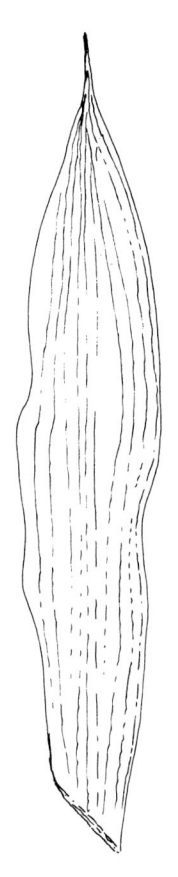

THE beautiful tumbleweed, much prized and respected medicinally, is a most fascinating garden plant, and easily grown from the small marble-like seeds that form at the end of each spike as the large flowering head ripens.

The bright deep pink flowers are a beautiful sight in spring, when the large umbel is eye-catching in the Transvaal Highveld on hillsides and in the open veld. After the flower head dries and blows away, tumbling over the veld like a spikey ball dropping its ripe seeds in its wake, the fan of strap-like leaves, grey green, sturdy and decorative, appears.

The huge bulb with silky layers of scales is extremely poisonous — the San or Bushmen used the bulb in the preparation of fish poison and the Khoikhoi (Hottentots) used it on their arrows for shooting game — and yet it is still used in some country districts as a remarkable painkiller. A moistened scale from the bulb is used to ease whitlows, boils, infected sores, septic scratches and cuts and inflammatory conditions. The drawing properties not only soothe but heal, and the old farming communities used the dried scale warmed and softened in water as a poultice over painful rheumatic joints, arthritic swellings and bruises, sprains and muscular strains. Several African tribes use the scales of the bulbs as a dressing for burns, and as a bandage for skin rashes, but one should approach the plant with caution.

In the Karoo people make a pillow or mattress stuffed with the scales of the bulb which they sleep on to calm hysteria and aid sleeplessness — but here a word of warning: the inhaled bulb may be narcotic!

The leaf is still used today over cuts and wounds to staunch bleeding, and is often dried for later use. The dried leaf could be moistened with milk or oil and used as a poultice for varicose ulcers, skin diseases, rashes, eczema and psoriasis.

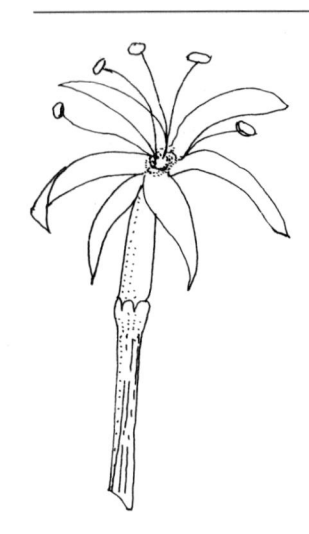

The parts of the tumbleweed above the ground are not toxic, as cattle browse the plant with no ill effect and the botanist and explorer Burtt Davy noticed that carrion birds eat the leaf, which he surmised was 'either to prevent harm from the ingestion of putrid flesh or to sharpen their vision'.

Inhaling the pollen or scent from the flowers often causes a headache or sore, red eyes (hence its Afrikaans common name seeroogblom), and giddiness and drowsiness sometimes follow. The plant is believed to have magical powers to ward off evil, bring rain, poison one's enemies, drive out sickness and weakness, keep the wearer free from evil influence, and to protect one's home. No wonder it is so sought after!

I grow the seeds in small pots to establish them before planting them out in the garden, and all the farm workers ask me for a plant to grow in their gardens to keep their homes safe. I have often screened the bulb with its familiar fan of leaves by growing bushes and grass around it in the hope of conserving it against witchdoctors and herdboys! When you do plant them in your garden be sure not to overwater them, as this will cause the bulb to rot. I have found the flowers give one a headache if brought inside, but the dried seed head is a beautiful decoration and has no toxic effect. Do not dig up the bulb from the veld — it does not transplant well — but rather collect fully ripened seeds.

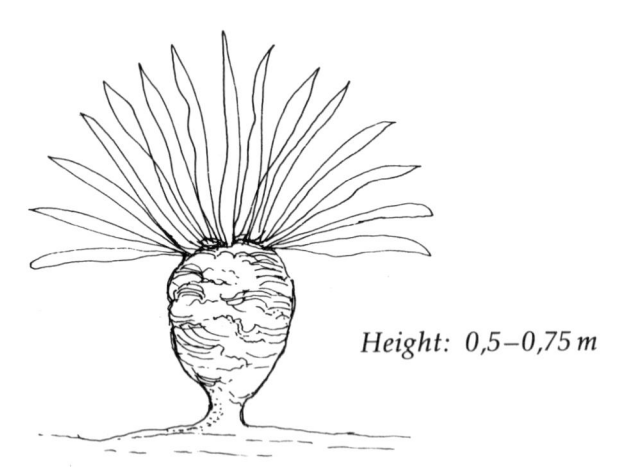

Height: 0,5–0,75 m

Waterblommetjie
Aponogeton distachyos

ENGLISH Cape pond weed, Cape asparagus, water onion
AFRIKAANS Vleikos, wateruintjie, watervygen, kleinwaterlelie

WATERBLOMMETJIES, the succulent and delicious traditional Cape delicacy, are now within the grasp of all the provinces, thanks to a fairly new Boland farming operation and our modern transport systems. In the spring many greengrocers in the Transvaal offer cleaned and neatly packaged waterblommetjies for sale which the Transvaalers pounce upon with delight.

The fragrant flowers have a scaly formation and must be washed carefully (preferably soaked) as sand lodges within the scales. These can be made into a succulent traditional Cape waterblommetjie bredie or chopped fresh into salads. I have tasted a salad of equal quantities of chopped celery stalks, cucumber and waterblommetjies sprinkled with chopped pecan nuts and lavishly tossed with mayonnaise, and found it wonderful!

The whole plant is high in mineral value and contains several vitamins, and the root is also edible, making it a valuable plant nutritionally.

The stems with their high juice content make a soothing treatment for burns and scrapes and take the pain out of sunburn if the juice is applied every hour until the redness fades. The leaf, well washed, makes a soothing poultice over sores and burns, and the flower petals applied over a pimple will help heal it quickly.

The San used the whole plant as a food, and the early Cape settlers were taught by the Khoikhoi to use it in soups and stews. They later developed a tasty waterblommetjie pickle recipe and also cooked and ate it like asparagus.

One fresh September day with heavy skies and the countryside rich and green after the winter rains, while travelling to Langebaan from Paarl, we came across on an isolated stretch of road a white-bearded farmer and his small coloured helper thigh-deep in a dam that was covered with masses of white-flowering waterblommetjies. We stopped to ask him about his remarkable crop and he told us he was going to make soup for his Sunday lunch (see recipe opposite). Under his arm he held a bunch of juicy stems — some a metre long — for his pigs and goats. He said he always fed them the stems in spring as the spring litter of piglets needed extra nourishment and the waterblommetjie leaves and stems helped lactation in nursing mothers. He also dug out the arum lilies — indicating the hill behind him, which was brilliant with green and white arum lilies flowering profusely — as these too increased the sow's milk flow, thus ensuring 'spekvet' piglets!

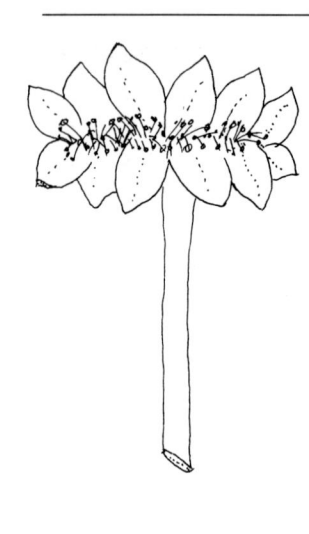

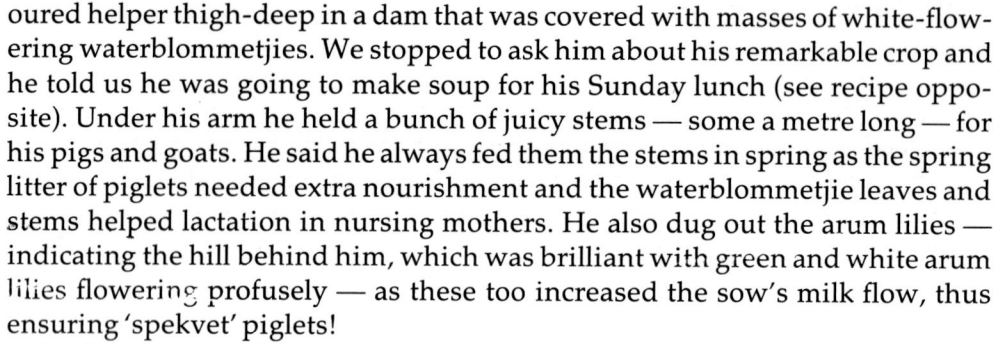

At that moment the sun came out from behind the clouds and I asked if I could take a photograph of him with his crop. He pulled his small helper up by his collar, blue with cold and with chattering teeth, and urged him to hold aloft his red and white OK Bazaars packet with its tangle of flowers and protruding stems. The farmer took his stance, with the bundle of long-stemmed lilies under one arm while he tugged at a growing stem with the other. I took the photograph, and with the dam, the arums and the white flecks of the waterblommetjies in that brilliant sunshine, it is one of the best memory stirrers I have — I can almost smell the flowers!

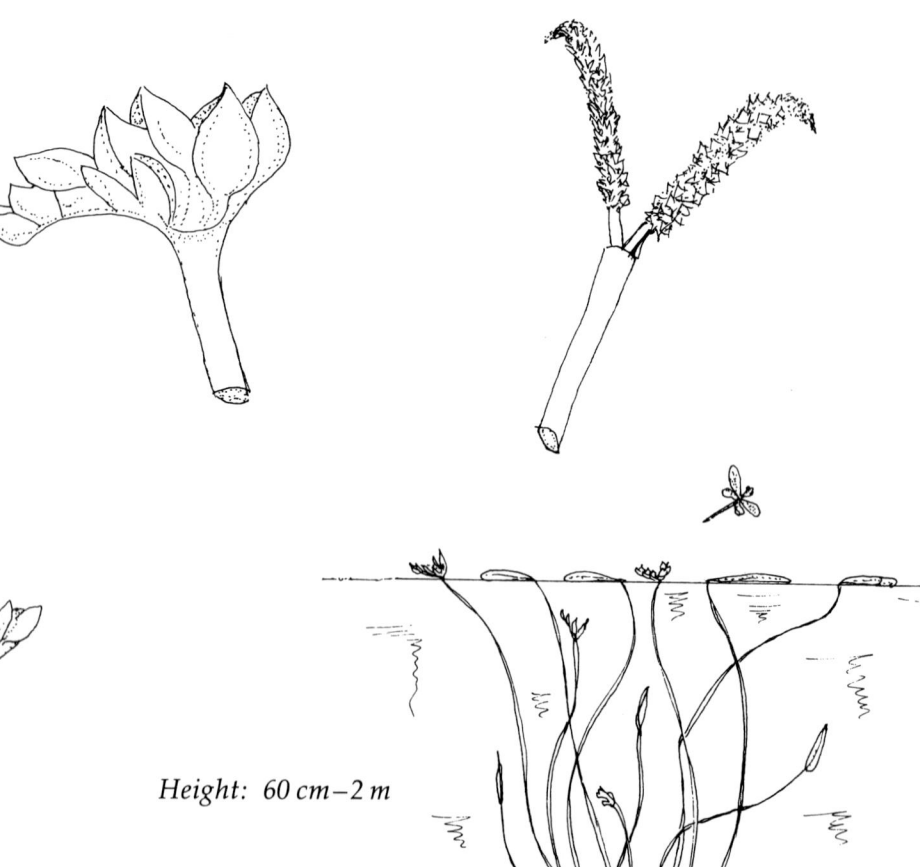

Height: 60 cm – 2 m

Waterblommetjie soup

Serves 6–8

2 large onions, finely chopped
2 tablespoons butter
2 cups celery, finely chopped
6 cups waterblommetjies, well washed and chopped
2 cups waterblommetjie stems, chopped
4 medium-sized potatoes, grated
salt and pepper to taste
1 litre milk
1 litre water

The quantities the farmer gave me were in handfuls and bunches, so these are the best measurements I can give, having come home and tried it out.

Fry the onions in the butter until golden. Add all the other ingredients and bring to the boil. Simmer slowly, stirring every now and then, for about 20 minutes. Add a little more water or milk if you prefer a thinner soup. Serve with croutons.

Traditional waterblommetjie bredie

Serves 8

little fat or cooking oil
1–2 kg mutton loin chops or rib or leg
honey
3 large onions, chopped
1 kg cleaned waterblommetjies
2 cups water
1 cup white wine
1–2 cups fresh sorrel leaves
salt and pepper
4 large potatoes, peeled and diced

Heat the fat or oil in a heavy bottomed saucepan. Brush the meat with honey and brown in the fat, turning every now and then. Add the onions and continue to brown. Add all the other ingredients, with the potatoes on top, and stew. Check the water every now and then, adding more if necessary, but do not stir too frequently.

Simmer gently until the meat is tender. Serve piping hot with boiled rice.

Water lily

Nymphaea caerulea

ENGLISH	Blue water lily, blue lotus
AFRIKAANS	Blou waterlelie
!KUNG BUSHMAN	//Ogbe
KWANGALI	/Numa
LOZI	Makwangala
THONGA	Matiba
ZULU	Iziba

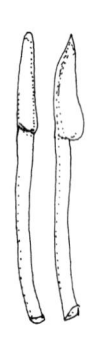

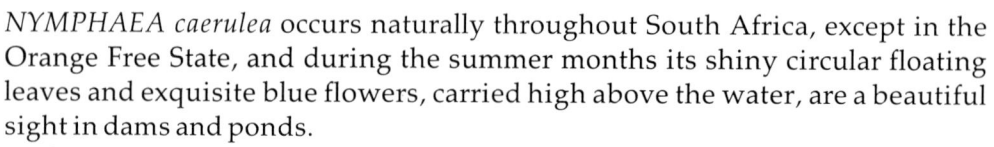

NYMPHAEA caerulea occurs naturally throughout South Africa, except in the Orange Free State, and during the summer months its shiny circular floating leaves and exquisite blue flowers, carried high above the water, are a beautiful sight in dams and ponds.

The genus name *Nymphaea* comes from the Greek goddess Nymphe, the goddess of springs and rivers, and a great variety of beautifully coloured hybrids have been cultivated, often with aromatic flowers.

The rhizomes and the seed are much sought after for medicinal use, and as a foodstuff. The seed is an old remedy for diabetes, eaten crushed with a little water once or twice a week, and was considered to be an effective treatment many decades before the benefit of modern drugs.

The root and stem are cooling, astringent and soothing, and possess antiseptic qualities. The leaves are a soothing dressing and are used to bring down inflammations and remove the pain of sunburn. Rashes and inflamed scratches are quickly soothed by a water lily leaf compress. On a long hike leaves placed in the shoes help sore feet and blisters. I have also placed a leaf in my hat when I was very hot and tired and found it refreshed and revived me, and squeezed juice from the stem onto my nose that was badly sunburned. It immediately soothed and protected it from further burn.

Both the root and stem are diuretic and have been used in the treatment of kidney and bladder disorders, and by some tribes for diarrhoea. The brew is also considered to be an excellent douche, skin wash and lotion for sunburn and heat rashes.

The flower boiled in water (usually 2 flowers boiled in 3 cups of water for 10

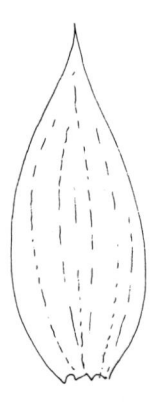

Wild camphor tree

Wilde als

Transvaal wild basil
Ocimum canum

Zulu basil
Ocimum urticifolium

White cat's whiskers

Wild asparagus

Spearmint
Mentha longifolia
polyadena

Balderjan
Mentha longifolia
capensis

Watermint
Mentha aquatica

Wild garlic

Wild foxglove

Wild cineraria

Tulbaghia

Wild gladiolus

Wild dagga
Leonotis leonurus

Klip dagga
Leonotis leonitus

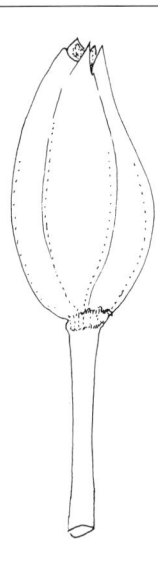

minutes, then strained) is used by the Zulu, the Tswana and white farmers for coughs and mucous on the chest (1 or 2 tablespoons are taken 3 times a day). Zulu farm workers boil up the stem and rhizome with wild garlic for coughs and colds and to rid the body of mucous.

The water lily can be easily grown in pots filled with soil and compost and special water lily tubs are available from nurseries and wholesalers. Plant the rhizomes in the rich soil and compost mixture, and gently lower the tub into the pond — they need to be sunk to a depth of approximately half a metre. In early August the rhizomes can be divided if they become too crowded for one pot.

I have grown a beautiful specimen of the blue water lily in a small pond in the garden where in the midday sun the blue flowers bloom in all their glory. I have had no trouble at all in growing the plant, and have now also established *N. capensis*, the blue Cape water lily. The plants are so easy to grow I urge you to do so too, for the joy one derives from these beautiful plants is unending.

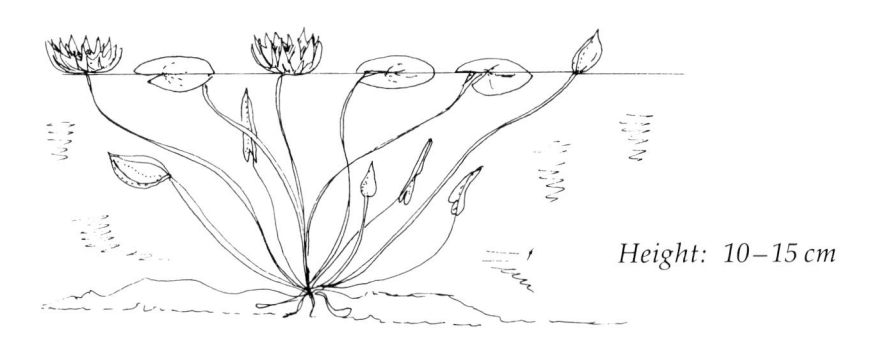

Height: 10–15 cm

White cat's whiskers
Clerodendrum glabrum

ENGLISH	Tinderwood tree, resin leaf
AFRIKAANS	Harpuisblaar, tontelhout, bitterblaar, kruitjie-roer-my-nie
SOTHO	Mohlokohloko
TSWANA	Umquaquane
XHOSA	Umqangazane
ZULU	Umqangazane, umquqongo

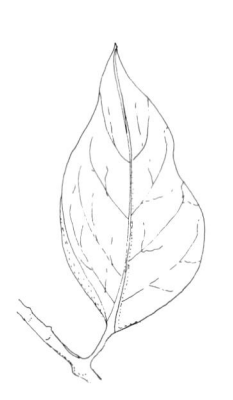

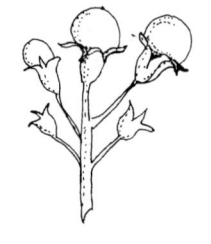

THE white cat's whiskers tree is small to medium in size, growing up to 10 m in height, and is found along river banks on rocky hillsides from the eastern Cape through Natal, the Orange Free State, Transvaal and Mozambique up into Zaire, and is often associated with termite mounds. It grows along sand dunes at the coast, where it usually flowers from July to about October, and inland it flowers around Christmas time and sometimes on into April.

The tree is rather inconspicuous when it is not in flower, but when it blooms it is a beautiful sight in the veld when little else is in flower with its masses of sweet smelling creamy white clusters of flowers. The flowers have long pink and white stamens which look exactly like cat's whiskers, hence its name. The small new buds in the cluster are mauvy pink and the pollen-laden cat's whiskers are pink and white, creating the overall impression of pale pink pom-poms.

Although the leaves are insect repellent, the sweet scent of the flowers attracts butterflies, moths and beetles, and it is an interesting experience to sit under the tree on a hot afternoon and witness the busy comings and goings of all kinds of winged creatures. Although the sweet smell of the flowers becomes sharp and almost unattractive as they age, if you pick and dry them when they are young they make a charming addition to pot-pourris.

Berries follow the flowers, and the fruit stays embedded in the calyx lobes for months afterwards, so even after the flowering and fruiting period is over, there is still much bird and insect activity in the tree.

The soft green leaves are velvety smooth, covered in the finest hair and grow opposite each other or in whorls of sometimes three or four. They emit a strong odour when crushed and have insect-repelling properties, and have thus been used for many decades by the indigenous people as a wash and lotion to prevent the development of blowflies, maggots and skin parasites in the wounds of animals and people, and are also taken internally for tapeworm, roundworm and threadworm.

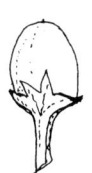

Farmers in the eastern Transvaal make a lotion by boiling a bucket of leaves and twigs in enough water to cover it for 20 minutes, leaving it to cool and then straining it, and use this as a drench or wash for tick-infested animals and for treating maggot-infested wounds on cattle.

The leaves boiled in water (¼ cup of leaves to 2 cups of water boiled for 15 minutes, then cooled for 10 minutes and strained) is a soothing cough and cold remedy. A little is taken at frequent intervals and has been successfully used particularly in Zimbabwe and Zaire for bringing down fever and as a treatment for colic. Combined with the root of the ginger bush (*Tetradenia riparia*, page 87) the white cat's whiskers tree is much favoured as an emetic in rheumatic ailments.

The leaf is believed to be soothing and sedative and this is probably why some tribes use pounded leaves under the armpits and on the neck of a child or adult suffering from convulsions. For old people, the Tswana make a weak tea of the leaves as a nightcap (usually 4 leaves in a cup of boiling water, stand for 5 minutes, then strain and sip sweetened with honey) to ensure a good night's sleep and to help night coughing and bad dreams.

The Tswana workers on my farm add the leaf infusion with bark scrapings to milk and give it to calves, farm dogs and donkeys to rid them of intestinal worms.

An infusion of the root (1 cup of cleaned, chopped root steeped in 2 cups of boiling water for 10 to 15 minutes, then strained) is drunk for snakebite by the Zulu, and the pounded root is bound over the area.

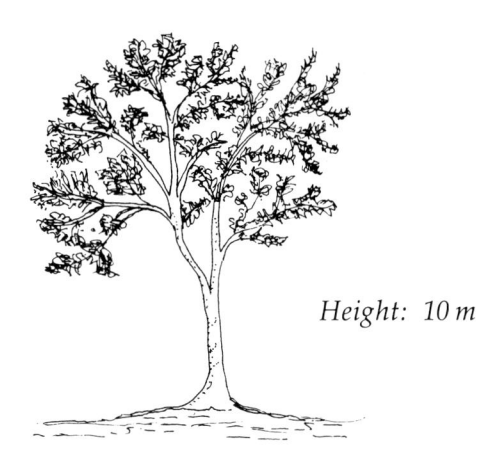

Height: 10 m

Wild asparagus
Asparagus africanus

AFRIKAANS	Haakdoring, wag-'n-bietjie
MPONDO	Kuboal
SOTHO	Lelala-tau-le lehola
TSWANA	Mositwane
XHOSA	Itali
ZULU	Isigobo

Height: 2 m

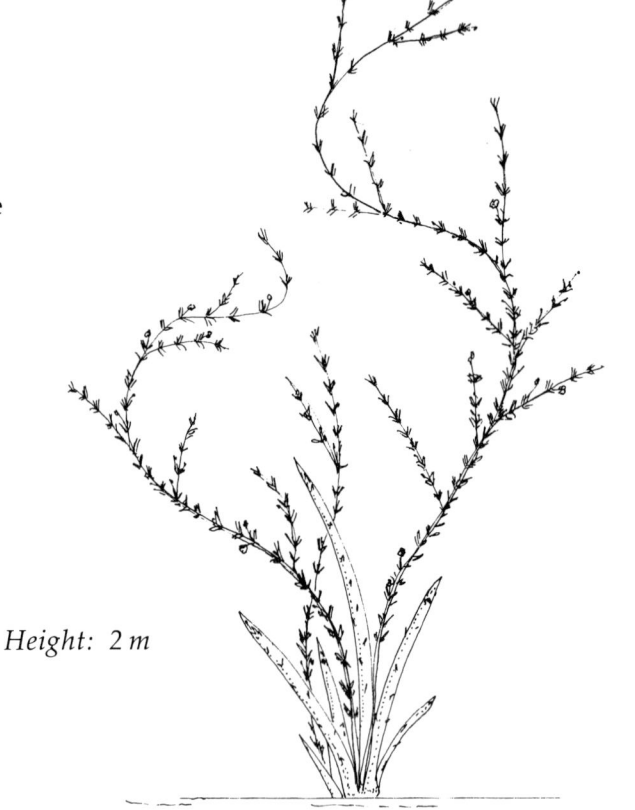

THE *Asparagus* species are a wide and fascinating group of plants, generally used in South Africa for the treatment of bladder and kidney ailments, lung conditions, tuberculosis and as a diuretic. The young shoots of several species are edible, and some species are much prized all over the world as pot plants.

A. africanus is a robust climber with a fine, frothy appearance. It is widespread throughout South Africa and can be seen climbing up bushes, fences and trees, its clear green colour, bright red berries and soft branches drawing the eye.

Van Riebeeck noticed the Khoikhoi digging up the young shoots of *A. africanus* and after trying them himself declared them to be as good as the asparagus in Holland! The early European settlers enjoyed the young shoots and cut back the growth of the long trailing branches frequently to encourage the new succulent young shoots. Several African tribes eat them as a vegetable too as well as using the plant medicinally.

The Southern Sotho rub the root which they have dried and pounded and crushed into a powder into scarifications on the back, stomach and legs of boys undergoing the circumcision rites and this is believed to give them strength and fearlessness.

A tea made by boiling the shoots in water (1 cup of chopped fresh shoots to 3 cups of boiling water) is used to clear pulmonary tuberculosis (a cup is drunk daily). A weaker brew is also used by both blacks and whites for bladder and kidney ailments, to cleanse the liver and for stomach complaints and cystitis. The Zulu use a tea of the shoots for nausea and colic to induce vomiting.

I always keep a sharp look out for the wild asparagus, and in spring it is always a race between me, the buck, the hares and the farm workers to collect those succulent shoots!

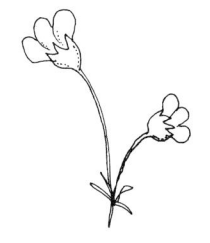

In early spring (the first week of August is best) crowns can be dug carefully from the ground and replanted in the garden. If you want a good crop of shoots prepare a deeply dug, richly composted bed and water the plant well until it is established. Cut back the plume-like growth in winter to encourage new shoots in spring.

Fresh wild asparagus

Boil the shoots or spears in enough water to cover them. As the water comes to the boil, throw that water off, refill the pot and boil the shoots in fresh water. When tender, drain, saving the water for soup or gravy, and serve with melted butter and a squeeze of lemon juice, and a little salt and pepper, or serve cold chopped into salads.

Wild asparagus tart   Serves 6

Pastry

1 cup cake flour
2 teaspoons baking powder
1 cup butter
1 cup grated cheese
about ½ cup water

Filling

2 tablespoons butter
2 tablespoons flour
1 cup milk
1 cup water in which the asparagus was cooked
2 beaten eggs
2 teaspoons mustard powder
little salt
paprika
2 tablespoons chopped onion
2 cups cooked wild asparagus, chopped (reserve water from cooking)
grated cheese

This is a tasty and easy dish that everyone loves! If you do not have enough wild asparagus, supplement with tinned asparagus to make up the correct quantity.

To make the pastry, coarsely grate the butter into the flour and baking powder. Add the other ingredients and mix together with enough water to make a fairly soft scone-like dough. Press into a flour-dusted baking dish.

To make the filling, melt the butter in a stainless steel pot and stir in the flour. Mix the milk with the asparagus water and slowly add to the mixture, stirring with a wooden spoon. Add the eggs and stir until it starts to thicken.

Turn the heat down low, then add the mustard powder, salt, paprika, onion and asparagus.

Pour into the pastry-lined baking dish, sprinkle with grated cheese, place in a medium oven (180 °C) and bake for about an hour, or until the crust starts to brown and the tart is set.

Serve hot or cold with a salad.

Wild basil
Ocimum species

AFRIKAANS   Boesmansboechoe, wilde basilikum

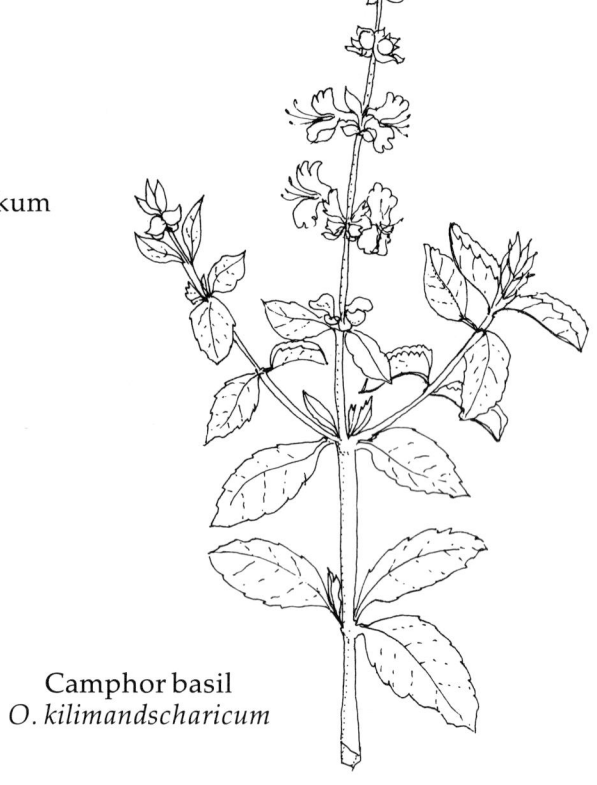

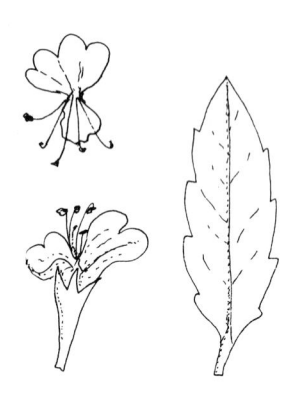

Camphor basil
O. kilimandscharicum

Camphor basil
O. kilimandscharicum

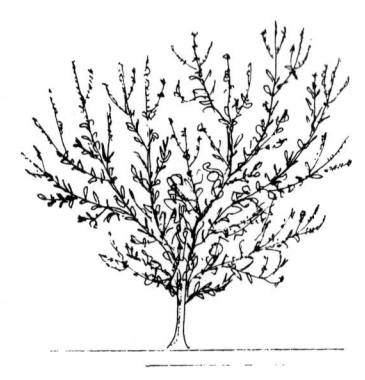

Height: 1–1,5 m

SEVERAL indigenous basils occur in the eastern Cape, Natal and the Transvaal, all of which have the most wonderful fragrances and can be used in much the same way as the garden basils. Through the years I have become fascinated with the great variety of basil species, from the well-loved sweet basil *(Ocimum basilicum)* and bush basil *(O. basilicum minimum)* to the beautiful purple ornamental basil *(O. basilicum purpurascens)*. I have grown tulsi, a large perennial basil *(O. gratissimum)* and sacred basil *(O. sanctum)* and variations on those old favourites like lemon scented basil *(O. basilicum citriodorum)*, lettuce leaf basil *(O. basilicum crispum)* and the large perennial camphor basil *(O. kilimandscharicum)*, which is an African basil named after Mount Kilimanjaro. It is also known as feverplant and is used as a febrifuge, and is also used by African tribes as a malaria treatment. A pot of this remarkable African basil on the patio is an excellent mosquito repellent, and I have used it as a lovely hedge in the herb garden. It grows about a metre high and the long flowering spikes in soft mauve or white make wonderful dried arrangements in the house.

In Central Africa people use a tea made of the leaves for infections (internally as a medicine and externally as a wash over the infected area) and for bringing down fevers. Cloths wrung out in the brew are applied over the head and chest. Leaves and seeds are placed as insect repellents among blankets and clothing.

A herdboy on my farm once showed me the compact and attractive small bushes of the Transvaal basil *(O. canum)* and told me that his mother smoked the dried leaves for chest coughs, colds and asthma and the Tswana people drank a tea of it for fevers and infections. (Pour 1 cup of boiling water over 4 basil leaves and a thumblength flowering or seed spike, stand, steep for 5 minutes, then strain and drink. Sweeten with a little honey if desired.) For a fever drink half a cup of warm or cooled tea frequently until the fever breaks, and use the cooled tea as a lotion over forehead and chest. This tea can be used as a mouth rinse too.

I tried wild basil tea when I had a feverish cold and so enjoyed the taste and the relief it gave that I have grown the plant in my herb garden ever since and always keep a sharp look out for it whenever I walk in the veld. It seems to be found all over South Africa from the eastern Cape through to the eastern Transvaal. I once found a long stretch of it growing abundantly in the hot summer sun at the roadside in the Free State — bone dry, yet vigorous.

I use the wild basil leaves in cooking instead of sweet basil, and I save the twigs and dry them for burning over braai fires — I love the taste the burning twigs give to chops! I sometimes use wild basil in salad dressings, but I always start cautiously as the leaves are strongly flavoured. Butter beans are delicious cooked with a sprig of wild basil, and boiled in their jackets potatoes take on a wonderful taste with a sprig of wild basil in the pot while they boil.

My Zulu studio assistant tells me that her village people burn leaves, stems and seed heads to repel mosquitoes and inhale the smoke for chest colds. She also uses another type of wild basil that grows into a large bush a metre high and wide known commonly as Zulu basil *(O. urticifolium)* which has lemon-scented leaves that are woolly and soft. These leaves are gathered by the handful and used as a rub for aching feet and calf muscles; and are packed into the shoes around the heels on long walks to soothe aches and pains.

I rub the leaves onto pillows and blankets to keep mosquitoes away, and the Zulu rub the leaves into their hair for the same reason!

I also use the wild basils in a most remarkable bathroom deodoriser. This keeps the bathroom and toilet fresh smelling and does away with the need to use aerosol sprays for deodorising, many of which do so much damage to the ozone layer.

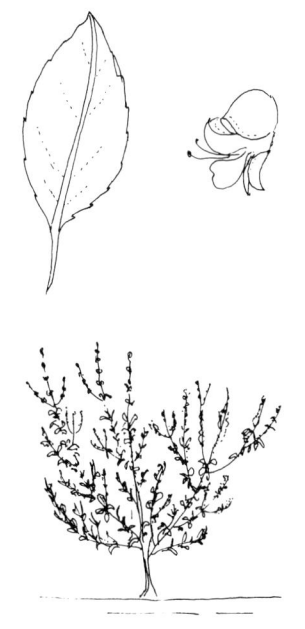

Transvaal basil
O. canum

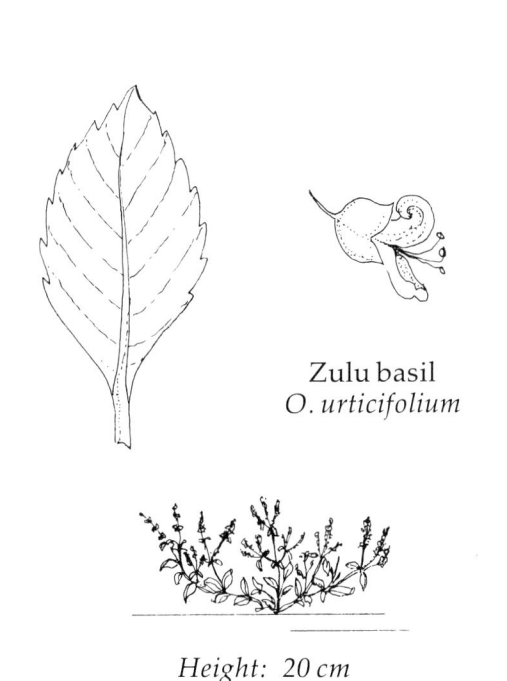

Zulu basil
O. urticifolium

Height: 30 cm

Height: 20 cm

Wild basil bathroom deodoriser

3 cups dried wild basil leaves, flowers and seed spikes
1 cup dried minced lemon peel
½ cup cloves
½ cup broken cinnamon pieces
½ cup crushed nutmeg
2 teaspoons clove oil

Mix all the ingredients, keep in a sealed jar and shake daily for 3 weeks. Add a little more oil from time to time if needed. Spoon into pot-pourri jars or bowls, cover loosely and stand in the bathroom and toilet.

This fragrant mixture will last you many years, and only needs to be revived every now and then with a little clove oil. I find the best way to do this is to turn the contents of the bowl into a jar, add the oil, screw on the lid and shake well, leaving overnight. The next day spoon into the bowls again.

As an insect repellent for fish moths and clothes moths, add 2 teaspoons of citronella oil to the above recipe and tie into sachets.

Wild basil and tomato sauce for pasta — Serves 6

2 large onions, finely chopped
1 green pepper, chopped
4 large tomatoes, skinned and chopped
6 fresh basil leaves, either Transvaal basil or Zulu basil, finely chopped
1 small cucumber, peeled and chopped
½ cup brown sugar
½ cup sunflower cooking oil
salt and pepper
little water

Fry the onions in the oil in a heavy-bottomed pot until golden brown. Add the chopped green pepper and stir fry with a wooden spoon. Then add all the other ingredients and stir well. Cook over medium heat, and add a little water if the mixture seems too dry. Cook for about 10 minutes, then remove from heat and stand covered while you prepare the pasta.

Pour the sauce over the pasta and serve piping hot.

Wild camphor tree
Tarchonanthus camphoratus

ENGLISH	Camphor bush, sage wood, wild cotton, wild sage wood, African flea bane
AFRIKAANS	Vaalbos, kanferbos, vaaibos, veldvaalbos
NDEBELE	Umnqebe
SOUTHERN SOTHO	Mofahlana
SWAHILI	Mkalambati
TSWANA	Mohata
XHOSA	Mathola
ZULU	Amathola

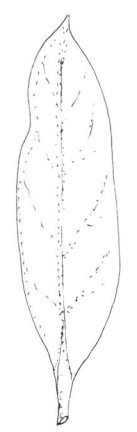

THE wild camphor tree grows into a shrub or small tree usually no more than 3 m in height, and is found in a great range of habitats — at sea level in coastal dunes, in mountainous areas, semi-desert, in high and low rainfall areas and in all soil types, all through the Cape, Natal, Orange Free State and the Transvaal up into Zimbabwe and Namibia.

The leaves are long, simple and narrow, grey-green with a grey-white back, and smell of camphor when crushed. The light grey-brown bark also smells strongly of camphor, and the wood is fragrant, fine grained and polishes to a beautiful sheen. This makes it a magnificent wood for musical instruments, fine cabinet making and carving. It is also much prized for boat building and strong branches are trimmed and used for fencing posts, or the smaller ones for basket struts and grain storage containers. The beautiful, fresh camphor fragrance remains in the wood for many years, and pieces of wood or twigs are used as an insect repellent among clothing, blankets and food.

The fruit is a small seed or nutlet covered in white, woolly hairs resembling a ball of cotton wool, and although the seeds are at their best in late winter and spring, they can be found on the tree at almost any time of the year. When burned the seeds give off a camphor smell and are used to fumigate huts, fresh leaves and twigs being added to the fire to make more smoke. This smoke inhaled is also believed to be good for rheumatism, headaches and sleeplessness.

A tea made of the crushed leaf (usually 6 or 7 leaves infused in 1 cup of boiling water for 5 minutes, then strained and drunk) is taken for stomach ailments, asthma, overanxiety and heartburn. The tree has been used for asthma by the San and Khoikhoi as a type of tobacco, and it produces a slightly narcotic effect. They also chewed the leaf for chest ailments, and the early settlers found the tea to be beneficial in the treatment of asthma, rheumatism and as a tonic for coughs, colds and flu.

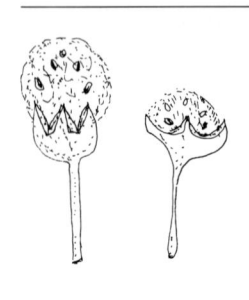

Several African tribes and some whites still use the plant as a treatment for bronchitis and chest ailments, placing a poultice of warmed leaves around the chest, and pour a strong infusion in a hot bath for paralysis and cerebral haemorrhage.

A soothing ointment made by macerating the leaves in Vaseline, aqueous cream or lard is a remarkable treatment for chilblains and sore feet. It can be rubbed onto the area several times a day to ease the discomfort. Some tribes use the ointment for anointing the body in religious ceremonies and as a massage for the legs, particularly before a long journey.

Several African tribes wear garlands of the leaves and rub the fresh leaves into their hair to keep it free of nits and dandruff, as well as for the fragrance it imparts. Leaves tucked under the pillow will ensure a peaceful night and if a leaf is chewed it will ease indigestion, prevent bad dreams and soothe a sore throat. A traveller will chew a leaf to protect him on his journey.

The Tswana stuff the leaves into their hats or wear a garland of leaves in the harsh midday sun, and also rub their feet with a handful of fresh leaves to give them strength on a long journey.

The Tswana and Venda use the woolly seeds to stuff pillows, which are considered to be excellent for headaches and sleeplessness.

The wild camphor tree is becoming a popular garden tree and some nurseries offer it for sale. It can be trimmed and shaped and will reward you with shade, fragrant firewood and a useful medicine chest!

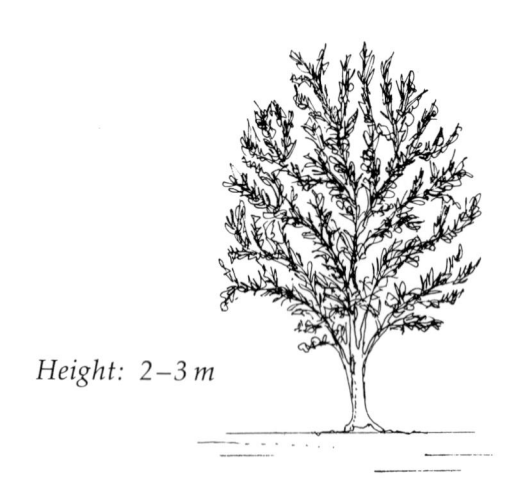

Height: 2–3 m

Wild cineraria
Senecio elegans

ENGLISH   Pink ragwort, American groundsel
AFRIKAANS   Strandblommetjie
XHOSA   Izuba

THE *Senecio* genus is one of the largest groups of flowering plants, many of which are indigenous to South Africa. Among them are *S. tamoides*, the canary creeper, a wonderfully showy garden creeper with brilliant yellow daisy clusters which is known to the Zulu as 'ihlozi-ikulu' and used by them to treat anthrax in cattle; and *S. purpureus, S. coronatus*, both of which are eaten as a vegetable to strengthen the blood and to protect against witchcraft. Many of the *Senecios* are poisonous, however, so the utmost care must be taken in identification.

The wild cineraria grows along the Cape coast. It is an annual, with an erect flowering head, fleshy divided leaves reminiscent of the 'ragworts', and their beautiful deep pink daisy flowers are a wonderful sight in spring and summer. The height, size and shape of the divided leaves vary according to its habitat.

The early settlers first noticed it growing in abundance on the hillsides near Alexandria, and seed was sent to Europe in about 1700. From then on the wild cineraria was used in hybridising with the cultivated cineraria species, as well as being used as an annual bedding plant in gardens overseas. The wild cineraria is just another of the many South African plants that has found a place in overseas markets.

The Khoikhoi were the first to use the plant as a remedy for chest ailments. The small leaves along the stem, and pieces of stem, are sucked and the saliva swallowed to help a tight chest, asthma and a tight cough. The other leaves can be eaten too, and are often made into a tea (¼ cup of fresh leaves and pieces of stem to 1 cup of boiling water, stand, steep for 5 minutes, then strain and drink) for kidney stones and painful kidneys, bladder infections, scant urine and cystitis. The plant has been much used in medicine, particularly by the Xhosa, and in some country districts it is still used and much respected today.

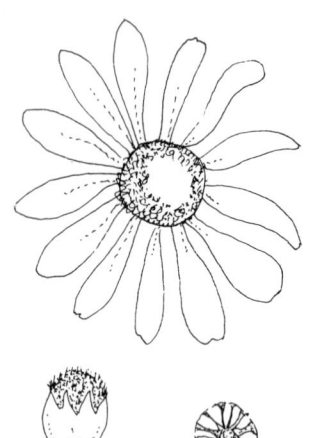

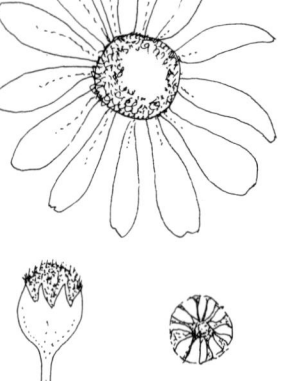

The beautiful deep pink flowers make a stunning garden display and seed and sometimes young plants are available at nurseries. Sow the seed in boxes for transplanting into their final position when they are big enough to handle.

The plant likes sandy, well-drained soil in full sun, and does not need a lot of water, but does well with a good weekly watering. I grow wild cineraria in massed planting for a spectacular spring display, planting it out in winter for the plants are able to withstand a little frost.

The flower heads are wonderful in pot-pourris as they retain their brilliant pink colouring when dry and the flowers last well in vases if they are picked, tied tightly into posies, and stood in a deep vase of water to which 2 to 3 tablespoons of sugar have been added.

Height: 0,5 m

Spring posy

Bunch flowers together in rings, starting with a large central flower — I usually start with a deep pink rose. Circle this with tulbaghias, then a circle of blue plumbago, then a circle of wild cineraria, and then a white fluffy circle of wild rosemary (kapokbossie), and ending in the leaves and flowers of the scented geraniums.

Tie with bright ribbons if you are giving this beautiful 'tussie-mussie' (the real name of the fragrant posy) away as a gift. It will last for over a week in water and then can be dried and added to pot-pourris.

Wild dagga
Leonotis leonurus

ENGLISH	Minaret flower, red dagga, lion's ear
AFRIKAANS	Wildedagga
XHOSA	Umfincafincane
ZULU	Umunyane

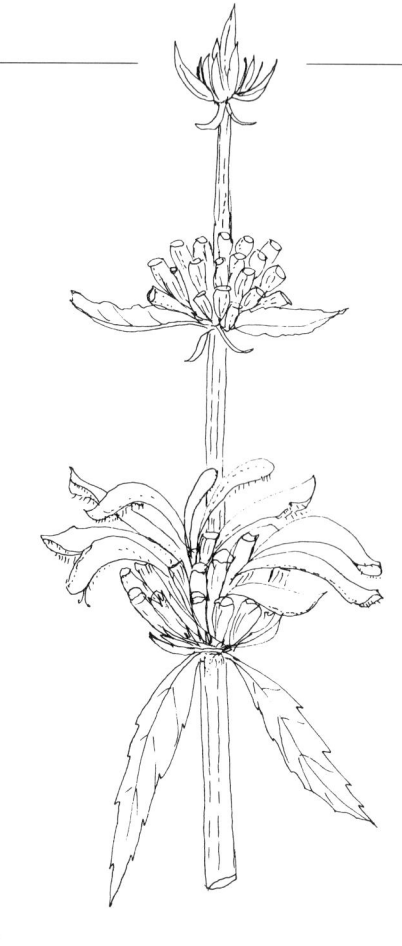

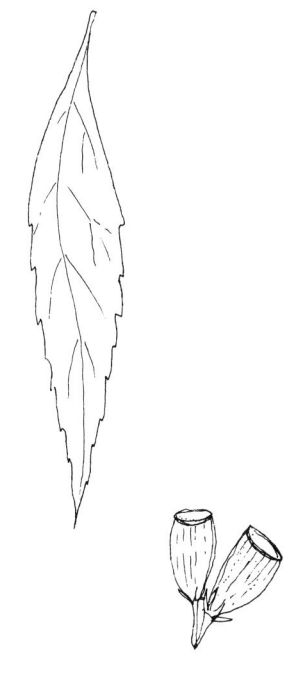

THIS beautiful and spectacular plant with its tiers of bright orange flowers grows all over South Africa. It was first used by the Khoikhoi as a tobacco and introduced to the settlers as a remarkable medicine chest. It is easily propagated from seed and cuttings and is often cultivated in gardens, where it blooms throughout summer and autumn. The flowers retain their orange colour when dried, and are good in a wild flower pot-pourri. I use the leaves as well, as they are pungent smelling and insect repelling.

The early colonists made an infusion of twigs, leaves and flowers for skin eruptions, including leprosy. Added to the bath the twigs give relief to muscular aches and pains, itchy skin and eczema. I have used a strong infusion (4 cups of leaves, stems and flowers, with 2 litres of boiling water poured over it — stand, steep and cool) in my bath water, and found it helped cramp and leg pains. This same brew can also be used to dab onto sores, bites, bee and wasp stings, and is said to help scorpion and snake bites.

The Zulu use the root for snakebite and sprinkle a decoction of the plant around their houses to keep snakes away. The Zulu and the Xhosa make a strong brew of the leaves boiled in water as an application for snakebite (a cloth is soaked in the brew and tied around the area), and often use a decoction or tincture of the root bark internally as a tea for snakebite.

The Zulu, Xhosa and many whites make a tea of the flowers (¼ cup of fresh flowers in 1 cup of boiling water, stand, steep for 5 minutes then strain, sweeten with honey if desired) for a wonderfully soothing cough and cold remedy. This same brew has been used effectively for the treatment of jaundice, cardiac asthma, haemorrhoids, headaches, chest ailments, bronchitis and epilepsy. The leaf is also smoked in the treatment of epilepsy and partial paralysis.

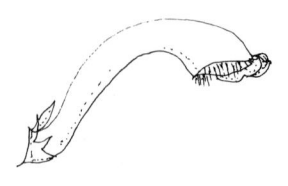

Interestingly, there are records of the tea being used as a diuretic 'to reduce corpulence'! This I have not tried, but many of the elderly people coming to my herb garden tell me that their parents and grandparents drank a tea of leaves and flowers daily for water retention, obesity and haemorrhoids. It is always fascinating to hear of these old remedies, and I am often startled by the strong dosages and treatments that were used. Those older generations seem to have been far tougher than we are today!

Wild dagga is also much respected in the treating of animals. The Tswana, Zulu and Sotho make a strong brew of leaves, flowers and stems and use this as an enema in sheep, goats and cattle, as well as, I might add, in human beings! This brew is also given to the animals for respiratory problems, and applied as a lotion to sores on stock and dogs, and as a wash for wounds, scratches, bites and stings. A few chopped leaves are tossed as food to chickens with diarrhoea and this seems to be very quick and effective.

I always marvel at the incredible medicinal values of this plant, and as I find it so easy to grow anywhere in the garden I encourage people to plant it. The bush needs clipping back after the winter to keep it attractive, otherwise bits of it seem to die off each season.

A lovely white variety, *L. leonurus* var. *albifolia*, is also used as a tea for coughs, colds, bronchitis, asthma, and as a wash for sores, rashes, bites and eczema.

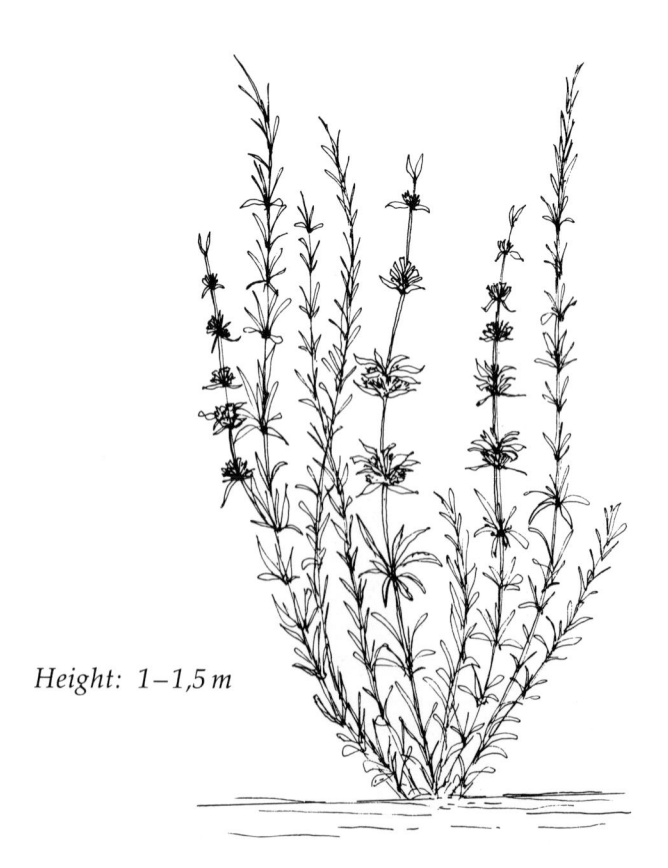

Height: 1–1,5 m

Klipdagga *(Leonotis leonitis)*

THE stately and unusual klipdagga often grows along the roadsides, where its tall stems (up to 2 m in some places) with round knobs of prickly calyxes and tier upon tier of bright orange flowers draw the eye. These make beautiful dried flower arrangements in winter when the flowers are over, and they seed themselves freely all over the garden. I treat this plant like an annual as it tends to become somewhat untidy and needs clipping back after the winter or to be replaced by new seedlings.

The klipdagga is also used as a tea (¼ cup of fresh leaves to 1 cup of boiling water — stand, steep for 5 minutes then strain) for coughs and colds, chest conditions, whooping cough, fevers, the aches and pains of flu, and as a gargle for laryngitis. The Xhosa also use this as a snakebite remedy, taken internally and applied externally to the fang punctures.

I have used the tea as a lotion to dab onto infected mosquito bites, bee stings and wasp stings and found it immediately soothing and helpful in reducing the swelling. Klipdagga also makes an excellent bath preparation for skin treatments, itchy skin and sunburn, and for aching muscles.

A sangoma from Louis Trichardt told me she grows klipdagga to give as a blood purifier, and to build strength in the young male during puberty. She crushes the seeds in milk and uses this as a tonic.

Overseas visitors find the klipdagga a most exciting plant to grow and it is much prized as an exotic African plant in their greenhouses! South Africans seem to take it for granted, as most do not appreciate its wonderful medicinal values.

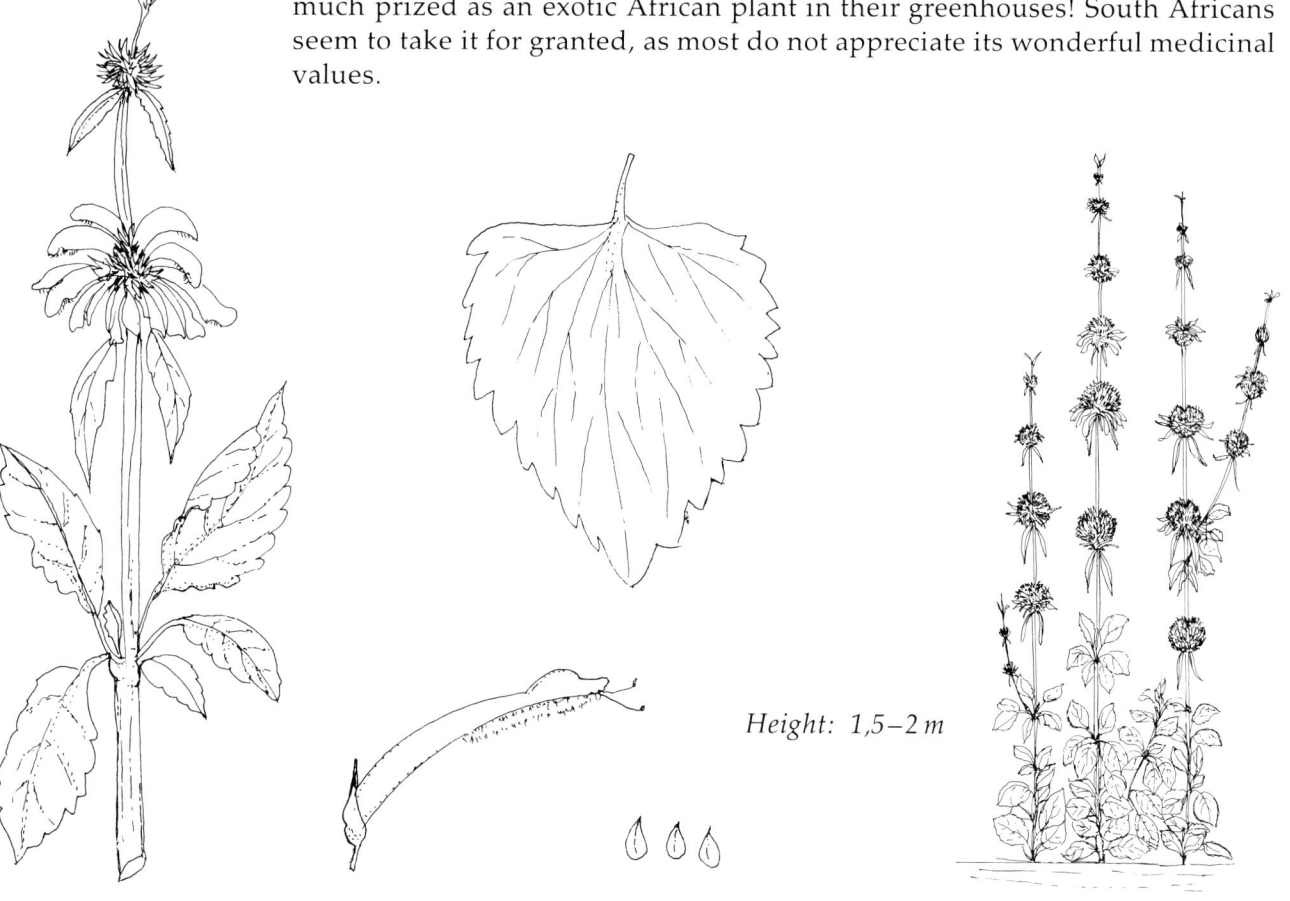

Height: 1,5–2 m

Wilde als

Artemisia afra

ENGLISH	Wild wormwood
SOUTHERN SOTHO	Zengana
TSWANA	Lengana
XHOSA	Umhlonyane
ZULU	Mhlonyane

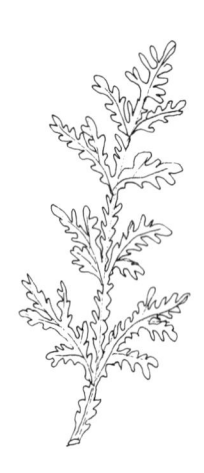

WILDE als is one of the oldest and best known of all the indigenous medicines. It has been used by African tribes for many centuries and was much used by the colonists, probably owing to its resemblance to European wormwood.

It is a warming, cleansing, disinfecting herb used for a great number of ailments. It is a well-known treatment for chest conditions, coughs, colds, colic, heartburn, flatulence, croup, whooping cough and gout, and is still used effectively today.

It grows wild in most of South Africa and is easy to grow and propagate. Many people remember their grandmothers using it, and it is fascinating to hear of the various treatments and remedies that they are familiar with. The usual preparation is in the form of a tea or decoction (¼ cup of fresh leaves to 1 cup of boiling water, stand and steep for 5 minutes, then strain, sweeten with honey and drink a little at a time) which is sipped to ease all the above ailments and is also a good gargle for a sore throat.

An infusion (pour 2 litres of boiling water over 1 cup of fresh leaves and stems, allow to draw for an hour, then strain) is used as a wash for haemorrhoids, and in the bath to bring out the rash in measles, to soothe fevers, to wash wounds, sores, rashes, bites and stings, and as an eyebath diluted with warm water to soothe red, smarting eyes.

A strong brew (½ cup of leaves to 1½ or 2 cups of boiling water — stand, steep and draw for 10 minutes then strain) is used as a mouthwash for gumboils (often a leaf taken from the brew is held in the mouth to draw the boil) and mouth ulcers, and can be dropped gently in the ear to relieve earache.

Wilde als brandy was an old standby kept in the medicine chest to treat

Wild pineapple

Wild olive

Wilde wingerd

Wild pear

Wild rosemary

Wild verbena

Yellow wandering Jew

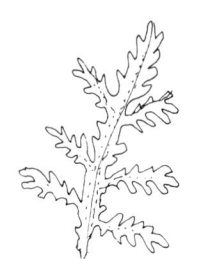

colds, coughs and chest ailments, as well as heartburn, indigestion and stomach cramps (see recipe overleaf). This 'magic potion' is still made today and is a much respected and popular medicine among rural people.

If the herb is boiled up in water the vapour or steam arising from the pot makes an excellent inhalent for bronchitis, blocked nose, tight chest, asthma and chest colds. I have found it to be excellent for blocked sinuses and a sinus headache. A quick way of making this vapour is to pack 2 or 3 cups of leaves into a bowl and pour over enough boiling water to cover it, then make a towel tent over your head, hold the bowl under your nose and inhale deeply until the brew cools.

I was taught by the Tswana on my farm to roll a leaf and insert it into the nose to clear a headache and a stuffy nose, and to pack an aching tooth with a leaf to help toothache until you can get to a dentist. Several African tribes smoke the leaves to help ease congestion, release phlegm, and soothe a sore throat and coughing at night.

The Zulu make an infusion by grinding up the leaves and adding hot water, and give this as an enema to children with worms and constipation. They also believe that this brew taken internally will cleanse the skin and the blood, and acne and boils are treated in this way, which makes good sense as the herb is disinfectant and deeply cleansing.

The Tswana and Venda make a wash of the plant for skin ailments, and use warmed leaves to draw out pimples and boils by applying them as a poultice. The warmed leaf is also an excellent and soothing poultice over a painful neuralgia, mumps swellings and on a sprain or strain, and bound over the stomach for babies with colic. The painkilling and relaxing properties of wilde als have made it a very important herb in country medicine — perhaps it should be brought more to our attention as it may be of use in our stressful world of today, replacing the more harmful drugs.

It is an excellent moth repellent and can be used in pot-pourris, natural insecticidal sprays and cupboard fresheners.

It grows easily from cuttings and is an attractive, feathery grey green shrub in the garden growing to about 1,5 m in height and spread. It benefits from quite harsh pruning and I find it needs a winter pruning to allow the lush soft spring growth to really come to its best. I have also pruned it as a pot plant in a large tub and find it most attractive and interesting as a patio plant, and a fascinating talking point.

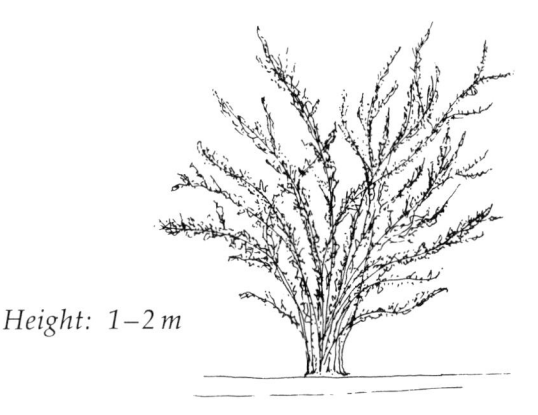

Height: 1–2 m

Wilde als brandy

1 bottle brandy
1 cup wilde als leaves
¼ cup thyme
½ cup mint leaves
1 cup sugar
1 thumblength piece ginger
¼ cup rosemary

Push all the ingredients into the bottle of brandy and shake well daily for a month. Then strain, or allow it to grow stronger with age by leaving the herbs in it. Take a tablespoon at a time in water.

Wilde als moth repellent

4 cups dried wilde als leaves
2 cups dried wild rosemary leaves
2 cups dried eucalyptus leaves
2 cups dried lavender leaves and flowers
1 cup mixed crushed nutmeg, cloves and cinnamon
1 cup crushed coriander seeds
1 cup coarse salt
2 cups minced dried lemon peel
lavender oil

Mix all the ingredients. Store in an airtight tin or crock, shake daily for 3 weeks, then add a little more oil. Add a few more cloves if desired, depending on your taste. Mix well, fill bowls or sachet bags with the mixture and place in cupboards and drawers. Revive with a little lavender oil from time to time.

Wilde als insecticide spray

1 bucket boiling water
¼ bucket fresh wilde als leaves and flowers
¼ bucket khakibos leaves and flowers
1 cup Jeyes fluid
1 cup soap powder

This is an effective spray for aphids, white fly and other plant pests.

Mix all the ingredients into the bucket of boiling water and allow to draw overnight. Next morning, strain. Dip a bunch of leafy twigs into the mixture and flick this over the plants. Pour the remainder into the ground around the stems of the infected plants.

Wilde wingerd
Cliffortia odorata

ENGLISH Wild grape
AFRIKAANS Wilde-vye-rank

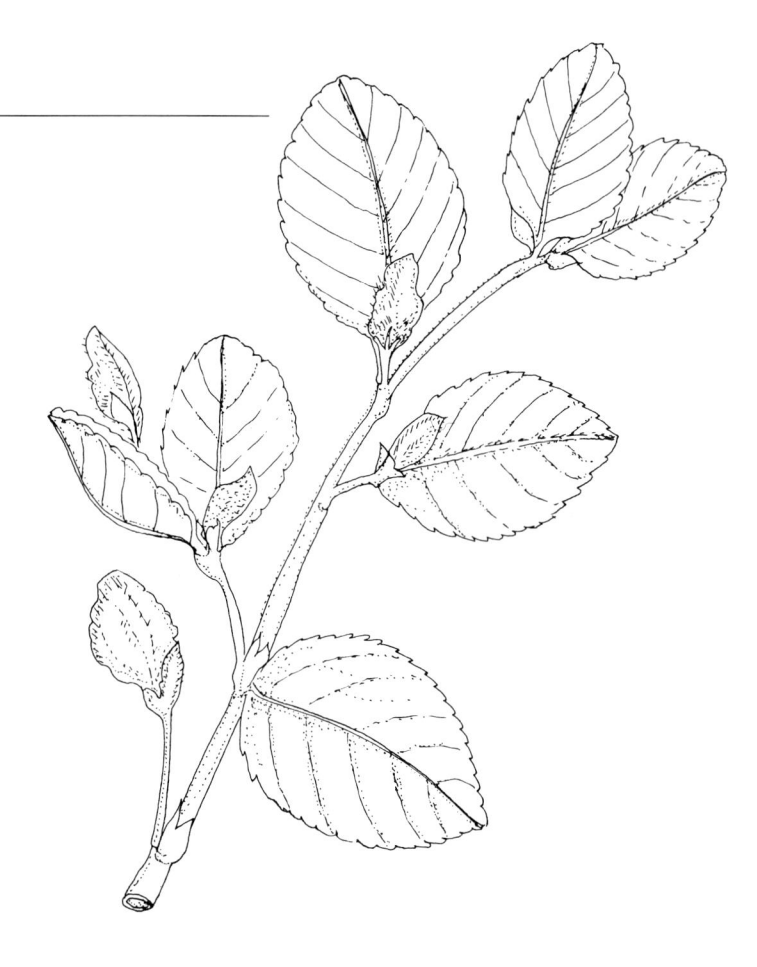

THE tough, shiny, large leaves of *Cliffortia odorata*, known throughout the Cape as wilde wingerd, are esteemed as an important medicine. The shrub is low growing, spreading and perennial. Its attractive leaves make it a good garden subject and although it occurs mainly in the Cape, it can be grown in the interior quite successfully. At first glance it looks like a grape vine with smaller, coarser, dark green leaves, and in cultivation can be trained up a fence or low wall, pruned, shaped and staked.

Its medicinal uses are many and varied and the young leaves and leaf tips on the stems are the parts usually used.

Farmers in the Western Province used wilde wingerd in the treatment of cattle diseases. A strong infusion was given internally, and a strong lotion was used — and on some farms is still used today — as a wash for wounds or infected sores, tick infections, blisters, abscesses and bites. The tea was used in miscarriage in cattle and also to prevent abortion. The Khoikhoi used the plant extensively for treating animals, particularly dogs and goats, and for themselves as well. They were probably the first people in Africa to discover the wonderful medicinal uses of the wilde wingerd, and taught the colonists who in turn taught their children. Today those early natural medicines still have their place in our modern, plastic world.

A tea made of the leaves (¼ cup of fresh chopped leaves to 1 cup of boiling water, stand, steep for 5 minutes, then strain) is used as a comforting treatment for colds, flu, sore throat (use as a gargle too), chills, fevers and threatened miscarriage.

The tea is also a most soothing treatment for haemorrhoids, and was one of the old remedies for piles used by the colonists. The tea can be used as a lotion or wash, or pads of cottonwool soaked in it can be used as a poultice at night.

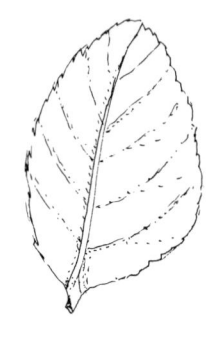

Boiling water is poured over the leaves and the steam inhaled to clear the nose and sinuses, and this is very effective in clearing a sinus headache.

For 'women's ailments' such as excessive, irregular or painful menstruation, a tea drunk daily during and a day or two before the period is remarkably soothing and helpful, the usual dose being ½ cup 2 or 3 times during the day. The diluted tea can also be used as a refreshing and cleansing wash during the period, and with a little apple cider vinegar as a douche.

The wilde wingerd is perhaps best known for its role in treating mouth and throat infections and diptheria — use the tea as a gargle and sip a little at frequent intervals.

On visits to my grandmother in Gordon's Bay as a child, her coloured housemaid would come in from her home in the mountains each morning bringing me a wild flower, or a spray of leaves, a seed pod or a grass head. I kept these precious things in fishpaste jars or Marmite bottles filled with water on my dressing table, and so got to know the Cape wild flowers through her. One day she brought me a leaf of wilde wingerd which I pressed in my note book, and told me how that morning her little brother had woken with a bad sore throat. Her mother quickly made a strong tea of wilde wingerd which he drank and gargled with immediately. I was duly impressed when the next day she arrived with a Hotnotsvy flower and leaf spray and announced that he was better, but that he'd have to chew a piece of this leaf, astringent and drying, frequently through the day, and drink a cup or two more of wilde wingerd tea. I too chewed a piece of the sour fig leaf, and imagined that little boy with his sore throat being cured by the wild plants. Little did I realise that this young girl would sow a seed that would live in me for ever — my interest in the fascinating world of herbal medicines.

Sometimes when I hear the South Easter and smell the Cape winter rain, and see the wild flowers wet with mist, I think back to those childhood days and my row of little vases each holding a wild flower, and the wilde wingerd and the sour fig that cured that little boy of what was most probably diptheria — it was the dreaded 'witseerkeel siekte' so prevalent in those times — and I am grateful that those country people used and appreciated the wild plants and could pass their knowledge on to others, so bringing it into all of our lives.

Height: 30–50 cm

Wilde wingerd cleansing douche

½ cup apple cider vinegar
1 cup wilde wingerd tea (made as above)
1 litre warm water

Mix well and use as a douche for cleansing between periods, and to clear any mild itch or infection. As in all home treatments don't forget to consult your doctor first!

Wild foxglove
Ceratotheca triloba

ENGLISH   Rhodesian foxglove
AFRIKAANS   Vingerhoedblom
NDEBELE   Inkunzanienkulu
ZULU   Udonqa batwa

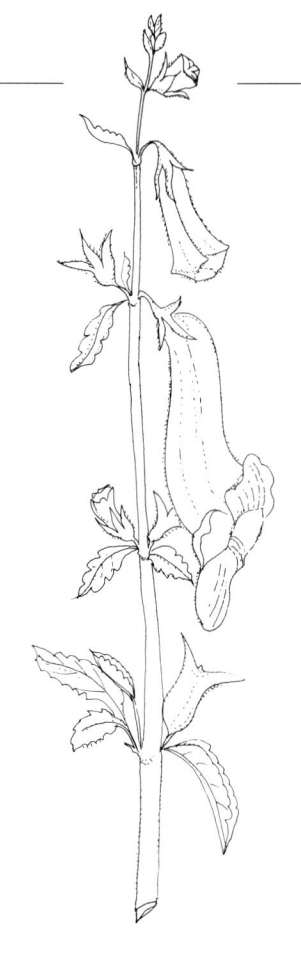

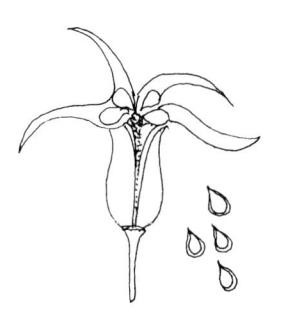

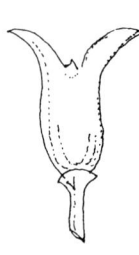

IN midsummer the tall, pale mauve spikes of the beautiful wild foxglove can be seen in the veld, on hillsides and along roadsides in the Transvaal, Orange Free State, eastern Cape and Natal and as far north as Zambia. At Kirstenbosch there is a huge and striking bed of them, showing how beautiful a group planting can be.

The plant is perennial and grows to about 1,5 m. The stem is erect, branched, and the leaves have a lobed appearance, hence the species name *triloba*. They have a strong, fetid odour, which is peculiar to this plant.

The flowers are beautiful in pressed flower pictures, retaining their pale colour for a long time. The seed capsule has two little horns, and the long spikes of dried seeds make interesting additions to dried flower arrangements.

A tea made of the leaves (¼ cup of chopped fresh leaves with 1 cup of boiling water poured over) is used to relieve painful menstruation and stomach ailments such as diarrhoea, cramps, flatulence, colic and nausea. The Zulu particularly appreciate this plant, and whites in Natal and the Transvaal use it for intestinal disorders.

I use the wild foxglove in an effective insect repelling spray, and as the seeds germinate well, I collect the seed from plants that are well established in the veld, and grow new plants every year for garden sprays.

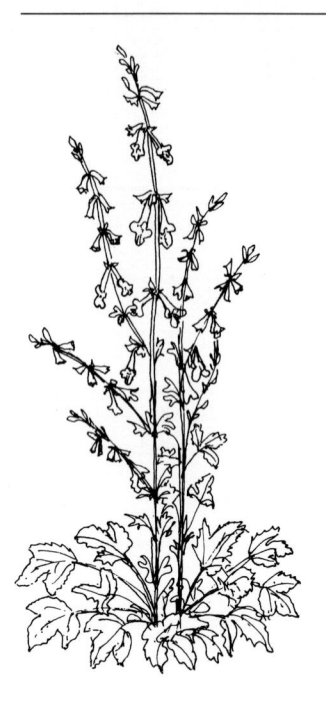

Height: 1,5 m

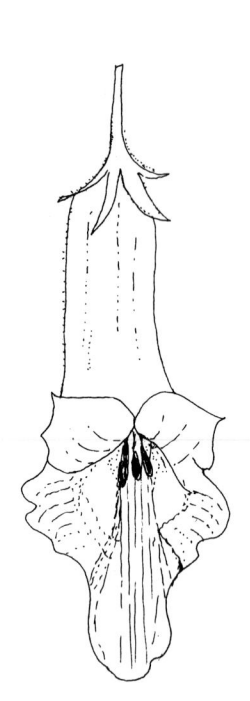

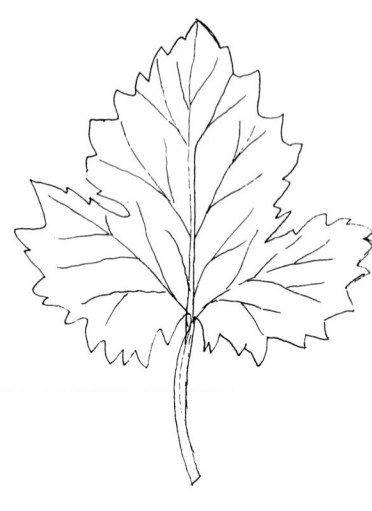

Wild foxglove aphid spray

¼ bucket fresh green wild foxglove leaves
¼ bucket khakibos leaves and seeds
¼ bucket chopped wild garlic leaves
1 cup soap powder
1 cup Jeyes fluid

Pour enough boiling water over the mixed leaves to fill the bucket. Leave to steep overnight. Next morning strain, add the washing powder and Jeyes fluid and mix well. Use in a spray or splash onto roses or pot plants. (Do not use this on vegetables — the taste is too strong!)

Wild garlic
Tulbaghia species

AFRIKAANS Wilde knoffel, wilde knoflok
ZULU Incinsini

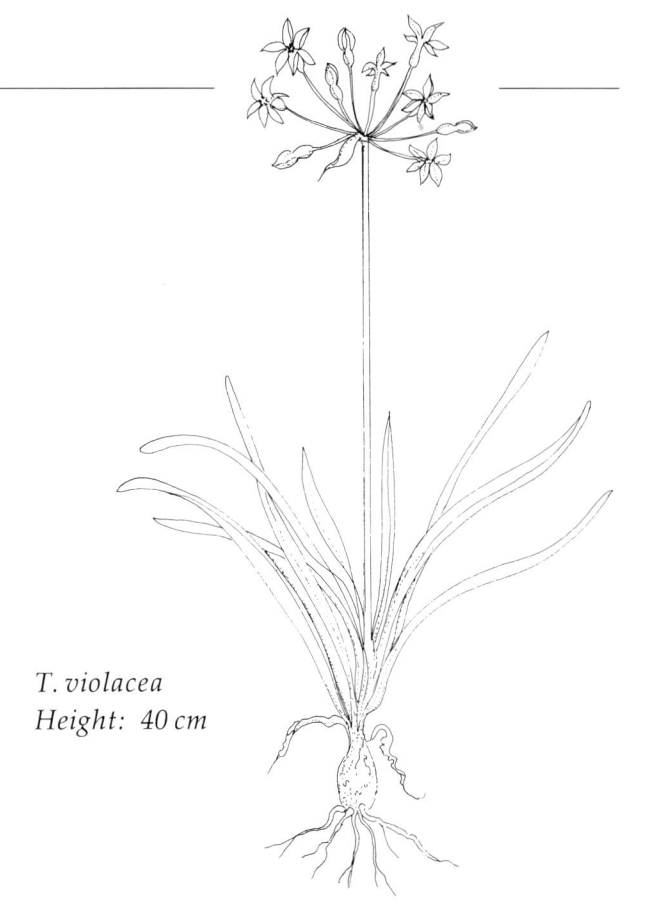

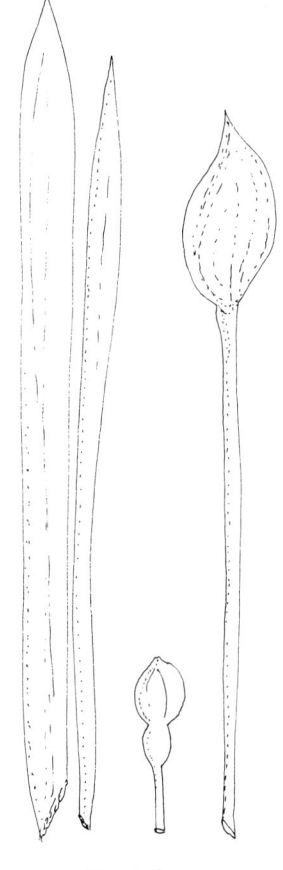

T. violacea
Height: 40 cm

T. violacea

TULBAGHIA *violacea* or wild garlic is a pretty garden plant which flourishes in many gardens in South Africa. It grows well in both sun and shade, and many believe it keeps snakes away — even today the Zulu encircle their huts with it, believing that no snake will pass through the dense, pungent smelling leaves. They also use the bulb in love potions.

Several African tribes use the leaves as a condiment with meat and potatoes, and one of the Zulu women working in my studio gave me the recipe for a spinach dish she makes using wild garlic. I made it and found it delicious (see recipe on p. 235).

The leaves and bulb are used to treat tuberculosis, and if eaten chopped and raw on bread or in a daily salad, wild garlic — like so many of the onion family — has remarkable antiseptic qualities. It will do much to clear colds, coughs and flu.

Another variety of wild garlic which I grow abundantly and with ease in my herb garden is *T. alliacea*, garlic chives or flat leaf chives. This has brownish green flowers quite different in colouring to the pretty mauve umbels of the wild garlic. Chopped into salads or added to stir fry dishes it has the delicious taste of garlic, but leaves none of the after-smell of real garlic.

Medicinally the roots can be tied into muslin and used in the bath to ease rheumatism, to help paralysis and to bring down a fever. A tea can be made of chopped bulbs and roots and used as a wash for the same illnesses (steep 3 to 6 bulbs and roots in 2 litres of boiling water, stand until cool, strain and use). This brew was also used by the Khoikhoi and the Basuto as a purgative.

Chopped into soups and stews the wild garlics give a delicious taste, and as the plants are so easy to grow they can be used lavishly. Adding the chopped leaves within the last few minutes of the cooking time seems to preserve their strong flavour. The plant is easy to grow — it needs full sun and a good weekly watering. It will form a dense cushion that remains green all year through. Pieces can be dug off the sides for propagation.

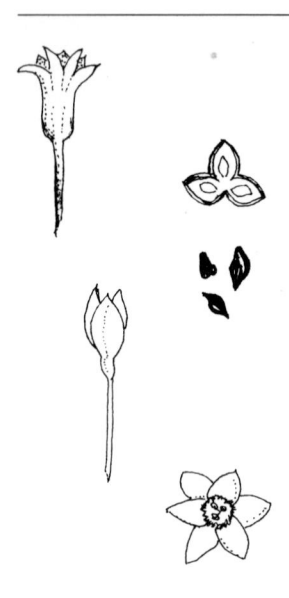

T. fragrans is a relative of the wild garlic but has coarser, thicker leaves, a thicker stem and juicier flowers with a rich lily-like fragrance. These are much loved by florists and they make excellent cut flowers. There is a white-flowered variety which is exquisite in the garden and in the vase; the scent is heady and even two or three flowers will fill a room with their fragrance.

I am particularly fond of *T. fragrans* for July wedding bouquets, as it blooms when few other flowers do and lasts beautifully. I also use it in pot-pourris, such as the springtime tulbaghia pot-pourri opposite. An all mauve pot-pourri is very beautiful with *T. fragrans*, violets, wisteria, lavender, sage and lemon verbena flowers, all of which bloom together. Don't use the wild garlic though as its strong garlicky smell will drown all else!

By early spring a border of tulbaghia makes a glorious ribbon of mauve or white and lasts for at least 6 weeks.

All the tulbaghias are well worth growing in the garden as they form attractive clumps and can be used to such advantage in the home. Divide the clumps in May or June to increase your plants — plant half a metre apart in well-composted soil.

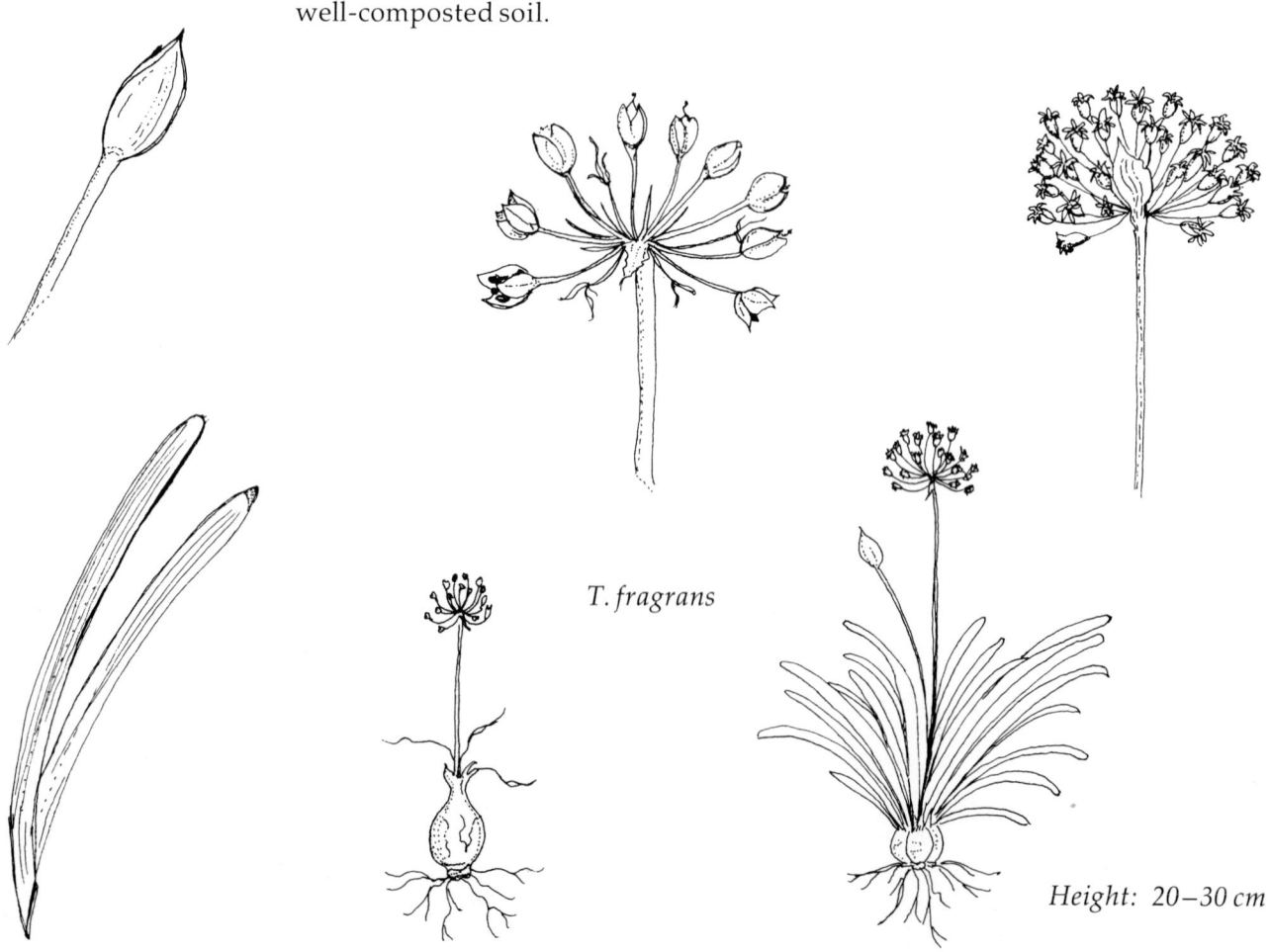

T. fragrans

T. fragrans

Height: 20–30 cm

Elizabeth's wild garlic dish *(T. violacea)*   Serves 6

4 large potatoes
4 cups spinach (she uses the wild spinach or 'morogo')
2 medium size onions, chopped
1 cup chopped wild garlic leaves
little oil or fat
little milk
salt and pepper

Peel and boil the potatoes in salted water. When done, drain and mash, adding the milk, salt and pepper.

Heat enough oil or fat to cover the bottom of a pot and brown the onions in it. Add the spinach and wild garlic and stir fry for 3 minutes (I add a little lemon juice here).

Toss this into the mashed potato and mix lightly with a fork. Serve as a vegetable. It goes particularly well with fish.

If there is any left over, mix in a beaten egg, form into flat cakes and fry these in oil. Cream cheese, grated cheese or bacon can also be added to make a delicious supper dish served with a salad.

Springtime tulbaghia pot-pourri

4 cups T. fragrans flowers
2 cups dried wild rosemary flowers and leaves
2 cups dried wilde als leaves
2 cups dried lavender flowers and leaves
2 cups dried wild pear flowers
2 cups minced dried lemon peel
1 cup cloves, nutmeg and cinnamon mixed
1 cup coarse sea salt
essential oil

Dry all the leaves and flowers on newspaper in the shade, turning daily. Then mix all the ingredients and store in an airtight container. Add the oil, and shake daily for 3 weeks. After 3 weeks adjust the fragrance by adding more oil or spices, according to your taste. Fill sachets and bowls, and place in cupboards.

This pot-pourri is made in springtime when the tulbaghia is at its best, so you can add other spring flowers like violets, jasmine and wisteria and then your essential oil would be predominantly one of those, e.g. violet oil, jasmine oil, lavender oil, etc.

Steamed garlic chives *(T. alliacea)*

This is a delicious and unusual dish and very easily made. Cut several garlic chive flowering stems into 3 cm long pieces. Place these with the flowers in a steamer and steam for half an hour or until tender. Serve as a vegetable with a little butter, salt and lemon juice.

Wild garlic spray for aphids and fruitfly

4 cups chopped leaves, stalks and flowers wild garlic (T. violaceae) or garlic chives (T. alliacea) or both
3 cups chopped wilde als leaves and stalks
3 cups wild basil leaves (you can substitute sweet basil or khakibos leaves here)
2 cups wood ash
1 cup soap powder

Place the herbs and the ash in a bucket and pour over half a bucket of boiling water. Leave overnight, then strain. Add the soap powder. Mix well, then dip a bunch of leafy twigs into the mixture and flick this all over the plants that are aphid infected. Do this daily for 3 or 4 days or pour it round the plants. I use it for red spider and white fly as well, and find it very effective.

Wild gladiolus
Gladiolus dalenii (formerly *G. natalensis*)

ENGLISH	Parrot lily, Natal lily
AFRIKAANS	Rooilelie
PEDI	Moxoxa-leleme
SOTHO	Kxahla-e-kholo

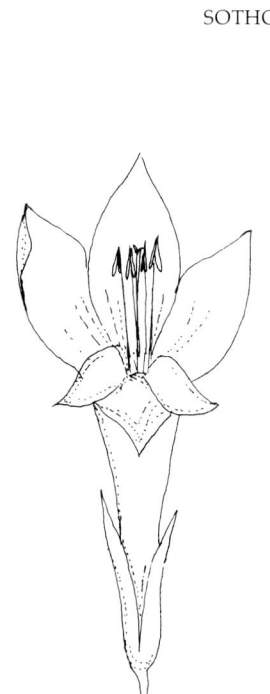

THE beautiful wild gladiolus, strikingly beautiful in its brilliant autumn reds and oranges, is believed to be the ancestor of several horticultural varieties of gladiolus. It was taken to England in the early 1700s where it was cultivated in greenhouses and new hybrids were developed.

The common name 'parrot lily' is derived from the colour and shape of the flowers, and in Natal, its home province, it is known as the Natal lily.

The Zulu and the Sotho have for many generations ground the corm down to a fine meal and taken this mixed with warm water in small quantities to help dysentery, diarrhoea and stomach upsets.

A tea made of the chopped fresh corm and the lower portion of the leaves (1 dessertspoon to 1 large cup of boiling water) is used for coughs and colds by the Sotho, Zulu and whites in Natal. The corms and leaves are burnt and the smoke from this is inhaled to clear a stuffy nose.

I started a patch of wild gladiolus in my herb garden with 3 corms, and by the following season they had multiplied enormously. Two years later I have a wonderful display of long, striking heads of bright orangy red flowers touched with yellow for 2 or 3 months during late summer and autumn.

The flowers can be eaten fresh in a salad or added to spinach or other cooked vegetables, or mixed into soups and stews.

The long strap-like leaves are attractive most of the year, but do need to be tidied up every now and then and the old corms replaced with new. Once the flowering period is over, cut off the dead spikes and leave the plant to rest. Lift the corms every second or third spring and replant the new ones in deeply dug, well-composted soil about 30 cm apart. Cover with compost and leaf mould, and water twice weekly until they are established. The arm-length flowering spikes will be at their best in their second year, and I have found they benefit from being picked and last well as cut flowers in the vase.

I have found them unfussy and very easy to grow and they seem to do well in most positions. I have seen great swathes of beautiful specimens in full sun on a hot dry hillside as well as a lush stand in a shady spot near a dam, with their feet constantly moist.

I use the flowers in wild flower pot-pourris as they add colour and substance to the pot-pourri and do not lose their colour over the years. Several nurseries offer corms for sale, so keep a look out for them.

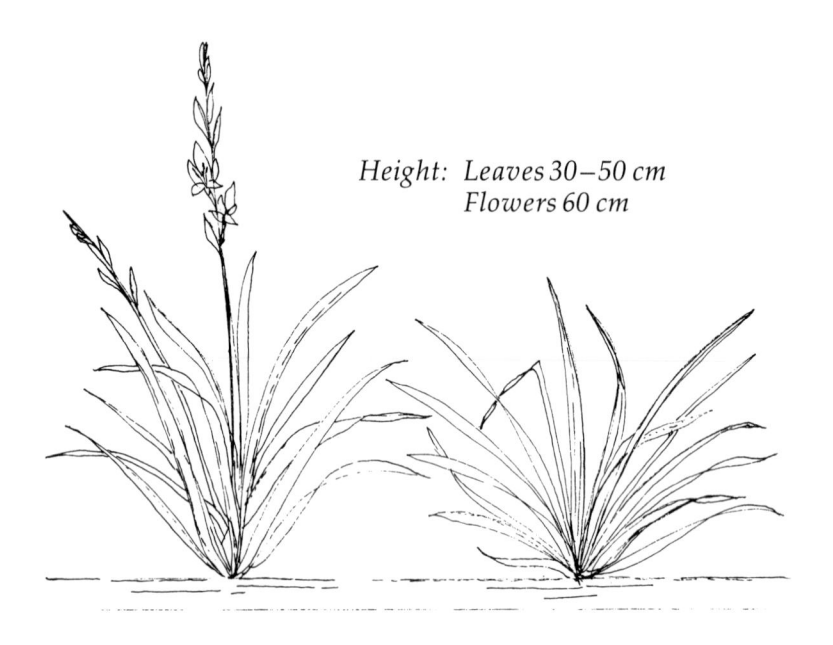

Height: Leaves 30–50 cm
Flowers 60 cm

Wild gladiolus cupboard freshener

4 cups dried wild gladiolus flowers
3 cups dried red hot poker flowers
2 cups minced dried lemon peel
3 cups dried eucalyptus leaves
3 cups dried lemon leaves
3 cups dried flowering eucalyptus pods
2 cups dried khakibos leaves and seeds
citronella oil

Dry the flowers and leaves on newspaper in the shade, turning them daily. When they are completely dry, mix in the other ingredients and keep in a sealed jar or bucket for 1 month. Shake daily. Then add more oil if necessary. Put into bowls or sachets and place in cupboards and drawers in the kitchen to keep them fresh smelling and insect free. It is also excellent to use in store rooms and kennels. Revive from time to time with citronella oil.

The wild gladiolus flowers absorb the oil in a remarkable way, and so they form an important ingredient in this unusual pot-pourri.

Wild medlar
Vangueria infausta

ENGLISH	Medlar
AFRIKAANS	Mispel, wilde mispel
!KUNG BUSHMAN	Mukatakassa, nuiri
NDEBELE	Umbizo, umtofu, umviyo
PEDI	Mmilo
SHANGAAN	Mpfilwa
SHONA	Mutufu, munzwiru, nyakanawenge, muzoza, muziringombe, munziro
SOTHO	Mmilo, mmilwa, moelwa
SWATI	Santulwan, umvile, umfilwa, amantulwane, imadnulu, infaylo, mavelo
THONGA	Mavilwa, mpfilu, pfilua
TSWANA	Mothwane, mmilo, mmilorotsane
VENDA	Mavelo, muzwilu
ZULU	Umvilo, umtulwa, iviyo, imivyo

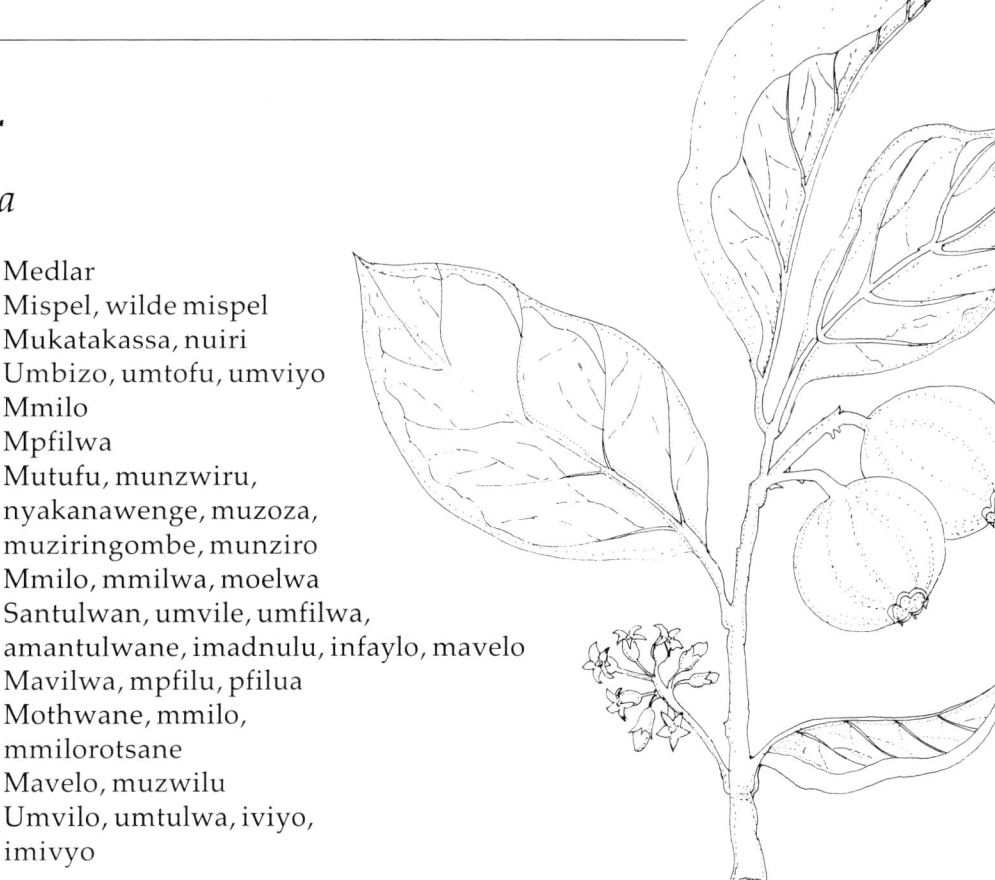

THE wild medlar is an attractive, small, deciduous tree with large, bright leaves. It is found wild in Natal and the Transvaal but is easily cultivated in gardens elsewhere. The stem is rough and knobbly and twists into interesting shapes and I have seen the tree pruned into an umbrella shape as a focal point on a lawn, where in midsummer it offers dense shade. The small insignificant flowers turn into delicious small, glossy fruits, which are greenish to begin with and turn light brown when ripe. The fruits have two large seeds inside them, and are covered by a tough skin. Many blacks cherish the berries, often drying them for times of food scarcity. They also eat the seed kernel, which contains protein, and use the fruit, which is high in vitamin C, as an important and healthy addition to the diet and as a medicine.

The root, scraped and boiled in water, is considered a beneficial treatment for menstrual problems in women and a tea made of the leaves and root is also used for treating coughs and colds and chest ailments. Some African tribes use the pounded root as a remedy for roundworm in humans and animals, as it is purgative. Dosages vary from 1 tablespoon to 1 teaspoon softened in water, which is then drunk. The somewhat bitter root is also chewed, the saliva swallowed, and the root discarded. It seems that the root is considered to be a pain killer as well as having magical properties.

In Zimbabwe some tribes make a mixture of wild medlar with certain other plants to treat and prevent malaria attacks, fever and pneumonia. The pounded leaves and twigs mixed with a little water are used as a poultice to bring down swellings provided the skin is not broken, and boiling water poured over the pounded leaves and twigs (usually 1 cup of twigs to 4 cups of boiling water)

and left to draw until pleasantly warm, is used as a soothing lotion for swellings, sprains and aches in the legs. The leaf well pulped is often used to ease toothache by packing it into and around the aching tooth.

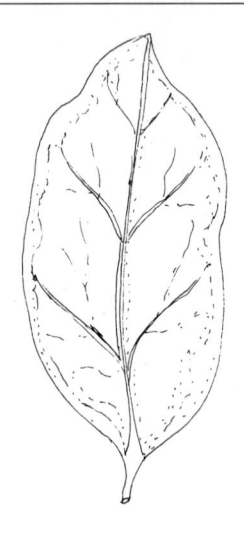

I once tasted a most delicious jam made of wild medlar, and if you are lucky enough to have a tree or to find one laden with fruit in the veld, it is well worth making.

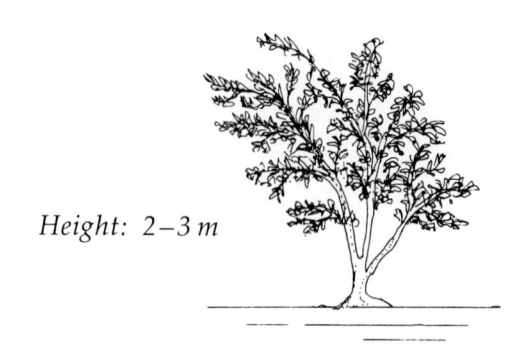

Height: 2–3 m

Wild medlar jam

1 kg ripe wild medlar berries
4 peeled and roughly chopped Granny Smith apples
1 kg brown sugar
3 cinnamon sticks
10 allspice berries
juice of 4 lemons
little grated lemon peel
little water (about ½ to 1 cup)

Cut open the berries and squeeze out the pips. Then mince the fruit, skin and all, being careful not to lose any juice.

Stir all the ingredients together in a heavy bottomed saucepan. Bring to the boil, then turn down the heat.

Simmer gently for about 1 hour or longer until the jam starts to thicken. When a drop 'sets' on a cold saucer, remove from heat, discard the cinnamon sticks and allspice berries and pour into hot sterilised bottles. Seal well and label.

Wild mint
Mentha species

AFRIKAANS	Kruisement
SOTHO	Kwena-ya-libida
XHOSA	Inxina
ZULU	Ufuthanelonghlange

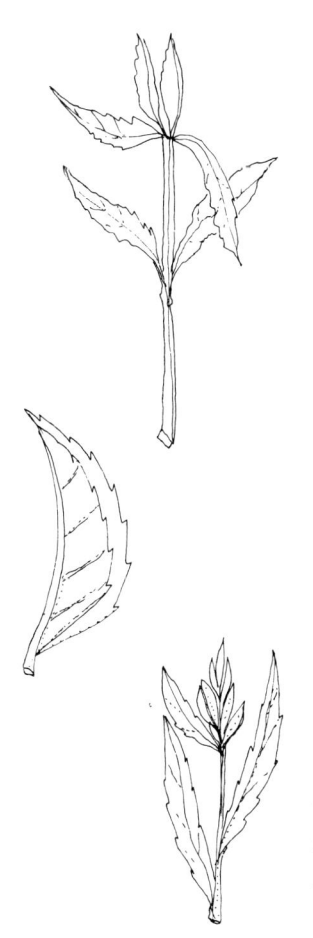

OF the many species of mint there are four indigenous varieties that grow all over South Africa in moist cool places along streams and in vleis. Mints cross pollinate and so there is often much confusion as to their proper identification.

Many years ago I had a mint garden — 12 long, parallel beds of about 20 m in length, enclosed by hedges and fed by a furrow from the river. Botanists and plant lovers spent long hours in that garden with me arguing about my 24 different mints. We rarely agreed — but we did so enjoy the arguing! I started with about 12 well-known mints, and then added to this collection the wild mints I had found: *Mentha aquatica, M. longifolia* (called balderjan, which I had got in the Cape) and a type of spearmint from Natal, probably *M. longifolia polyadena*. Then the confusion started! My conclusion was that everyone seems to have their own common names for the numerous varieties of this precious plant. Here I have attempted to be as botanically correct as possible, and at the time of writing the species names are to the best of my knowledge correct.

The Khoikhoi were the first to use all the varieties of mint medicinally and subsequently the mints found wide use in colonial medications, both taken internally and applied externally. When planting mints remember they are heavy feeders and often deplete the soil, then moving on to new ground by roots and runners.

Studying the mints is a fascinating hobby and if you have enough water in your garden and partial shade, the mints will give you endless pleasure. They make good potplants too, but need to be repotted every season or two into fresh compost and soil. Cut them back frequently to encourage new growth.

Wild water mint (Mentha aquatica)

AFRIKAANS	Kruisement
SOTHO	Kwena-ya-libida, kwena-e-nyenyane
XHOSA	Ityaleba
ZULU	Umayime, umnukani

THE wild water mint is commonly found in moist areas such as vleis, streams, and at the edge of dams and in furrows throughout South Africa. It is low growing with fairly broad egg-shaped leaves and spreads by runners, ever seeking new ground. The leaves are alternatively hairy or smooth and the flowers white or pale mauve in whorls.

A tea made of the leaves (¼ cup of fresh leaves steeped in 1 cup of boiling water for 3 to 5 minutes) is drunk for colic, stomach upsets, colds, coughs, and as a blood cleanser and diuretic. It has also been known to ease gall bladder attacks and is taken after dinner to aid digestion.

Fresh sprigs tied into a piece of muslin or pantihose make an invigorating scrub for tired muscles in the bath, and add a refreshing fragrance to the bath water.

Added to salads and vegetable dishes, this mint gives a most delicious taste.

Height: 30 cm

Minted onions Serves 6

4–6 large onions, peeled and thickly sliced
6 sprigs water mint
1 litre milk
3 tablespoons flour or maizena
1 tablespoon butter
2 beaten eggs
salt and pepper
1 cup grated cheese

This recipe was given to me by an old lady in the eastern Cape who said her grandparents ate this every day during winter and never got colds. Whenever I make this dish I think of Mrs Adams and her faith in mint and onions!

Boil the onions in water with a sprig of mint until tender. Strain, and arrange in a dish. Make a white sauce by melting the butter in a pot, stirring into it the flour or maizena with a wooden spoon, then gradually adding the milk, stirring all the time. Add the beaten egg, salt and pepper and stir until it thickens.

Chop the sprigs of mint finely and add to the sauce. Stir well and pour over the onions. Sprinkle with cheese. Place under the grill for 5 minutes to brown and melt the cheese. Serve piping hot as a vegetable dish with meat or fish.

Horsemint (Mentha longifolia)

THERE are three subspecies of *Mentha longifolia,* which can be distinguished by a careful look at the leaves and flowers.

Mentha longifolia var. *capensis*

AFRIKAANS Balderjan

This subspecies has long, sharp leaves, grey green in colour and hairy, with a lighter grey surface underneath. The flowers are fairly small and compact and are mauve or mauvy pink in colour. Its common name in the Cape is balderjan, of which ballerja is a corruption.

The flavour is strong and the smell pungent, and fresh leaves rubbed onto the arms and legs will keep mosquitoes away. Farm workers pick bunches on their way home from work to rub onto pillows and blankets to keep mosquitoes away during the summer months.

A tea made from this mint (¼ cup of fresh leaves to 1 cup of boiling water, stand for 3 minutes then drink) is excellent for stomach upsets, croup, painful menstruation, colds, colic, indigestion and flatulence. A leaf chewed also aids digestion, heartburn and colic.

An old Cape remedy for all the above ailments, particularly colds and stomach disorders, is a mixture containing 3 sprigs of mint about 10 cm in length, 1 sprig of buchu about 10 cm in length or 1 sprig of rosemary about 10 cm in length, all steeped in 1 litre of boiling water for 5 minutes, then strained and a dash of brandy added. Half a cup of brew is sipped at intervals until the condition eases (warm the brew before drinking — keep the rest covered in the refrigerator).

Height: 50–70 cm

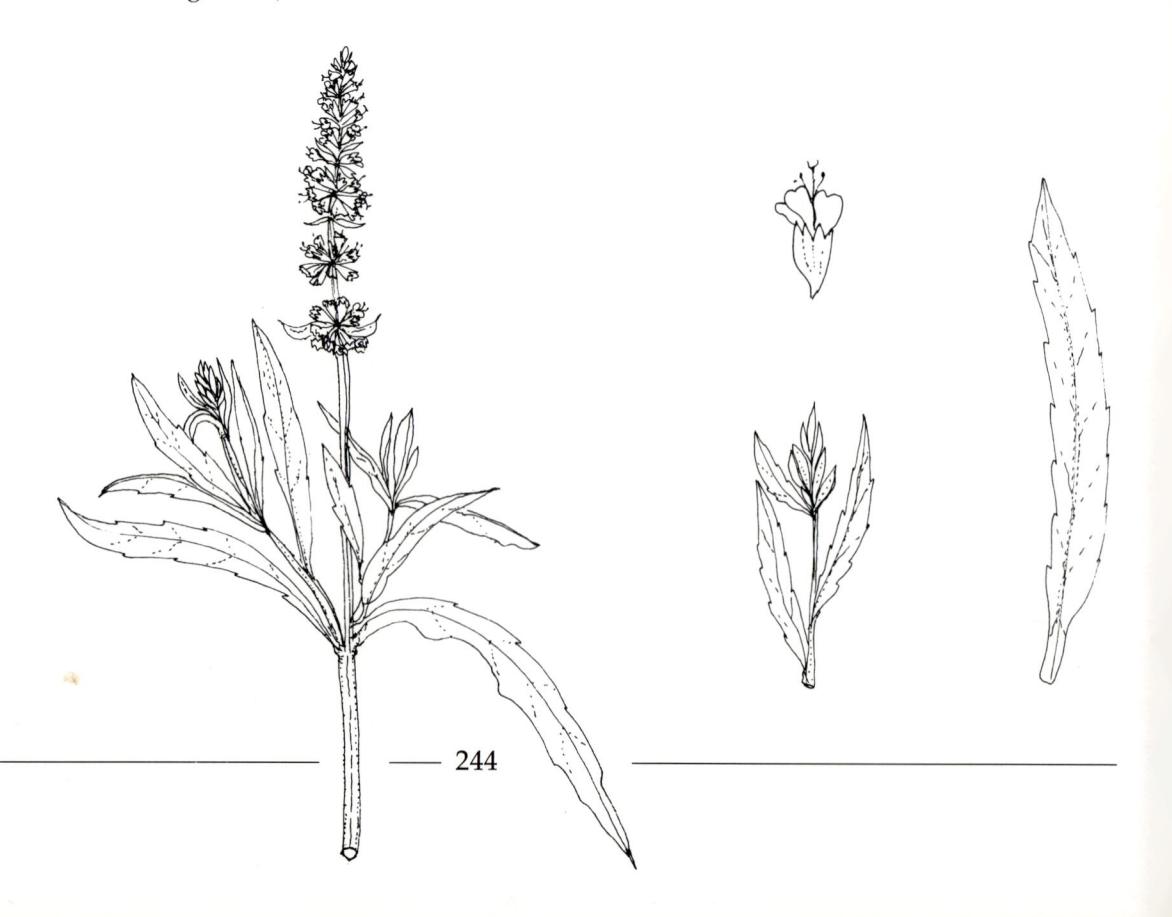

Mentha longifolia var. *polyadena*

ENGLISH   Spearmint

Height: 0,5–1 m

Although this is not the real spearmint, *M. spicata*, it has been called 'spearmint' throughout South Africa owing to its long spear-like leaves and its typical spearmint flavour. The leaves are lime green, hairless and soft and although the plant is similar to *M. capensis*, the flowers grow in spikes well above the leaves and are white in colour.

Untidy in growth, this is the tallest of the mints, and needs severe cutting down during the winter to encourage new growth in spring.

A tea of the leaves (¼ cup of leaves to 1 cup of boiling water, stand for 3 minutes, then strain) is drunk for colds, coughs, stomach cramps, flatulence, indigestion, headaches (the steam from the tea is inhaled as well to ease sinus and blocked nose) to hasten confinements, ease cramps, and as a poultice to soothe muscular pains and sprains and swellings.

Mentha longifolia var. *wissii*

ENGLISH	Woolly long-leaf mint, Cape velvet mint
SOUTHERN SOTHO	Kwena-ya-thaba
XHOSA	Inxina, inzinziniba
ZULU	Ufuthanelonhlanga

This mint has soft, grey green, velvety leaves, narrow and pointed with the finest hairs, and white to mauve flowers. It is more low growing and not quite as vigorous and prolific as its close cousins, and occurs mainly in Namibia, Namaqualand and the western Cape.

This mint is used like the others for colds and coughs, cramps, colic, indigestion, headaches and stomach ailments, and is also used to wrap meat and game to keep it free of flies and the flesh fresh.

Height: 30 cm

Spearmint cool drink   Serves 10

3 sprigs spearmint, well pressed down
2 litres boiling water
1 litre fresh fruit juice (e.g. orange, pineapple, grape, apple, apricot, peach, granadilla)
sugar or honey to sweeten

This cool drink is wonderfully refreshing and I make it all through the summer, varying the fruit juice.

Pour the boiling water over the leaves and allow to draw for at least 1 hour but preferably overnight, then strain. Discard the mint, mix in the sugar or honey and the fruit juice, and chill. Serve with ice in tall glasses decorated with a lemon slice and a sprig of mint. Be sure to grow enough as you'll be making it all summer through!

Wild olive
Olea europaea subsp. *africana*

ENGLISH	Olive wood, brown olive
AFRIKAANS	Olienhout, wilde olienhout, swartolienhout
NDEBELE	Umquma
SHONA	Muguma, mutoba
SOTHO	Mohlware, motholoari
SWATI	Umnquma
TSWANA	Motlheware
XHOSA	Umquma
ZULU	Umhlwati, umsityane, umquma

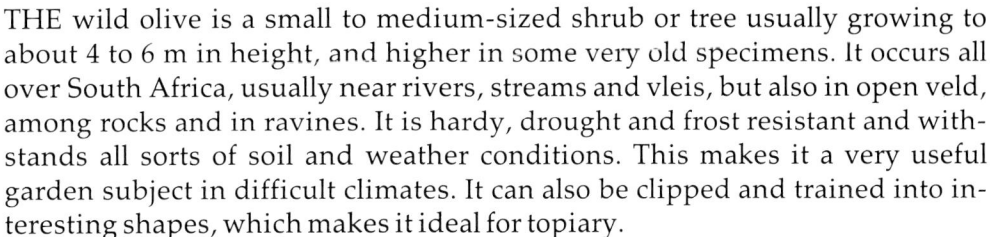

THE wild olive is a small to medium-sized shrub or tree usually growing to about 4 to 6 m in height, and higher in some very old specimens. It occurs all over South Africa, usually near rivers, streams and vleis, but also in open veld, among rocks and in ravines. It is hardy, drought and frost resistant and withstands all sorts of soil and weather conditions. This makes it a very useful garden subject in difficult climates. It can also be clipped and trained into interesting shapes, which makes it ideal for topiary.

The evergreen leaves are greyish green, giving the tree an attractive silvery look in the wind. The insignificant flowers produce edible purplish black fruits much enjoyed by birds, baboons and monkeys and the seed sows itself everywhere, so propagation by seed is easy. The wild olive is a close relation of the European olive and I have often wondered if it would be possible to turn the wild fruits into a commercial enterprise — perhaps if treated properly the wild olive could produce oil and commercial olives. The commercial olive tree has been successfully grafted on to the wild olive.

The timber produced by mature wild olive trees is much sought after and has a hardy, finely grained heavy texture that is golden brown in colour with patches of darker brown figuring, and once it is properly worked it is satiny smooth and very beautiful, making it excellent for furniture and cabinet making. It is also a long lasting, durable wood, so it is ideal for fencing posts and makes a long-burning, pleasant-smelling fuel. Walking sticks, knobkieries and spearhandles are considered valuable and of superior strength if they are made from wild olive, and they carve easily when freshly cut. All the pieces of bark and wood that are worked away are saved for kindling, and some tribes believe that inhaling the smoke of a fire built with wild olive will clear the head and the blood after over imbibing!

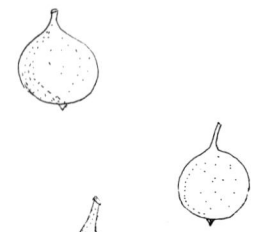

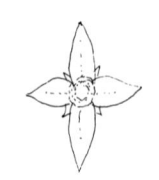

Medicinally the tree has many uses, and is used by many African tribes as well as many whites. An infusion of the fresh leaf is an excellent eye bath (usually 10 leaves to 1 cup of boiling water — leave to draw for 10 minutes then strain) for eye infections, inflammation and opthalmia in man and stock. A pad of cotton wool wrung out in the tea and placed over the closed eye is also soothing, relaxing and healing for tired, bloodshot eyes.

The leaves boiled in water (1 cup of leaves to 4 cups of water for 5 minutes, then cooled and strained) is an effective gargle for sore throats and diptheria — a little is used frequently. For bladder and kidney ailments a strong decoction of the root is drunk first thing in the morning before breakfast, and last thing at night. Gently boil 2 cups of chopped, washed roots in 6 cups of water for 15 to 20 minutes in a closed stainless steel pot. Allow to stand overnight and next morning draw the tea off, discard the root pieces and keep in a closed container in the fridge. Take half a cup morning and evening and sometimes twice during the day to clear infection. This brew can also be taken as a headache remedy and is used by several tribes for influenza and fevers, and rheumatism.

Long ago the dried powdered leaf was considered to be a styptic, and some tribes still use it as such.

An old folk remedy sometimes still used today is to dry the fruits and pound them into a coarse powder, which is then mixed into oil — usually sweet oil — and left for a fortnight before being used as a rub for aching joints or rheumatism. On some skins this causes irritation, however, so test a sensitive area first.

The Tswana use a large bunch of leaves twisted into a wad to wash with, believing that in this way all impurities will be released from the body. I have found this to be very refreshing, particularly over the feet. Perhaps the astringency of the leaves tightens the skin to a small degree.

In spite of their astringency the leaves are much grazed by stock, and the tree is valuable for scarce winter grazing in arid areas. As it is evergreen, many farmers plant it in grazing camps for their cattle as a valuable winter fodder. It is much favoured in the Orange Free State where there is heavy frost and the wild olive is the tree emblem of that province.

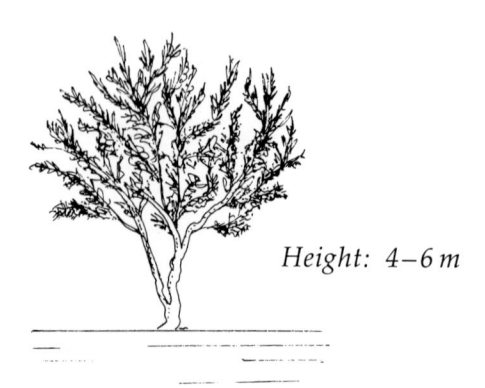

Height: 4–6 m

Wild pear
Dombeya rotundifolia

ENGLISH	Plum blossom tree
AFRIKAANS	Dikbaspeer, vaalbas, gewone drolpeer
NDEBELE	Umwane
SHANGAAN	Siluvari
SOTHO	Mokhuba
TSWANA	Motubani
VENDA	Mulanga
XHOSA	Mutokwe, mutokswa
ZULU	Hlebehlo, inhlizya enkulu

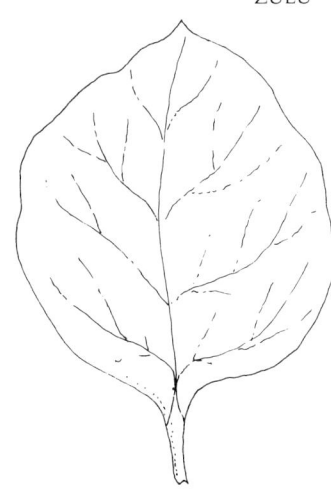

THE wild pear is a familiar sight in spring with its clusters of sweetly scented white blossoms and is much sought after for its wonderful medicinal uses. As well as being an attractive garden subject, this beautiful tree is one of the most useful of the wild trees. It has a remarkable wood which makes excellent wagon wheels, spokes, handles, yokes, naves and gate or fencing posts. When it is seasoned the wood is termite proof, which makes it particularly valuable for farmers.

The bark contains a tough, flexible fibre which is pounded and worked into ropes and string by several tribes. The bark is carefully stripped from a branch that has been soaked in water for 2 days to make it supple, then on flat stones it is pounded with round stones until it is smooth and soft. These fibres are twisted into strings and ropes by rolling between the hands, a job often given to the old women, who sit patiently in the sun working at the rope, the end piece held firmly around their big toe! The bark is also used as a binding or splint support for broken limbs in humans and stock. The Tswana use it to hold a bandage or dressing in place, believing it to have healing powers in restoring injured bone and muscle.

The bark is also used in a tea (boil approximately 1 cup of bark in 8 cups of water for 2 hours, cool, then strain) for delayed menstruation (some tribes believe it will induce abortion), to bring on labour if the birth of the baby is delayed, and to treat internal ulcers, stomach ailments and excessive diarrhoea.

The wild pear has long been used by many whites as a treatment for haemorrhoids, diarrhoea, stomach cramps and vomiting (boil 1 cup of twigs and leaves in 4 cups of water for half an hour, strain and take half a cup at a time).

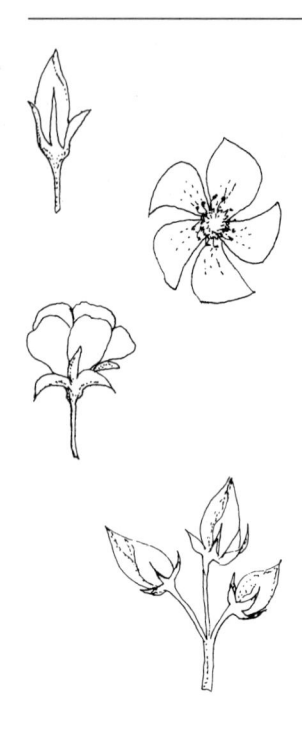

The brew is also used as a lotion applied externally for haemorrhoids and varicose veins.

The root boiled in water is an ancient colic remedy and sometimes pieces of root are chewed and the saliva swallowed, the rest spat out, to relieve gripes, stomach aches and cramps. In some rural areas today a popular folk medicine is made by boiling the twigs and pieces of root in twice the amount of water for 1 hour with a little piece of ginger and a few cloves. This is then strained and an equal quantity of brandy mixed into it. The mixture is poured into warmed bottles and kept tightly corked. A tablespoonful in a little hot water sweetened with honey is slowly sipped to soothe a stomach ache, leg cramps, nausea, flatulence or menstrual pains.

When the wild pear is in full blossom I carefully pick the exquisite white sprays before they start to go brown, hang them upside down in the garage until they are dry and brittle, and use them for dried flower arrangements. Although they brown with age they are simply charming in a vase and I derive much pleasure from them all year round.

Several nurseries offer the trees for sale, and I urge you to find a place in your garden for one. The tree is beautiful in every season — breathtaking in spring in its white blossom, shady and lovely in summer with its round pale green leaves, and stark and interesting in winter with its black graceful branches. It grows in any soil, is undemanding and will survive ably on rainfall and a little compost dug in around it from time to time.

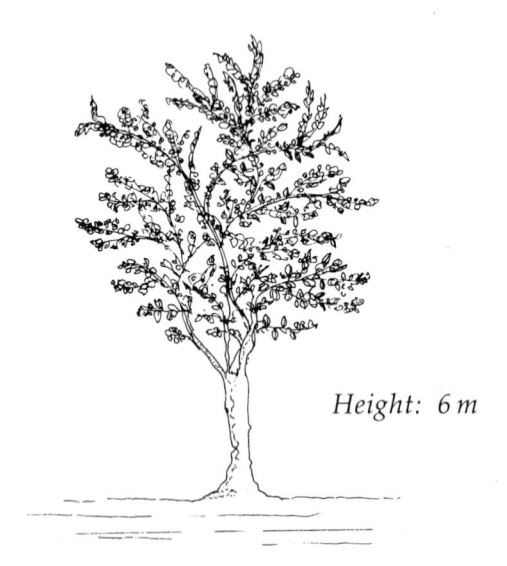

Height: 6 m

Wild pineapple
Eucomis undulata

ENGLISH	Eucomis
AFRIKAANS	Wilde pynappel, krulkop
PEDI	Maphuma-difala
SOUTHERN SOTHO	Kxapumpu
TSWANA	Mothuba-difala
ZULU	Umakhunda

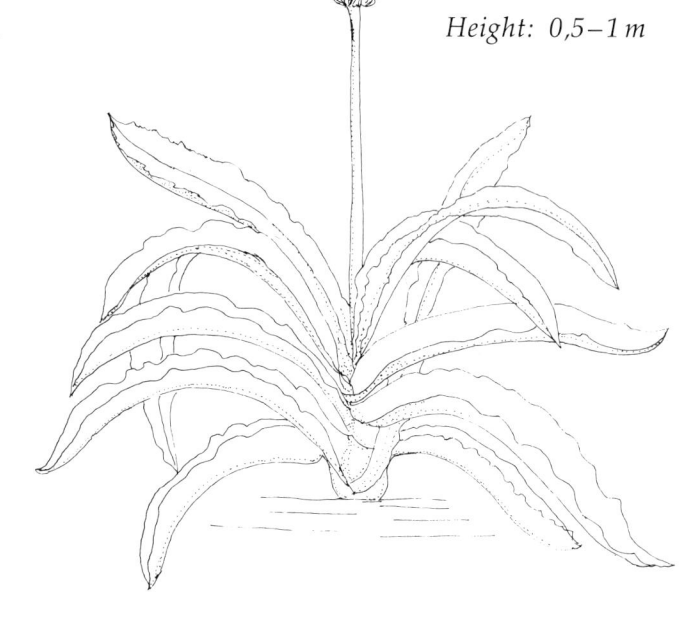

Height: 0,5–1 m

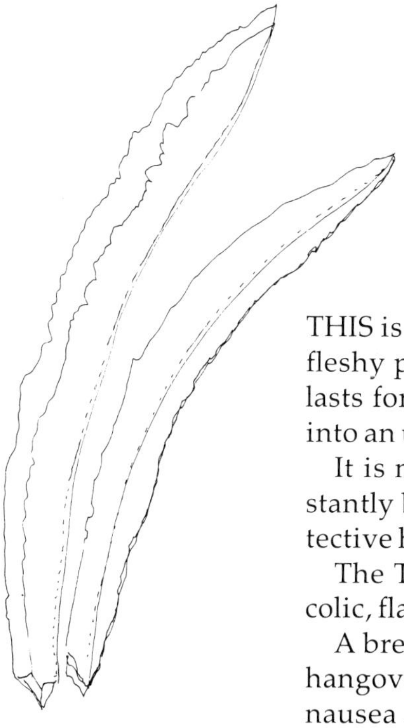

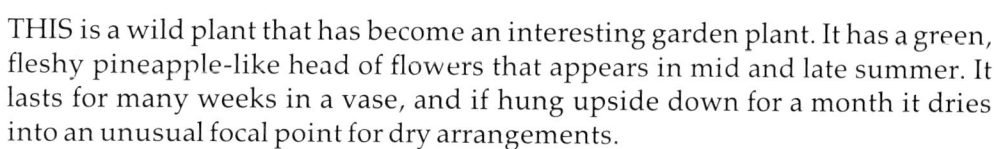

THIS is a wild plant that has become an interesting garden plant. It has a green, fleshy pineapple-like head of flowers that appears in mid and late summer. It lasts for many weeks in a vase, and if hung upside down for a month it dries into an unusual focal point for dry arrangements.

It is much prized by the African people as a medicinal plant and I am constantly begged for plants by the witchdoctors, who believe *Eucomis* to be a protective herb, as well as a treatment for various ailments.

The Tswana use shaved bulbs and roots boiled up in milk or water to ease colic, flatulence and abdominal distention.

A brew from the bulb is also used by both the Tswana and the Sotho to ease a hangover, and the Zulu use a decoction to treat kidney and bladder ailments, nausea and coughs. Some African tribes use the root and bulb in a brew as an enema, and mixed with ash and sheep or goat fat they smear this all over the body to protect themselves against evil or illness.

The long, strap-like leaf acts as a poultice on suppurating sores and boils, and wound round the wrists is believed to bring down fever. The juice of the stem is also used to soothe scratches and rashes, but I have found it not nearly as effective as bulbinella. The leaves and the bulb can be added to lucerne or mealie leaves and given to cattle to treat gall sickness and other diseases.

There are several species of *Eucomis*, all of which make fascinating garden plants. I grow *Eucomis* in both sun and shade, and find the more water it gets the lusher the growth and the bigger the flower head. It dies down completely in winter, but quickly comes into leaf and flower again as soon as the soil warms up. Several nurseries offer the plants for sale.

Wild rosemary
Eriocephalus species

ENGLISH	Snow bush
AFRIKAANS	Kapokbossie, wilde roosmaryn

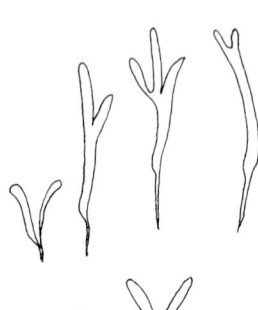

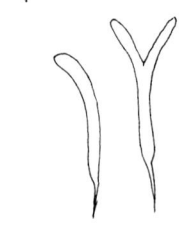

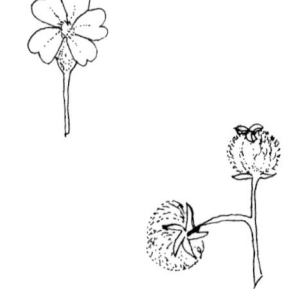

KAPOKBOSSIE is the common name applied to all species of *Eriocephalus* because of their white, woolly, hairy seeds. There are many fascinating varieties, such as kleinkapok, veerkapok, grootbergkapok and kapkappie. The Kapokberg in the Cape gets its name from the abundance of kapokbossie growing on its slopes — when the plants are in seed and flower the mountain looks white with snow.

The *Eriocephalus* species are widespread throughout South Africa, and are much used as teas for coughs and colds, flatulence and colic, as a diuretic and a diaphoric, for cosmetic purposes like bath preparations and hair rinses, for foot baths for delayed menstruation and oedema of the legs and for menstrual cramps. (Pour 1 cup of boiling water over a thumblength sprig, stand for 5 minutes, then strain. Drink sweetened with honey, and add a little lemon juice for a cough and cold, or drink it plain for other ailments.)

I grow *E. africanus* in my herb garden, and find it propagates easily from both cuttings and seed. It likes dry feet so I grow it on a slope or in well-drained soil. *E. ericoides* or Karookapok has a most beautiful seed head and fragrance, and was used by the Khoikhoi and Griqua to stuff pillows and is still used by farmers in the Cape today.

I use twigs or fresh leaves for flavouring bean dishes, fish dishes and poultry stuffings and as a substitute for ordinary rosemary, and find it gives a fresh and unusual taste.

Many nurseries grow indigenous plants now, and I have found wild rosemary in several as well-established plants.

Wild rosemary makes a most refreshing bath addition, as well as a superb hair rinse. Wild rosemary seems to have similar qualities to ordinary rosemary as both have an invigorating effect on the skin and hair.

You can boil up sprigs of wild rosemary (1 measure of twigs and flowers to 2 measures of water, boil for 15 minutes then stand to cool) and add it to the bath or use as is for a hair growth stimulant and conditioner. Use wild rosemary in sachets and pot-pourris too — (see page 268 for wild flower pot-pourri).

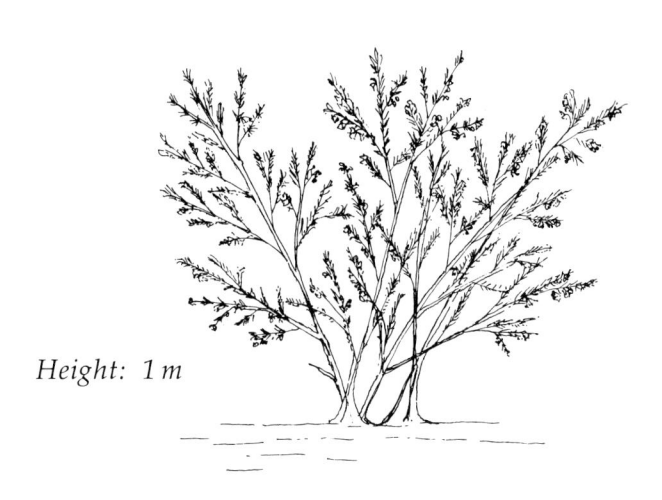

Height: 1 m

Wild rosemary bean bake   Serves 6

500 g Haricot beans
1 cup sultanas or raisins
2 large onions, chopped
little oil
4 large tomatoes, skinned and chopped
1½ cups brown sugar
2 cups vinegar
salt and pepper to taste
water
4 thumblength sprigs wild rosemary

Soak the beans and the raisins overnight. Lightly brown the onions in the oil. Add the tomatoes, sugar, vinegar, salt and pepper. Stir well, then add the soaked beans, raisins, and enough water to cover them. Push the wild rosemary twigs deep into the bean mixture. Cover and cook slowly, stirring every now and then. Add more water if it dries out.

I usually cook this dish overnight in a slow cooker, or in an unglazed clay pot in a very slow oven. (Be sure to add enough water to keep the beans moist and succulent.)

When tender, remove the wild rosemary sprigs and serve either hot or cold. It is delicious on toast with a simple salad, or mashed with mayonnaise and cucumber slices as a sandwich filling.

For a variation you can add bacon, ham, dried peaches or pecan nuts.

Wild rosemary bath vinegar or hair rinse

1 bottle white grape vinegar
about 12 sprigs wild rosemary (flowers and seeds included)

Pour out a little of the vinegar to make room for the herb and push enough twigs into the vinegar to fill the bottle. Stand in the sun for 3 weeks, shaking every day. Once or twice during those 3 weeks replace the wild rosemary twigs with fresh ones to strengthen the vinegar. Then strain it off, discarding the herb, and pour the vinegar into a clean bottle. Push a pretty fresh spring into it for identification.

Use this vinegar in the bath about ½ cup at a time, and as a hair rinse for dandruff. I use ½ cup in a basin of warm water after shampooing my hair.

Wild sage
Labiatae family

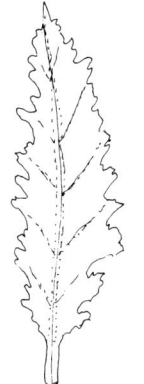

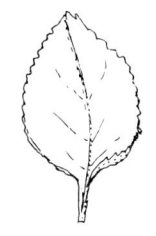

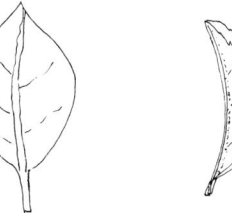

THE sages are a widespread group of fragrant plants found all over the world. Most species contain strong oils and possess remarkable healing qualities and have thus been much used as a food flavouring and medicine. In South Africa there are about 20 indigenous sages which were used by the African people and the colonists in the last century with great effect, and are still much used today. For plant collectors this group of wild sages could become a fascinating study. The few listed below are ones that I am familiar with and which I have used with positive results in cooking and medicine. They are all similar to the well-known *Salvia officinalis*, or garden sage, which is used throughout the world. They have a typical labiate, two-lipped flower and their pungent sage-like scents of varying intensities can be quickly recognised in the veld.

Several African tribes use the sages as washes, lotions and disinfectants, often burning the leaves in their houses. The leaves of any of the wild species of sage are chewed to ease a sore throat and soothe indigestion and flatulence.

Afrikaanse salie *(Salvia chamelaeagnea, S. panticulata)*

ENGLISH   Aromatic sage
AFRIKAANS   Afrikaanse blou salie, wilde salie

Height: 60 cm – 1 m

THIS blue-flowered sage has smooth, strong-smelling leaves and grows into a large, dense bush, particularly in the southern and south-western Cape. Like its relatives it is used in the form of a tea as a treatment for coughs, colds, whooping cough, diarrhoea, colic, heartburn, flatulence and female ailments (pour 1 cup of boiling water over 1 tablespoon of fresh leaves and flowers, stand for 5 minutes then strain). Half a cup of the tea can be sipped slowly at intervals 3 to 6 times a day to ease the condition.

An infusion of the dried leaf (usually 1 teaspoon of dried leaves to 1 cup of boiling water, stand for 5 to 8 minutes, then strain) was once taken as a treatment for epilepsy and convulsions and is still valued by many country people today.

Blue sage *(Salvia africana coerulea)*

ENGLISH Wild sage
AFRIKAANS Blousalie, bloublomsalie, Afrikaanse salie, wilde salie

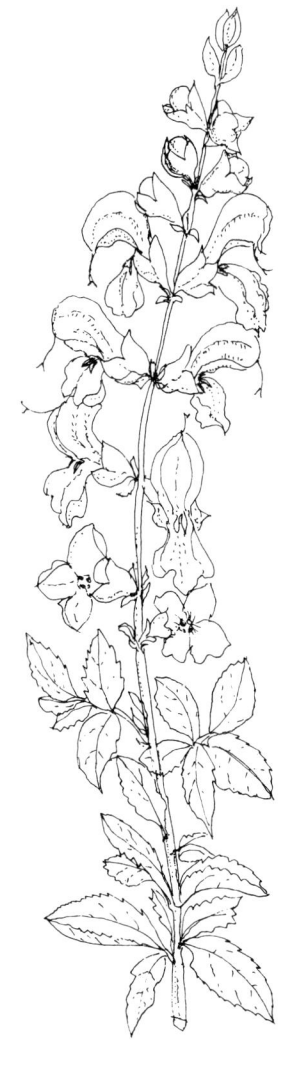

BLUE sage tea is popular as a treatment for coughs and colds, flu and chest ailments. Pour 1 cup of boiling water over 1 tablespoon of fresh leaves, draw for 5 minutes, sweeten with honey and add a slice of lemon if desired. Sip a little frequently to ease a cough and whooping cough or drink half a cup 4 times a day for chest ailments, and for painful or excessive menstruation.

Blue sage was one of the first plants to be used by the Dutch at the Cape and an old remedy for stomach ailments, diarrhoea, colic, flatulence, heartburn, gripes and indigestion was a fresh twig of sage about 12 cm long infused for 10 minutes in 2 cups of boiling water, strained, with 3 tablespoons of Epsom salts and 3 tablespoons of lemon juice added. This mixture was stored in a screw-top bottle and a small dose taken at a time.

The same mixture was, and still is, given to a cow after calving to assist the expulsion of the afterbirth. Some farmers still believe this to be the best household medicine, and this sage is an excellent substitute for the ordinary cultivated sage, *Salvia officinalis*. A leaf chewed will ease a sore throat and help voice loss, and sage tea made as above is an excellent gargle for sore throats and night coughing. I have used the fresh leaves chopped and mixed into honey and lemon juice (1 tablespoon leaves, 1 tablespoon honey, 2 tablespoons lemon juice) as a cough mixture for a persistent cough, and found it to be very soothing. I take a teaspoon every half hour until the cough eases.

The same tea is drunk to clear up female ailments and used externally as a wash, and as a douche (dilute 1 litre blue sage tea to 1 litre warm water with 2 tablespoons apple cider vinegar) for vaginal thrush and irritation.

The plant propagates by cuttings and in the garden if you keep the bush neatly trimmed it will reward you with blue flowers almost all year round. Take care not to overwater as, like all sages, it does not care for waterlogged soil.

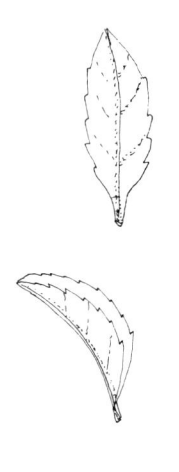

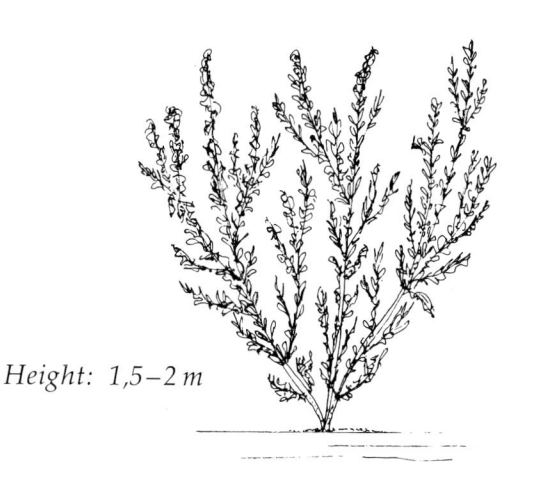

Height: 1,5–2 m

Brown sage *(Salvia africana-lutea, S. aurea)*

ENGLISH   Dune sage, sand sage
AFRIKAANS   Bruin salie, sand salie, strand salie, geelblomsalie

GREY-leaved with large brown flowers, this sage grows easily all along the Cape coast from the west coast and Namaqualand through to the eastern coast, and can be easily recognised. It makes large dense aromatic dune coverings and the small leaves are fragrant and crisp.

The sweet nectar in the flowers attracts sunbirds and moths, making brown sage a rewarding plant to have in the garden.

I love using the leaves dried in pot-pourris as they retain their shape, colour and much of their fragrance, and mix well with the other ingredients. Surprisingly, I have managed to grow it in my Transvaal herb garden and although it is not as lush or as prolific as its Cape relatives, it does reasonably well — even with a little frost and very cold winter winds.

It grows well from slip, and several nurseries offer the plants for sale. In the garden it can be trimmed and clipped and its grey foliage makes it an attractive garden subject. Apart from attracting nectar-seeking birds, brown sage makes an excellent tea for coughs, colds, bronchitis and female ailments (pour 1 cup of boiling water over 1 thumblength sprig of leaves, stand for 5 minutes then strain and drink sweetened with honey), making it a most worthwhile addition to the garden.

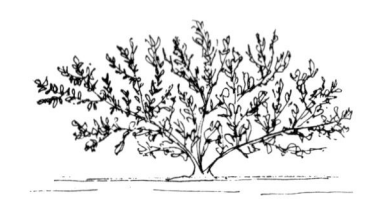

Height: 1,5 – 2 m

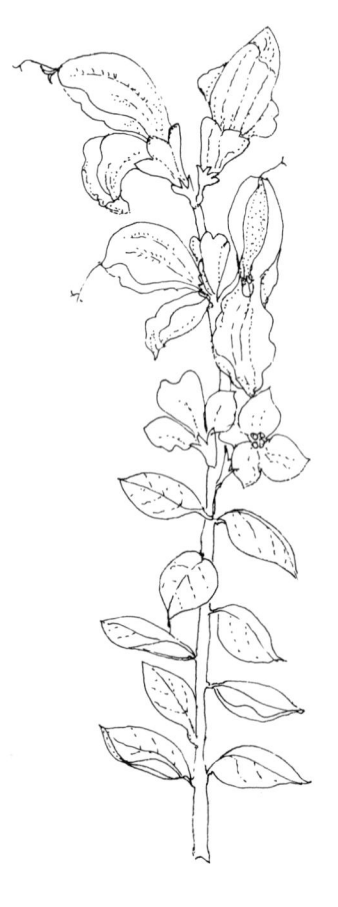

Creeping sage (Salvia repens)

ENGLISH	Small sage
AFRIKAANS	Saliebossie, kruip salie, klein salie
SOUTHERN SOTHO	Mosisili-oa-loti
TSWANA	Mosisili
XHOSA	Usikiki

THIS low-growing, creeping sage is different from the other sages in that it has a spreading root or rhizome. It has large mauvy blue flowers like most of the other sages, and is strong smelling.

It is found all over South Africa and is used differently in each province. In the Orange Free State it is a diarrhoea and stomach ailment remedy; in the Transvaal it is used as a tea for coughs and colds; and in the Cape it is used for fevers, flu, bronchitis and flatulence. It is also used as a disinfectant and the Sotho and the Tswana burn it in their huts to drive out insects and bedbugs, and to clear the air after an illness.

The tea for the above ailments seems to be standard: ¼ cup (or a little less) of fresh leaves and stems steeped in 1 cup of boiling water for 3 to 5 minutes, then strained. Drink sweetened with honey if desired as this is a strong-tasting herb.

It is a popular skin treatment (pour 1 litre of hot water over 1 cup of leaves, stems and flowers, draw for 20 minutes, then strain) used as a wash for sores, rash, scratches and infected bites.

Height: 10–15 cm

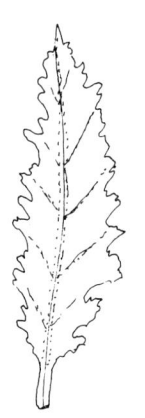

Groot salie (Salvia rugosa (formerly S. disermas))

ENGLISH Wild giant sage, Transvaal sage
AFRIKAANS Wilde salie, Transvaal salie
TSWANA Mogasane

Height: 60 cm

TALL, stiff, and hairy, this large sage is found in the western Transvaal, Orange Free State, Karoo, and the northern and western Cape. It is strong-smelling, pungent, and is used in cooking — very sparingly — and as a lotion (steep 1 cup of fresh leaves and stems in 2 litres of hot water for 15 minutes) for skin rashes, stings, bites, pimples, infected scrapes, scratches and sores.

A little of this brew, or a tea made by steeping 1 tablespoon of fresh leaves and flowers in 1 cup of boiling water for 5 minutes, then strained, is taken as a tonic, to re-energise the body after a long illness and to bring down fevers. A tablespoon at a time is a good heartburn remedy.

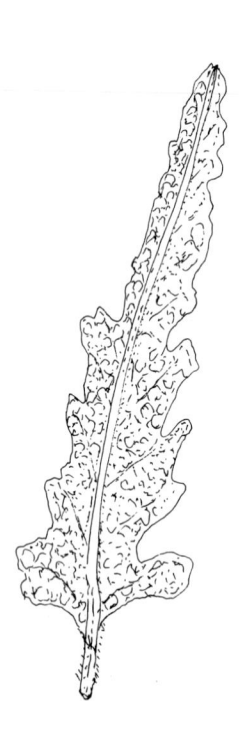

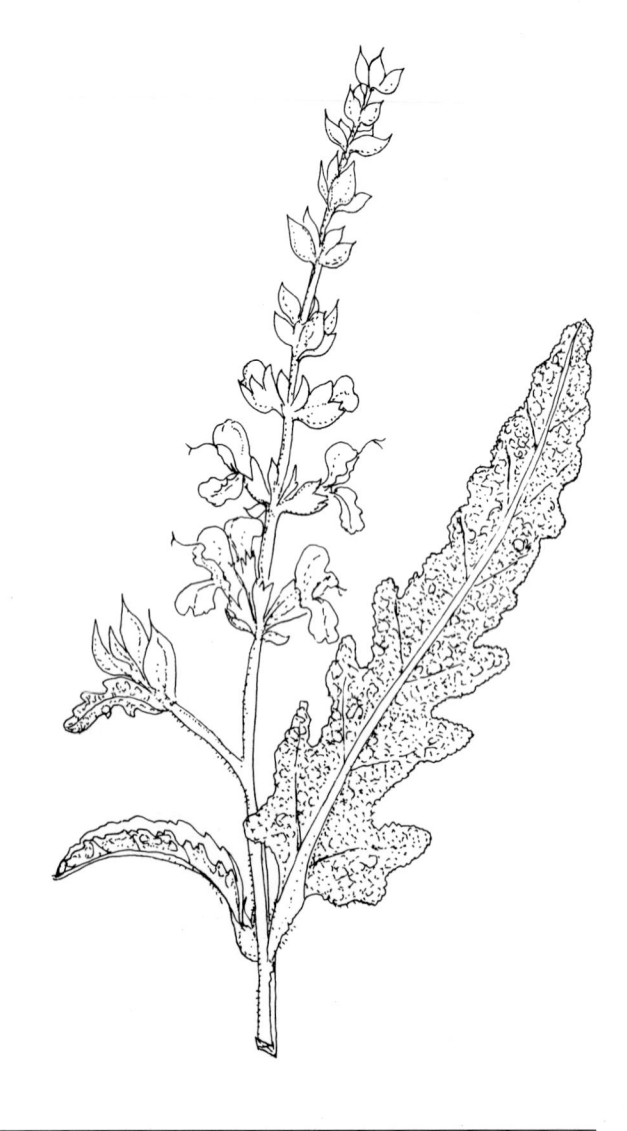

Narrow leaf sage *(Salvia stenophylla)*

AFRIKAANS Fynblaar salie, maagpynbos, fynblaar wilde salie
SOTHO Mosisili

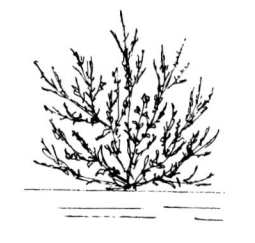

Height: 10–20 cm

THIS variety of wild sage has tough, strongly scented, finely shaped leaves that are rough and deeply lobed. It is low growing and unobtrusive with small, pale mauve flowers. It grows along the roadsides and in the veld in the Transvaal and Orange Free State and in parts of the eastern Cape through to Namibia.

A brew made by pouring 1 cup of boiling water over 1 tablespoon of fresh leaves and flowers and left to draw for 5 minutes, is taken as an antispasmodic tea for cramp, colic, flatulence, heartburn and chronic indigestion. It is also an excellent wash for sores and infected bites and scratches, and can be dabbed on as a lotion or used as a wash. The tea also makes a soothing sore throat gargle, and as a drink for colds and flu it will ease the discomfort and help bring down a fever.

I have grown it easily from seed.

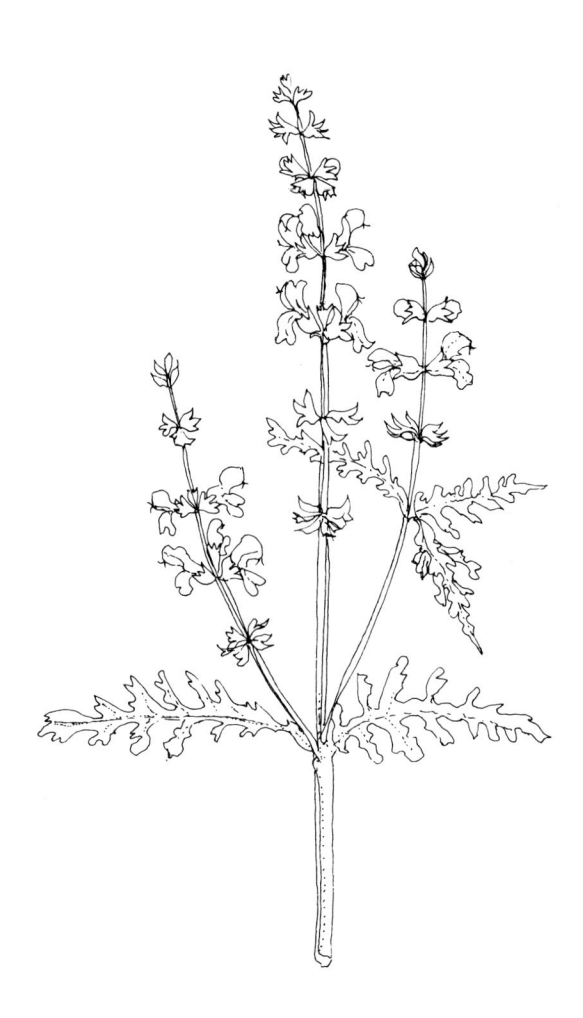

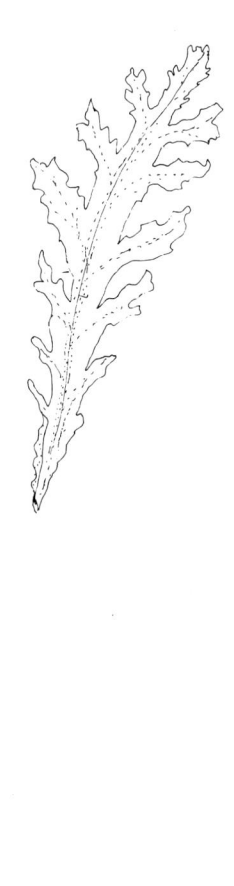

Vrystaat salie (Salvia verbenaca)

ENGLISH Free State sage
AFRIKAANS Wilde salie

THIS is a small sage growing from a rosette of deeply notched leaves, which distinguishes it from the other sages. It grows in the Orange Free State and the Karoo and in the northern Cape up to Namibia.

The leaves are used in a tea for heartburn and colic, and the usual dose is 1 tablespoon of fresh leaves steeped in 1 cup of boiling water for 5 minutes, then strained. Drink sweetened with honey for colds, colic, indigestion, flatulence and fever.

A few nurseries offer it for sale, and it grows easily from seed. Once when I stopped for tea on the road near Steynsburg on the way to Hendrik Verwoerd Dam, I found this sage growing thickly under the concrete table and all over the cleared ground. I collected eight seeds, and every one grew!

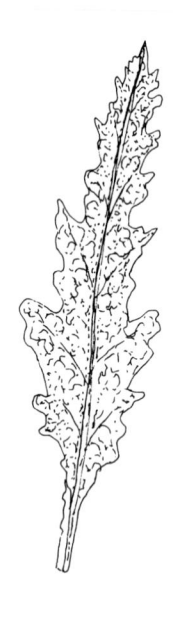

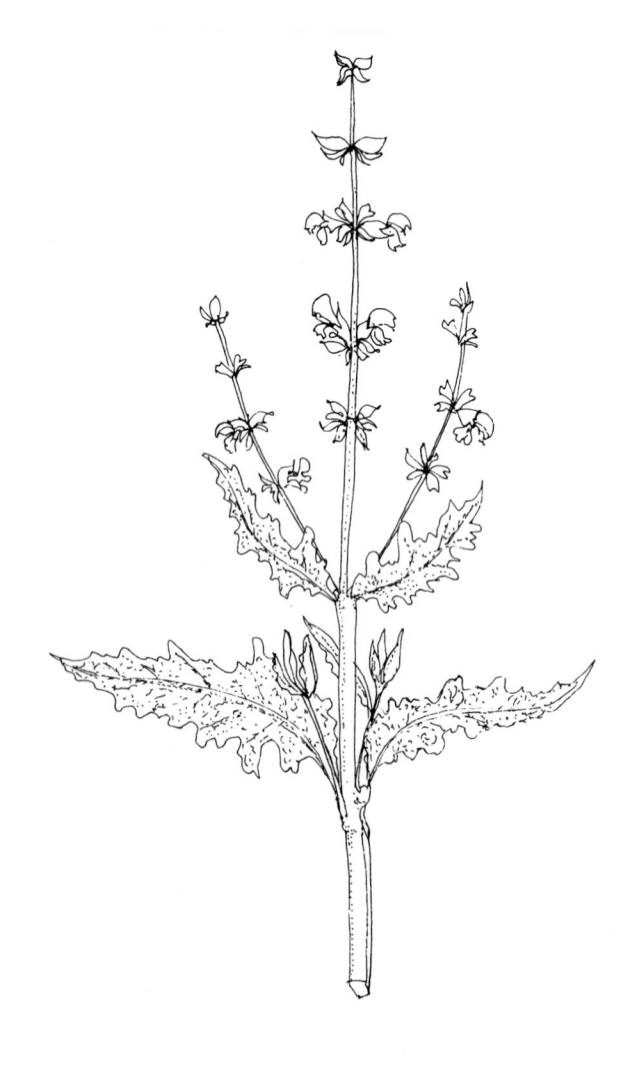

Height: Leaves 10 cm
Flowers 30 cm

Wild verbena
Pentanisia prunelloides

AFRIKAANS	Sooibrandbossie
LOBEDU	Amatunga
SOUTHERN SOTHO	Setimo-mollo, ngelile
TSWANA	Setimo-mollo
XHOSA	Irubuxa
ZULU	Icimamlilo, icishamliolo-omncane

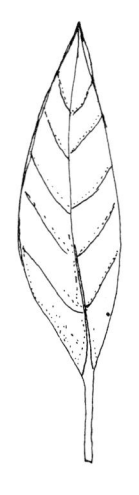

THIS pretty little blue flowered perennial grows in the open veld and on mountain slopes in the Transvaal, Orange Free State, eastern Cape and Natal. It is low growing with a strong, woody rootstock which withstands veld fires. It flowers profusely from early November to March, and can often be seen in the veld along the roadsides after the spring rains. It has small, oval, hairy leaves in pairs, and the flowering head is similar to that of a garden verbena with tiny pale mauvy blue flowers, and sometimes darker blue flowers.

This precious herb has been used by all the indigenous people to counter many ailments and illnesses. It is much respected and revered, and is believed to have magical properties: burning the root at night will ensure the finding of a good job; having a bunch of leaves and flowers in the house will protect against lightning; and the Sotho believe that using the plant as one of the ingredients in a magic brew will ensure that no sorcerer will find the door of his hut!

The wide variety of ills that this little plant cures is quite amazing. The leaves and stems warmed in water make an excellent poultice for sprains, swellings, sores, rheumatoid arthritis and ordinary arthritis.

The root is pounded in hot water and applied as a poultice to haemorrhoids, left on overnight, and is also used to draw boils, abscesses and suppurating wounds, as it seems to have the quality of drawing pus. The washed, pounded root is applied directly to the area, and is usually left on overnight. A fresh poultice is applied daily until the condition clears.

A decoction or tea made of the root (usually 1 tablespoon of chopped root to 1 cup of boiling water, stand, draw for 5 minutes, then strain) is used in the treatment of chest ailments, colds, blood impurities (possibly even venereal diseases), fever, flatulence and colic, tuberculosis (used particularly by the Xhosa) and influenza. During the devastating flu epidemic in 1918 it was reported that a group of Zulus who had made a tea of the root and drunk some daily were the only ones not to be affected. From this report tests for antibiotic properties in the plant were conducted, with positive results. So this pretty little wild plant is a natural antibiotic!

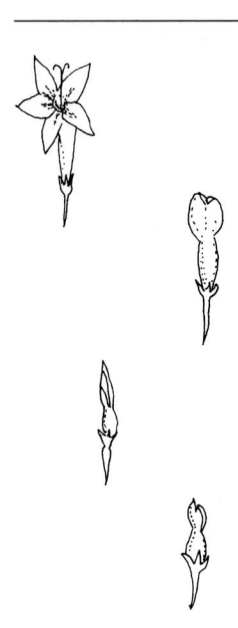

An infusion made of leaves, stems and even some flowers is a refreshing wash and tea used by some African tribes to assist the milk flow in a new mother, to help expel the afterbirth in humans and cattle, and to bring down a fever. The Zulu place a leaf poultice over the lower abdomen to help expel the afterbirth and also give a tea to the patient to heal her internally.

A well-known cure for heartburn is to chew a piece of root and this is a favourite remedy among whites as well.

A warm weak tea made of the leaves (1 cup of leaves to 2 litres of boiling water, stand for 10 minutes then strain) is a soothing wash for skin infections, burns, rashes and scratches, and some tribes use it as an enema. The leaves and flowers can apparently be cooked with vinegar and spices and eaten as a relish, and dried leaves and root are ground to a powder and dusted onto wounds by some tribes.

The only way of propagating wild verbena is by seed carefully sown in separate bags, as it does not transplant easily. When the seedling is big enough tear away the bag and try to disturb the roots as little as possible. Place in a well-prepared moist hole and water well until it is established. Thereafter it thrives on no attention.

Height: 20 cm

Yellow wandering Jew
Commelina africana

ENGLISH	Wild commelina, wild tradescantia, mouse ears, day flower
SOTHO	Lekxopswana
TSWANA	Kxopo-enyenyane
ZULU	Indangabana

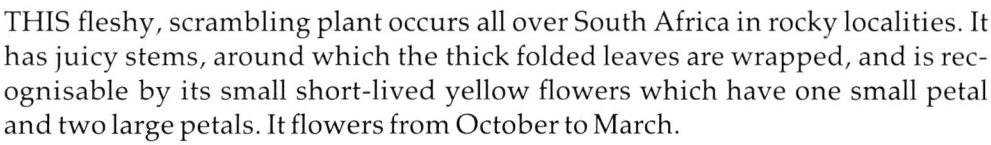

THIS fleshy, scrambling plant occurs all over South Africa in rocky localities. It has juicy stems, around which the thick folded leaves are wrapped, and is recognisable by its small short-lived yellow flowers which have one small petal and two large petals. It flowers from October to March.

It is a common species in the Transvaal and hybridises with its relatives (like *Commelina benghalensis*) and so is variable.

It is much sought after for tribal medicines in the treatment of hysteria, nervousness, barrenness in young women and venereal diseases. A crushed, bruised leaf is used by the Tswana as a burn dressing, and the juice from the stem is used by several African tribes as a salve, squeezed directly onto a sore or burn.

The Sotho dry the leaves and stems and burn these for their ash, which is mixed into a medicine and applied externally to the loins for sterility. A tea is also made of the ash and a small dose taken daily as a medicine.

I was taught by a Tswana kitchen worker to make a tea of the whole plant and use this as a bath or wash for a restless child — she told me it was especially good for restless grandchildren! The leaf is also rubbed over the pillow to calm the child.

The plant is edible and can be mixed with other vegetables as a nourishing dish and is very important in times of drought and food scarcity.

The Ndebele and the Tswana make a decoction of the root for difficult menstruation, particularly in young girls, and also use this tea to treat venereal diseases, bladder ailments and hip pains.

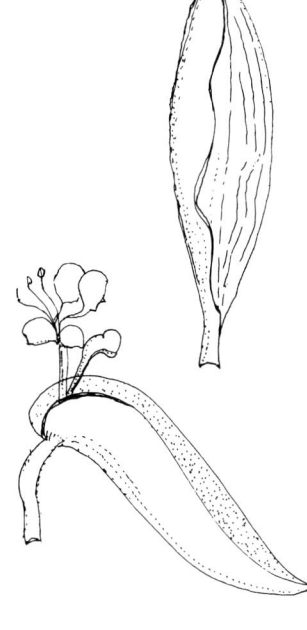

The plant grows from cuttings — the noded stems pressed into wet sand will root easily — and it is useful in the garden. It can either be burned and used as an ash dug in around plants as a fertiliser or mixed in with other plants as a foliar feed (see below), or the plant can be placed in layers in the compost heap to accelerate decomposition.

Height: 10–15 cm

Foliar feed

2 cups ash from dried yellow wandering Jew
1 cup finely minced fennel leaves
1 cup finely minced cabbage leaves
1 cup finely minced fresh yellow wandering Jew leaves
½ cup soap powder

Pour 4 litres of boiling water over all the ingredients. Leave overnight to draw. Splash the mixture onto plants and water some in around the roots. The soap powder acts as an insect repellent, as does the fennel, and will help the 'tea' to stick to the leaves onto which it is splashed.

Epilogue

AS you turn these last pages my greatest hope is that there will have stirred within you a desire to know more about our precious wild plants and to grow them. Each one of us, in our own small corner, can make South Africa more beautiful by sowing, propagating and protecting the indigenous plants of our country — and by teaching others to appreciate and value them too.

Wild flower pot-pourri

10 cups mixed dried wild flowers
4 extra cups scented geranium leaves
4 cups dried mixed lemon and orange peel, and pips
1 cup roughly broken cloves, cinnamon and nutmeg mixed
1 cup lightly crushed coriander seeds
1 cup coarse sea salt (omit at the coast)
essential oil of your choice e.g. scented geranium oil, lavender oil, jasmine oil

After many long years of using and growing our indigenous plants, this one recipe best sums up for me the beauty and fragrance of the wild countryside.

Use any of the flowers listed below. Gather the flowers when fully open and dry them on newspaper in the shade, turning them daily.

wild rosemary	wild dagga
tulbaghia	buddleja
jasmine	plectranthus
scented geranium	plumbago
agt-dae-geneesbossie	pompon tree
lavender tree	Hottentots' tea
sausage tree	wild camphor tree
heather	buchu
weeping boer-bean	agapanthus
cancer bush	doll's protea
ginger bush	harebell
knobwood	lemon bush
melkbos	wilde als
parsley tree	pincushion
raasblaar seeds	renosterbos
sand olive	silverleafed vernonia
sweet thorn	traveller's joy
wild basil	wild cineraria
wild foxglove	wild gladiolus
wild mint	wild sage
Cape honeysuckle	

Mix all the ingredients. Place in a closed jar or bucket and shake daily for 2 weeks. Add a little more oil if necessary, then when beautifully fragrant, fill bowls, sachets or baskets lined with cellophane, and place around the house. Every now and then revive with a few drops of essential oil — empty the pot-pourri into a jar or bucket, add the oil and a few freshly dried flowers, seal, shake well and leave for 2 days, then return to the containers.

Where to obtain indigenous plants and seeds

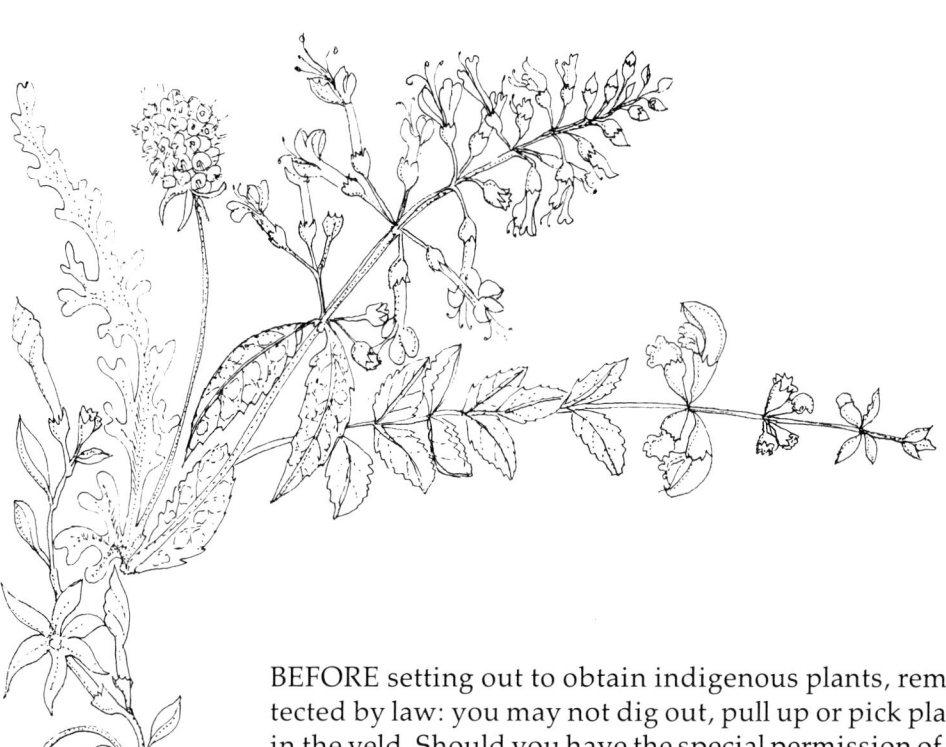

BEFORE setting out to obtain indigenous plants, remember that they are protected by law: you may not dig out, pull up or pick plants along the roadside or in the veld. Should you have the special permission of the landowner on whose land you find a specimen to take a seed or cutting, always bear the Golden Rule in mind: take only one seed or one cutting. Also, wild plants as a rule do not transplant easily, so don't be tempted to try it — rather develop a seed sowing system.

To obtain free seeds and free entry to the botanical gardens countrywide, I urge you to join the Botanical Society of South Africa, Kirstenbosch. For a nominal yearly subscription you will also receive their fascinating little quarterly publication *Veld and Flora*. Some of the botanical gardens have nurseries where you will find a wealth of indigenous plants for sale, such as Kirstenbosch's utilisation nursery where rare and unusual indigenous plants with horticultural potential are propagated for sale. Many ordinary nurseries in South Africa also offer indigenous plants for sale.

The following is a list of nurseries and societies where indigenous plants and seeds and information are obtainable:

CAPE

Kirstenbosch Nursery and Plant Utilisation Nursery, Botanical Gardens, Kirstenbosch, Claremont 7700, Cape Town

Botanical Society, Kirstenbosch, Claremont 7735, Cape Town

Rust-en-Vrede Indigenous Nursery, PO Box 231, Constantia 7848, Cape Town

Von Lyncker Bulb Nursery, PO Box 18200, Wynberg 7824, Cape Town

Honingklip Nurseries and Book Sales, 13 Lady Anne Avenue, Newlands 7700, Cape Town

Feathers Wild Flower Seeds, PO Box 13, Constantia 7848, Cape Town

Parsley's Cape Seeds, PO Box 1375, Somerset West 7130 or 1 Woodlands Road, Somerset West 7130

TRANSVAAL

Afriflora Indigenous Plant Nursery, PO Box 2076, Honeydew 2040, Transvaal

The South African Pelargonium and Geranium Society, The Treasurer, PO Box 8714, Johannesburg 2000

Eloffs Indigenous Seeds, PO Box 75, Delmas 2210, Transvaal

Patryshoek Kwekery, PO Box 17078, Pretoria North 0116

The Dendrological Society, PO Box 104, Pretoria 0001

Department of Environment Affairs, Private Bag X447, Pretoria 0001

Seed of trees: The Responsible officer, Seed Section, PO Box 727, Pretoria 0001

South African Nurserymen's Association, Volkskas Building, Market Street, Johannesburg

Protea Seed and Nursery Suppliers, PO Box 98229, Sloanepark 2152, Johannesburg

Charles Craib, PO Box 67142, Bryanston, Sandton 2021

Witkoppen Wild Flower Nursery, PO Box 67036, Bryanston 2021 or Cedar/Rietvallei Road, Witkoppen.

Forestry and Environmental Conservation Branch for Information:

Bloemfontein (051) 47-3191
Bloemhof (018022) 456
Cradock (0481) 4701
Knysna (0445) 46-7010
Nelspruit (02311) 23244
Pietermaritzburg (0331) 28101
Pretoria (012) 310-3911

Bibliography

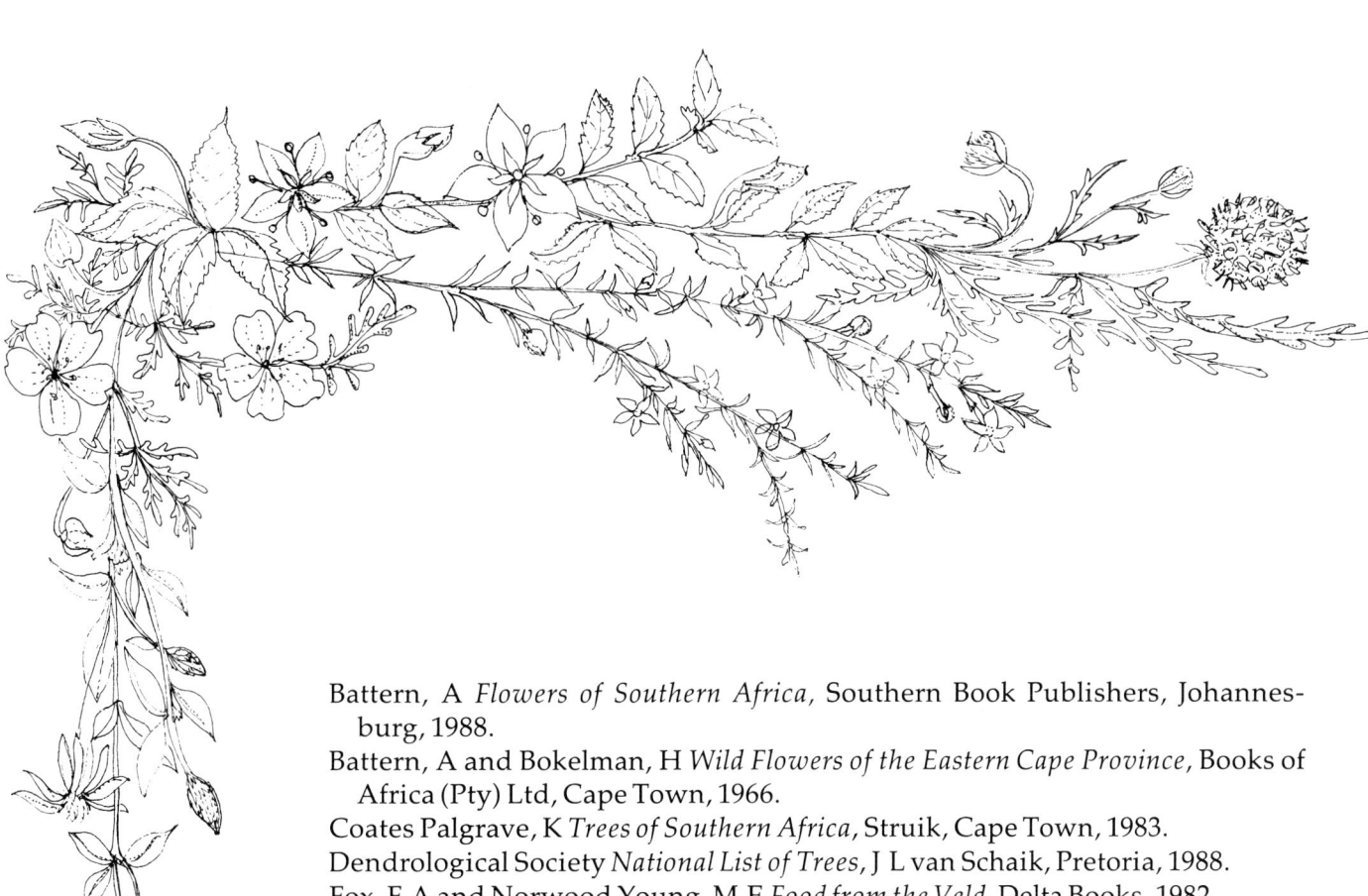

Battern, A *Flowers of Southern Africa,* Southern Book Publishers, Johannesburg, 1988.
Battern, A and Bokelman, H *Wild Flowers of the Eastern Cape Province*, Books of Africa (Pty) Ltd, Cape Town, 1966.
Coates Palgrave, K *Trees of Southern Africa*, Struik, Cape Town, 1983.
Dendrological Society *National List of Trees*, J L van Schaik, Pretoria, 1988.
Fox, F A and Norwood Young, M E *Food from the Veld*, Delta Books, 1982.
Hancock, F *Ferns of the Witwatersrand*, Illustrated by A Lucas, Witwatersrand University Press, Johannesburg, 1973.
Letty, C *Wild Flowers of the Transvaal*, Wild Flowers of the Transvaal Book Fund, 1962.
Lucas, A and Pike, B *Wild Flowers of the Witwatersrand,* Purnell, Cape Town, 1971.
Mitchell, J and Breyer Brandwyk, M *The Medicinal and Poisonous Plants of Southern and Eastern Africa*, E & S Livingstone Publishers, Edinburgh and London, 1962.
Smith, C A *Common Names of South African Plants*, Government Printer, Pretoria, 1966.
Van der Walt, J J A and Vorster, P J *Pelargoniums,* Vols I, II, III, Illustrated by E Ward-Hillhorst, National Botanic Gardens, Kirstenbosch, 1988.

Index

Aambeibos 80
Aambeibossie 66
Abdomen
   – ailments 132
   – distention 251
   – flatulence 91
   – pain 60
   – upsets 141, 204
Abortion 249
Abscesses 33, 66, 67, 73, 89, 102, 126
   170, 184, 188, 229, 263
   – internal 185
   – *see also* Poultices
Acacia 199
   – *karroo* 199
Aches, to ease 23, 44, 126, 161, 203
Aching joints 170, 248
Acne 51, 54, 66, 114, 227
Adansonia digitata 19
Adelaarsvaren 37
Adelaarsvaring 37
Adenandra uniflora 44
Adiantum capillus-veneris 120
African flea bane 219
African walnut 104
Afrikaanse blou salie 256
Afrikaanse salie 256, 257
Afterbirth
   – to expel 49, 61, 264

   – to expel in cattle 257
Afterlife 34, 71
After-dinner digestive 108
Agapanthus 111, 268
Agathosma
   – *betulina* 42
   – *cerefolium* 43
   – *ciliaris* 43
   – *crenulata* 42
   – *dielsana* 43
   – *pulchellum* 43
Agt-dae-geneesbossie 7, 268
Aloe
   – *davyana* 9
   – *ferox* 10
      konfyt 12
      insect repellent 11
   – *marlothii* 13
Aloysia triphylla 118
Alsbossie 81
Amarabossie 186
Amaranth 132
Amatungula 14
   – jam 15
   – syrup 16
American groundsel 221
Anaemia 79
Analgesic 164
Angel's fishing rod 91

Anosterbos 159
Anthericum
   – *ciliatum* 100
   – *revolutum* 100
Anthrax 28, 81, 114, 133, 137, 221
Anthrax-infected meat, to disinfect
   114, 119
Antibiotic 264
Antiparasitic 85
Antispasmodic 261
Antiseptic 43, 137, 148, 171, 190, 193,
   210, 233
Ants 44, 127
Anxiety 43, 63, 78, 93, 94
Anysboegoe 44
Aphids 127, 228
   – wild foxglove spray 232
   – wild garlic spray 236
Aphrodisiac 185
Aponogeton distachyos 207
Appetiser 59
Appetite, lack of 94, 159
Apple of Sodom 28
Arabian numnum 14
Aromatic sage 256
Aronskelk 17
Artemisia afra 113, 226
Arthritic swellings 205
Arthritis 40, 263

Arum lily 17
Asclepiadaceae family 30
Asclepias
   – *fruticosa* 128
   – *physocarpa* 129
Asparagus
   – Cape 207
   – fresh wild 215
   – tart 215
Asparagus africanus 214
Aspirin 59
Assegaaibos 81
Assegai wood 74
Asthma 20, 22, 35, 89, 109, 110, 118, 137 142, 149, 204, 216, 219, 220, 221, 224, 227
   – cardiac 223
Astringent 77, 81, 90, 96, 170, 171, 210
Athrixia phylicoides 51

Backache 17, 22, 33, 40, 43, 55, 73, 76, 77, 79, 126, 156, 161, 164, 201
Bad dreams, *see* Nightmares
Balderjan 244
Ballerja 224
Balsem kopieva 47
Baobab 19
Barosma betulina 42
Barrenness 74, 265
Basboom 153
Basil
   – bush 216
   – camphor 216
   – lemon-scented 216
   – lettuce leaf 216
   – purple ornamental 216
   – sacred 216
   – sweet 216
   – tulsi 216
   – wild 216
   – wild, and tomato sauce for pasta 218
   – wild, bathroom deodoriser 218
Basterboegoe 44
Bath, relaxing 56, 65
Bath freshener 117, 119, 149, 242, 252
Bath vinegar
   – hard fern 90
   – jasmine 108
   – wild rosemary 254
Bathroom deodoriser, wild basil 218
Bean bake, wild rosemary 253
Bean dishes, to flavour 252
Bedbugs 259
Bedding, stuffing 77
Bee stings 48, 223, 225
Beer, marula 124
Beetles, to attract 212

Bergaalwyn 13
Bergaartappel 182
Bergboegoe 42
Bergpatat 182
Bergtee 61
Berula thunbergii 201
Besembos 35
Biesie 139
Bietou 53
Birds
   – to attract 13, 32, 74, 104, 200
   – to attract nectar-seeking 10, 258
   – *see also* Sunbirds
Birdseed grass 26
   – peach chutney 27
Bird's brandy 24
Birth, *see* Childbirth
Bites 7, 9, 17, 24, 38, 91, 114, 118, 132, 170, 193, 223, 224, 226, 229, 259, 260
   – dressing for 7
   – infected 165, 261
   – itchy, to relieve 191
Bitter 15
Bitter aloe 10
Bitter apple 28
Bitteraalwijn 10
Bitterappel 28
Bitterblaar 55, 212
Bitterbos 66, 147
Bitterhout 30
Bitterwortel 30
Black nightshade 132
Black stick lily 35
Black water fever 152
Bladder
   – ailments 42, 50, 55, 178, 210, 214, 248, 251, 265
   – infections 28, 43, 61, 73, 98, 100, 221
   – pains 63
   – stones 98
Bleeding, to staunch 49, 98, 205
Bleeding gums 57
Blinkblaar 32
Blinkblaar wag-'n-bietjie 32
Blister leaf 40
Blisters 9, 29, 47, 184, 193, 195, 203, 210, 229
   – on feet 112
   – fever 47
Blocked nose 40, 227, 230, 237, 245
Blocked sinuses 227
Blood
   – to purify 50, 51, 53, 66, 132, 156, 186, 225, 227, 242, 264
   – to strengthen 53, 221
   – disorders 80, 114
   – poisoning 7
Blou waterlelie 210

Bloublomsalie 257
Bloukandelaar 5
Bloulelie 5
Blousalie 257
Blowflies 213
Blue bottle stings 190
Blue lotus 210
Blue sage 257
Blue water lily 210
Boat building 219
Bobbejaanbroodboom 19
Boegoe 42
Boer-bean
   – karoo 105
   – weeping 104
Boesmansboechoe 216
Boesmanstee 51
Boetabessie 53
Boils 17, 33, 61, 66, 67, 89, 102, 126, 145, 188, 200, 205, 227, 251, 263
Bokbessie 53
Bollota africana 109
Bones
   – broken 139, 153
   – injured 249
Boophane disticha 205
Boraginaceae family 7
Boslelie 68, 72
Bossietee 96
Bosysterhout 163
Bots 204
Bouquets 121
Bow wood 74
Bowel
   – complaints 187
   – haemorrhages 49
Bows, cross-berry 74
Bracken 37
   – in mustard sauce 39
Brake 37
Brandblaar 40
Brandblad 40
Brandewynbos 157
Brandy
   – buchu 43
   – wilde als 228
Brandy bush 157
Breast diseases 162
Breath freshener 136, 162
Bredies
   – Hotnotskool 101
   – tomato, with spekboom 194
   – waterblommetjie 209
Bronchitis 22, 26, 57, 63, 88, 109, 114, 120, 121, 142, 164, 185, 220, 223, 224, 227, 258, 259
Broodboom 19
Brown olive 247

Bruidslelie 5
Bruin salie 258
Bruises 22, 71, 85, 91, 103, 119, 132,
　　137, 156, 170, 174, 190, 205
Buchu 63, 127, 268
　– brandy 42
Buddleja 43, 268
Buffalo thorn 32
Buffelsdoring 32
Bulbine frutescens 47, 103
Bulbinella 47, 103
Bulbs, corns and tubers
　– *Boophane disticha* 205
　– *Crinua bulbispermum* 72
　– crinum lily 72
　– *Cyperus esculentus* 135
　– *Dierama pendulum* 91
　– *Dioscorea elephantipes* 83
　– elephant's foot 83
　– *Eucomis undulata* 251
　– *Gladiolus dalenii* 237
　– *Haemanthus coccineus* 137
　– harebell 91
　– nutgrass 135
　– *Oxalis pes-caprae* 188
　– paintbrush lily 137
　– sorrel 188
　– *Tulbaghia* species 233
　– tumbleweed 205
　– wild garlic 233
　– wild gladiolus 237
　– wild pineapple 251
Bulrush 49
Burn jelly plant 47
Burns 9, 11, 47, 100, 188, 195, 205, 207,
　　264, 265
Bush basil 216
Bush lily 68
Bushman's tea 51
Bush-tick berry 53
　– cordial 54
Butterflies, to attract 45, 104, 212
Butterfly bush 45

Cabbage palm 111
Cabbage tree, common 111
Cabinet making 219, 247
Caffer tea 102
Cakes, to decorate 62, 121
Calf muscles, aching 172, 217
Callouses 117, 178
Camphor basil 216
Camphor bush 219
Camphor leaf geranium 168
Canary creeper 221
Cancer 55
Cancer bush 55, 268
Candelabra flower 205

Canoes 166
Cape Aloes 10
Cape asparagus 207
Cape gum 199
Cape honeysuckle 268
Cape leadwort 151
Cape pond weed 207
Cape velvet mint 246
Cape water lily 211
Cape willow 59
　– hair growing oil 60
Capillaire 121
Cardiac asthma 223
Carissa edulis 14
Carpobrotus edulis 190
Carving, *see* Wood carving
Castor oil leaf 50, 52
Cat herb 109
Catarrh 96, 102, 120, 185
Cat's tail 47, 49
Cat's whiskers, white 212
Cattle
　– diarrhoea 200
　– diseases 229
　– to expel afterbirth 257
　– gall sickness 80, 88, 114, 251
　– opthalmic disorders 81
　– sores 147
Centella asiatica 143
Ceratotheca triloba 231
Cerebral haemorrhage 220
Chafes, on horses 195
Chameleon's berry 24
Chamomile, wild 63
　– and hottentotskooigoed pot-
　　pourri 64
　– massage oil 64
Charm 152
Chenopodium album 133
Chest
　– ailments 6, 20, 22, 24, 33, 35, 57, 87,
　　89, 102, 110, 118, 121, 149, 164, 167,
　　168, 184, 204, 220, 221, 223, 225, 226,
　　227, 239, 257, 264
　– colds 89, 217
　– poultice 20
Chicken pox 55
Chickens, diarrhoea in 224
Chilblains 220
Childbirth 6, 184
　– difficult 79
　– to ease 69, 74, 130
　– wash 81
Children
　– frail 195
　– frightened 89, 142
　– grazes and cuts 48
　– over-excited 35, 265

Chills 43, 229
China flower 44
Chinese lantern 184
Chironia baccifera 66
Chives
　– flat leaf 233
　– garlic 233
Cholera 43
Christmas berry 66
Chrysanthemoides monilifera 53
Chutney, birdseed grass peach 27
Cider tree 123
Cineraria, wild 221
Circulation, to assist 80
Circumcision 131, 214
　– wounds 185
Citronella plectranthus 149
Clematis
　– *brachiata* 203
　– *vitalba* 203
Clerodendrum glabrum 212
Cliffortia odorata 229
Climbers
　– canary creeper 221
　– Cape honeysuckle 57
　– *Clematis brachiata* 203
　– flame lily 85
　– *Gloriosa superba* forma *superba* 85
　– jasmine 107
　– *Jasminum multipartitum* 107
　– *Tecomaria capensis* 57
　– traveller's joy 203
　– wild asparagus 214
Clivia 68
Clivia miniata 68
Cliviine 69
Clothes moths 218
Cocoa substitute 135
Coffee substitute 34, 104, 135, 199
Cold sores 47
Colds 6, 20, 30, 40, 43, 46, 51, 56, 63, 73,
　　78, 80, 87, 88, 89, 94, 96, 98, 102, 109,
　　110, 113, 116, 120, 121, 133, 142, 149,
　　159, 161, 164, 168, 184, 187, 200, 204,
　　211, 213, 216, 220, 223, 224, 225, 226,
　　227, 229, 233, 237, 239, 242, 244, 245,
　　246, 252, 256, 257, 258, 259, 261, 262,
　　264
　– in children 121, 199
Coleonema album 43
Colic 28, 30, 43, 46, 61, 63, 78, 80, 87, 98,
　　105, 109, 110, 112, 113, 116, 136, 141,
　　147, 163, 168, 170, 176, 184, 197, 199,
　　213, 226, 227, 231, 242, 244, 246, 250,
　　251, 252, 256, 257, 261, 262, 264
Colitis, ulcerative 187
Combretum zeyheri 155
Commelina

– *africana* 265
– *benghalensis* 265
Common cabbage tree 111
Common coral tree 70
Common nightshade 132
Common sand olive 163
Compost, to accelerate decomposition 266
Compress, *see* Poultice
Condiment 26, 233
Confectionary 200
Confetti bush 43
Confinement, *see* Labour
Congestion 227
 – in lungs 161
 – nasal 22, 40
Constipation 91, 141, 187
 – in children 227
Contused veins 130
Contusions 91, 156
Convulsions 132, 159, 213, 256
Cool drink
 – to decorate 151
 – peppermint geranium 174
 – spearmint 246
Copaiba 47
Coral tree 70
Cordial, bush-tick berry 54
Corms, *see* Bulbs, corms and tubers
Corns 30, 193
Coronary thrombosis 77
Cortizone 84
Cortizone plant 83
Cotine 128
Cotyledon leucophylla 145
Cotyledon orbiculata 145
Coughs 6, 14, 20, 23, 26, 28, 30, 33, 43, 46, 51, 56, 63, 73, 78, 80, 87, 89, 96, 102, 109, 110, 113, 118, 120, 121, 133, 142, 149, 159, 164, 168, 187, 195, 200, 204, 211, 213, 216, 220, 221, 223, 224, 225, 226, 227, 233, 237, 239, 242, 245, 246, 251, 252, 256, 257, 258, 259
 – in cattle and dogs 200
 – in children 199
 – chronic 110, 114, 118
 – night 213, 257
Cough syrup
 – heather 94
 – kattekruie 110
 – nastergal 133
Cracked nails 48
Cracked skin 124, 178
 – fingers 48
 – heels 48, 116, 170
 – lips 48
Cramps 43, 63, 78, 112, 168, 223, 245, 246, 261

 – legs 113, 250
 – menstrual 252
 – stomach 113, 178, 231, 250
Crane's bill 81
Cream of tartar tree 19
Creeping sage 259
Crinum bulbispermum 72
Crinum lily 72
Crocodiles 20
Cross-berry 74, 158
Croup 63, 226, 244
Cupboard freshener 46, 64, 117, 118, 119, 227
 – wild gladiolus 238
Curry, nutgrass 136
Cussonia spicata 111
Cuts 25, 47, 51, 190
Cyanella lutea 7
Cyclopia genistoides 96
Cyperus papyrus 139
Cystitis 214, 221

Dagga
 – red 223
 – wild 223
Dais cotinifolia 153
Dandruff 148, 220, 254
Davy, Burtt 206
Day flower 265
Deadly nightshade 132
Death rites 157
Decorations, cake, pudding 121
Dehydration 193
Deodoriser 44
 – wild basil bathroom 218
Depression 93, 94
Diabetes 128, 186, 210
Diaphoric 252
Diarrhoea 30, 33, 38, 46, 49, 55, 57, 61, 63, 67, 80, 81, 87, 88, 98, 105, 113, 124, 132, 159, 163, 167, 169, 170, 176, 178, 183, 184, 186, 190, 197, 199, 200, 210, 231, 237, 249, 256, 257, 259
 – in calves and lambs 81
 – in cattle and dogs 200
 – in chickens 224
 – in children 128, 159
Dichrostachys cinerea subsp. *africana* 184
Dicoma anomala 80
Dierama pendulum 91
Digestion 242, 244
 – to aid 56, 61, 108, 136
Digestive complaints 81, 116, 169, 183
Dikbaspeer 249
Dioscorea 83
Dioscorea elephantipes 83
Dioscorine 84

Diosphenol 43
Diptheria 230, 248
Disinfectant 165, 187, 226, 227, 255, 259
Distemper 128
Diuretic 28, 43, 50, 137, 143, 191, 210, 214, 224, 242, 252
Dodonaea angustifolia purpurea 164
Dodonaea viscosa 163
Dogs
 – diarrhoea 200
 – to prevent scratching 191
 – sores 149
Doll's protea 268
Dombeya rotundifolia 71, 249
Doringboom 199
Douche 98, 210, 257
 – sour fig 191
 – wilde wingerd cleansing 230
Douwurmbos 7
Dragon lily 85
Dressings 7, 10, 210
Dried flower arrangements 6, 27, 37, 50, 56, 78, 89, 103, 104, 112, 121, 155, 187, 225, 231, 250, 251
Drill grass 98
Dropsy 88
Dry skin 116
Dull geranium 197
Dune sage 258
Dusting powder, *see* Talc
Dwellings 20
Dye, ivy-leaved geranium 171
Dysentery 30, 33, 49, 57, 61, 80, 81, 98, 124, 132, 142, 167, 169, 176, 183, 185, 186, 190, 197, 199, 237
Dysentery geranium 169
Dysentery herb 81
Dyspepsia 81, 147, 159

Eagle fern 37
Earache 71, 73, 98, 130, 145, 226
Earth almond 135
Eczema 7, 142, 206, 223, 224
 – infantile 190
 – on puppies 48
Edible pelargoniums 179
Eight-day healing bush 7
Elephantitis 185
Elephant's foot 83, 193
Elizabeth's wild garlic dish 235
Emetic 112, 213
Enema 91, 124, 132, 149, 170, 204, 224, 251, 264
 – for children 227
Energiser 93, 132, 150, 157, 260
Energy, lack of 94
Enteritis 49

Epilepsy 114, 146, 185, 224, 256
Equisetum
   – *arvense* 98
   – *ramosissimum* 98
Erica 93
Erica hirtifolia 93
Eriocephalus
   – *africanus* 252
   – *ericoides* 252
Erythrina lysistemon 70
Eucomis 111, 251
Eucomis undulata 251
Everlasting, wild 102
Everlastings 76
Evil 32, 131, 206, 251
Eyes
   – ailments 55, 133
   – infections 81, 248
     in cattle 24
   – sore 24, 46, 148, 185, 226, 248
Eyewash 46, 55, 147, 185, 226, 248

Fagara capensis 88, 113
Famine food
   – for humans 34, 37, 106, 165, 196
   – for stock 50, 112
Fat hen 132
Fatigue 23
Fear 93, 103
Febrifuge 216
Febrile conditions 69
Feet
   – aching 52, 71, 114, 156, 210, 217, 220
   – cracked 124
   – hard skin 52
   – to refresh 204, 248
Female ailments 55, 256, 257, 258
Ferns
   – *Adiantum capillus-veneris* 120
   – bracken 37
   – hard fern 89
   – maidenhair fern 120
   – *Pellaea calomelanos* 89
   – *Pteridium aquilinum* 37
Fertiliser 139, 266
Fertility 6, 28
Fever 19, 20, 55, 56, 57, 59, 63, 80, 87, 88,
   103, 110, 112, 113, 118, 132, 143, 149,
   163, 170, 186, 197, 213, 216, 217, 225,
   226, 229, 233, 239, 248, 251, 259, 260,
   261, 262, 264
Fever blisters 63
Fever tea 118
Fever tree 113, 118
Feverplant 216
Fiddleheads 37, 38
Fine-leaf maidenhair 120
Fish, grilled, garnish 188

Fish dishes, to flavour 252
Fish moths 218
Fish poison 112, 164, 205
Fishing baskets 155
Fishing nets 131
Fixative 104, 166, 175
   – pompon tree 154
Flame lily 85
Flat leaf chives 233
Flatulence 28, 63, 78, 81, 87, 91, 113, 116,
   136, 141, 147, 178, 199, 226, 231, 244
   245, 250, 251, 252, 255, 256, 257, 259,
   261, 262, 264
Flatulent colic 159, 168
Fleas 11
Flies 150
   – to keep off meat 246
Flower arrangements, dried 6, 27, 37,
   56, 78, 89, 103, 104, 112, 121, 155,
   187, 225, 231, 250, 251
Flu 40, 55, 56, 57, 63, 109, 110, 113, 118,
   159, 161, 220, 225, 229, 233, 248, 257,
   259, 261, 264
   – preventative 114
Flute, bulrush 50
Foliar feed 266
Food scarcity, *see* Famine food
Fractures 42, 152, 153
   – to bind 131
Free State sage 262
Fruit salads, to decorate 62, 151
Fruitfly 127
   – wild garlic spray 236
Fuel 159, 247
Fumigant 65, 78, 219
Furniture making 104, 247
Fynblaar salie 261
Fynblaar wilde salie 261

Galbessie 132
Gall bladder attacks, to ease 242
Gall sickness, in cattle 80, 88, 114, 251
Gansiebos 163
Gansies 55, 128
Garden nightshade 132
Gargle 26, 46, 51, 102, 109, 113, 133, 134,
   164, 190, 225, 226, 229, 230, 248, 257,
   261
Garlic chives 233
   – steamed 235
Gastric ulcers 14
Gastritis 53
Geel sterretjie 195
Geelblomsalie 258
Geelboslelie 85
Geelkatstert 47
Geelsuring 188
Geranium

   – camphor leaf 168
   – carpet 61
   – dull 197
   – dysentery 169
   – hooded 170
   – ivy-leaved 171
   – nutmeg 172
   – oak-leaved 173
   – peppermint 174
   – raspleaved 175
   – rose-scented 176
   – salmon 197
   – scented 167
   – stork's bill 197
   – wild rose 178
Gewone drolpeer 249
Gewone kiepersol 111
Gewone koraalboom 70
Ghaukum 190
Gifappel 28
Gifbol 205
Ginger bush 87, 114, 213, 268
Gladiolus, wild 237
Gladiolus
   – *dalenii* 237
   – *natalensis* 237
Glandular swellings 98
Glass making 38
Gloriosa lily 85
Gloriosa superba forma *superba* 85
Gonorrhoea 124
Good fortune 69
Gotu kola 143
Gouna 190
Gout 17, 43, 63, 226
Grasklokkies 91
Grass aloe 47
Grazes 25, 46, 114, 161, 171, 188, 190,
   207, 260
Grazing, scarce 248
Greasy skin, *see* Oily skin
Grewia
   – *flava* 157
   – *occidentalis* 74, 158
Gripes 163, 250, 257
Groot salie 260
Grootbergkapok 252
Groundcovers
   – *Berula thunbergii* 201
   – Cape honeysuckle 57
   – carpet geranium 61
   – *Carpobrotus edulis* 190
   – *Centella asiatica* 143
   – *Commelina africana* 265
   – creeping sage 259
   – curry bush 76
   – *Geranium incanum* 61
   – *Helichrysum peteolatum* 78

– *Helichrysum* species 76
– pennywort 143
– peppermint geranium 174
– *Salvia repens* 259
– sour fig 190
– *Tecomaria capensis* 57
– toothache root 201
– yellow wandering Jew 265
Gryshout 80
Gum arabic tree 199
Gum boils 226
Gum infections 117
Gums, bleeding 57, 117, 149

Haakdoring 214
Haemanthus
– *coccineus* 137
– *hirsutus* 137
– *rotundifolius* 137
Haemorrhages
– bowel
– cerebral
– intestinal
Haemorrhoids 55, 66, 67, 73, 80, 81, 85, 103, 109, 130, 132, 156, 161, 164, 223, 224, 226, 229, 249, 250, 263
Hail 186
Hair
– conditioner 59, 253
– to keep fairness 63
– rinse 63, 150, 252
  wild rosemary 254
– to stimulate growth 59, 63, 253
  Cape willow hair growing oil 60
Hands, cracked 124
Hands, sore 71
Handskoentjie 57
Hanging baskets 174
Hangover 105, 251
Hard fern 89
– bath vinegar 90
– tea 90
Harebell 91, 268
– dried 91
Harness sores 28
Harpuisblaar 212
Head cold 89
Headaches 17, 30, 40, 59, 105, 132, 142, 152, 156, 184, 195, 204, 219, 220, 223, 227, 245, 246, 248
– sinus 227, 230
Heart 80, 110
– ailments 6, 76, 78, 133, 173
– racing 77
– treatment for animals 77
Heart lung ailments 110
Heartburn 78, 81, 89, 105, 116, 147, 159, 184, 199, 219, 226, 227, 244, 256, 257,

260, 261, 262, 264
Heat exhaustion 59
Heat rash 59, 66, 118, 163, 210
Heath 93
Heather 93, 268
– cough syrup 94
– pot-pourri 95
Heatstroke 156, 159, 163, 193
– to prevent 203
Heels
– aching 172
– cracked 48, 71, 116, 170
Heide 93
Helichrysum
– *auriculatum* 76
– *crispum* 77
– *foetidum* 77
– *nudifolium* 102
– *odoratissimum* 78
– *peteolatum* 78
Heteromorpha arborescens 141
Heteropyxis natalensis 116
Heuningblomtee 96
Heuningbostee 96
Heuningtee 96
High blood pressure 173
Hikers, to refresh and revive 204
Hip pains 265
Hoarseness 110
Hoenderuintjie 135
Honey
– to flavour 108
– relaxant 108
Honey tea 96
Honeybush tea 96
– jelly 97
Honingtee 96
Hooded geranium 170
Horehound 109
Hormonal disturbances 195
Horse sickness 13
Horsemint 244
Horsetail 98
Hot toddy, maidenhair fern 122
Hotentots' bread 18
Hotnotsbrood 83
Hotnotskooigoed 76
Hotnotskool 100
– bredie 101
Hotnotsvy 190
Hottentots bean 106
Hottentots bean tree 104
Hottentots' cabbage 100
Hottentots' fig 190
Hottentots' tea 102, 268
Hottentotsbedding 76
Hottentotskooigoed 76, 127
– and wild chamomile pot-pourri 64

Hound's berry 132
Hypertension 77, 78
Hypoxis
– *argentea* 195
– *nyasitica* 195
– *oliqua* 195
– *rigidula* 195
– *rooperi* 195
Hysteria 100, 109, 205, 265

Iboza 87
Iboza riparia 87
Impotence 28, 53, 74
Indigestion 42, 63, 220, 227, 244, 245, 246, 255, 257, 261, 262
– in old people 199
Infections 7, 63, 216
– to clear 98
– vaginal 230
Inflammation 106, 163, 210
– of eyes 248
– of vagina 98
Influenza, see Flu
Inhalent 117, 227
Injuries 139
Inkbol 195
Insect bites, see Bites
Insect stings, see Stings
Insect repellent 44, 117, 127, 150, 212, 213, 216, 218, 219, 223, 231, 259
Insect-repelling pot-pourri 115, 142
– plectranthus 150
Insecticide 11, 124, 236
– spray, wilde als 228
Insects, to attract nectar-loving 165
Insomnia 93
Intestinal disorders 231
Intestinal ulcers 98
Invalids, jelly for 97
Itch, vaginal 230
Itchy skin 47, 223, 225
Ivy-leaved geranium 171

Jack-in-the-pulpit 17
Jakkalsbos 79
Jam
– *Aloe ferox* 11
– amatungula 15
– nastergal 134
– sour fig 192
– wild medlar 240
Jasmine 107, 268
– bath vinegar 108
– pot-pourri 108
– tea 108
Jasminum multipartitum 107
Jelly
– honeybush tea 97

– marula 125
– peppermint geranium 174
Joints
– aching 170, 248
– painful 41, 43

Kaapse sandolien 163
Kaapse wilger 59
Kafferbessie 157
Kafferboom 70
Kafferdruiwe 22
Kaffertee 51
Kaffir tea 51
Kalahari Christmas tree 184
Kalkoenbos 55
Kamferblaar 168
Kaneelbol 183
Kanferbos 219
Kankerbos 55
Kannabos 153
Kapkappie 252
Kapokbossie 252
Kardomom 113
Kareedoring 199
Karoo boer-bean 105
Karoo botanic gardens 84
Karoodoring 199
Karookapok 252
Karroo thorn 199
Katjiedrieblaar 40
Katstert 47
Kattekruid 109
Kattekruie 109
– cough syrup 110
Kei-apple 28
Keiserkroon 5
Kerriebos 76
Kew 84
Khakibos 11, 73, 127
Kidneys
– ailments 42, 50, 76, 158, 161, 170, 178, 185, 210, 214, 248, 251
– infections 43, 73
– pains 60
– stones 76, 221
Kiepersol 111, 147
Kierie 114
Kigelia africana 165
Kindling 93
Klawersuring 188
Klein salie 259
Kleinkapok 252
Kleinperdepram 113
Kleinwaterlelie 207
Klimop 203
Klipdagga 127, 225
Knobwood 113, 268
Knopdoring 113

Knoppieshout 74
Knowltonia
– *vesicatora* 40
– *transvaalensis* 40
Kolsuring 171
Kommetjie teewater 44
Konfyt, Aloe ferox 12
Koorsbossie 80
Koringblom 147
Korsbossie 80
Kouterie 145
Kraaibos 141
Kremetartboom 19
Kruidjie-roer-my-nie 126, 212
Kruip salie 259
Kruisbessie 74
Kruisement 241, 242
Krulkop 251

Labiatae family 255
Labour
– to bring on 249
– difficult 147
– to ease 71
– to hasten 245
– prolonged 116
– retarded 74
– to strengthen contractions 49
Lack of appetite 94, 159
Lack of energy 94
Lactation 69
– to assist 58, 162, 165, 264
– in pigs 208
Lamiaceae 87
Lantana rugosa 24
Lapaalwyn 10
Large-fruited bushwillow 155
Large-fruited combretum 155
Laryngitis 225
Lasiospermum bipinnatum 65
Lavender 160
Lavender tree 116, 268
– bath freshener 117
– pot-pourri 117
Laventelboom 116
Leather
– making 38
– to soften 124
Legs
– aching 114, 175, 201, 240
– cramps 113, 250
– massage for 220
– pains 22, 223
– sores 7, 100
– ulcers 143
Lemon bush 118, 268
– tea 118
Lemon-scented basil 216

Lemon verbena, wild 116, 118
Lemonade tree 19
Lemonspur flower 149
Leonitis
– *leonitis* 225
– *leonurus* 223
var. *albifolia* 224
Lepidium africanum 26
Leprosy 66, 156, 185, 223
Lettuce leaf basil 216
Lice 118
Lightning 32, 131, 152, 157, 263
Liliaceae family 9, 10, 13
Lily-of-the-Nile 17
Limbs, weakness of 79
Linen freshener 119
Ling 93
Linnaeus 82
Lion's ear 223
Lippia javanica 118
Lips, cracked 47, 124
Liqueur, buchu brandy 43
Liver
– ailments 55, 133
– to cleanse 214
Lobostemon fructicosus 7
Loss of voice 51
Love potions 233
Lucky beans 70, 71
Lucky-bean tree 70
Luibos 7
Lumbago 33, 63, 57, 114
Lungs
– ailments 214
– congestion 161
Lycorine 69

Maagbossie 80
Maagpynbossie 261
Maagtee 186
Maagwortel 80
Maartblom 137
Maggots 213
Maidenhair fern 120
– hot toddy 122
Malaria 19, 59, 73, 87, 88, 112, 118, 124, 132, 204, 239
Mange 28
Mannetjie rabassam 61
Many-petalled jasmine 107
Maroela 123
Marrubium
– *africana* 109
– *vulgare* 109
Marsh pepperwort 143
Marula 123
– jelly 125
Massage, for legs 220

Massage oil, wild chamomile 64
Matjiesgoed 49
Matricaria
  – *africana* 63
  – *glabrata* 63
Mayonnaise, sweet and sour yellow sorrel 189
Measles 33, 118, 226
Meat
  – to disinfect 114
  – to keep fresh 119, 246
Medicinal Plant Index 40
Medlar 239
Meidjiejanwillemse 61
Meidjie willemse 66
Melianthus 126
  – *cosmosus* 7
  – *major* 7, 126
  – *minor* 126
Melkbos 30, 128, 268
Menstruation 136
  – absorbent pad for 49
  – cleansing wash 230
  – cramps 252
  – delayed 249, 252
  – difficult 265
  – excessive 61, 257
  – irregular 7, 61, 230
  – painful 148, 230, 231, 244, 250, 257
  – problems 239
  – wash 81
Mental disorders 141
Mentha
  – *aquatica* 242
  – *capensis* 245
  – *longifolia* 244
    var. *wissii* 246
    var. *capensis* 244
    var. *polyadena* 245
  – *spicata* 245
Mildew 54, 99, 139
Milk bush 30
Milk flow, *see* Lactation
Milk wort 30
Milkweed 128
Mimosa 199
Mimosa thorn 199
Minaret flower 223
Mint
  – Cape velvet 246
  – wild 241
  – wild water 242
  – woolly long-leaf 246
Minted onions 243
Miscarriage
  – to prevent 130
  – threatened 229
Mispel 239

Misty plume bush 87
Mobile 155
Moles 129
Monkey bread tree 19
Monkeys, to attract 32
Monkey's tail 35
Monson, Lady Anne 82
Monsonia
  – *angustifolia* 81
  – *burkeana* 82
Morning bride 147
Moroggo 26, 133
Mosquito bites 9, 47, 142, 225
  – to relieve itch 191
Mosquito repellent 44, 216, 217, 244
Moth repellent 160, 173, 218
  – wilde als 228
Mother-in-law's tongue 130
Moths, clothes 218
Moths, to attract 104, 212, 258
Mountain aloe 13
Mountain buchu 42
Mouse ears 265
Mouth 126
  – infections 117, 143, 149, 190, 193, 200, 230
  – ulcers 48, 226
Mouthwash 117, 226
Mucous 211
Muishondbos 173
Mumps 227
Muscles 119
  – aching 63, 114, 162, 170, 201, 204, 223, 225
  – injured 249
  – painful 245
  – relaxant 44, 52
  – spasm 63, 65, 112
  – sprained 164
  – strained 205
  – tension 35, 63, 78, 167
  – tired 242
Musical instruments 219
Myrothamnus flabellifolius 161

Naaldbossie 81
Nagskaal 132
Nagskade 132
Nails 69
Narrow leaf sage 261
Nasal congestion 22, 40
Nastergal 132
  – jam 134
  – sore throat gargle 134
Natal lily 237
Natal plum 14
Nausea 43, 59, 87, 105, 112, 163, 167, 169, 176, 178, 183, 197, 214, 231, 250, 251
Neck stiffness 59
Nerve paralysis 125
Nervousness 93, 141, 167, 265
Nettle 132
Neuralgia 227
Nicholas klapper 155
Night coughing 110, 213, 227, 257
Nightcap 63, 89, 94, 213
Nightmares 43, 63, 94, 103, 213, 220
  – in children 142
Nightshade 132
Nits 220
Noem-noem 14
Nooiensboom 111
Nooienshaarvaring 120
Norra 182
Norretjie 182
Nose
  – bleeding 117, 118
  – blocked 227, 230, 237, 245
Num-num 14
Nurseries, list of 269–70
Nutgrass curry 136
Nutmeg geranium 172
Nymphaea
  – *caerulea* 210
  – *capensis* 211

Oak-leaved geranium 173
Obesity 224
Ocimum
  – *basilicum* 216
  – *basilicum citriodorum* 216
  – *basilicum crispum* 216
  – *basilicum minimum* 216
  – *basilicum purpurascens* 216
  – *canum* 216, 217
  – *gratissimum* 216
  – *kilimandscharicum* 216
  – *sanctum* 216
  – *urticifolium* 217
Oedema 252
Oil of geranium 178
Oily skin 90, 112, 171
Ointment, soothing 78
Old man's beard 139, 203
Old people 22, 87, 97, 116, 195
Olea europaea subsp. *africana* 247
Olienhout 247
Olifantboom 193
Olifantstert 35
Olifantsvoet 83
Olive, wild 247
Onions, minted 243
Oorlams plakkie 145
Opthalmia 156, 200

– in cattle 11, 24, 81, 200
– in dogs 200
– in man 248
– in stock 248
Orange River lily 72
Osteomylitis 200
Otitis media 145
Over-imbibing 248
Over-anxiety 219
Over-excitement 78, 109
– in children 35
Over-exhaustion 193
Over-full feeling 168
Oxalis pes-caprae 188

Pain killer 44, 57, 70, 71, 114, 184, 201, 205, 227, 239
Paintbrush lily 137
Palmiet 49
Paper, papyrus 140
Paper making 20, 131, 140
Papirusriet 139
Papkuil 49
Pappe, Carl 102, 146
Papyrus 139
Papyrus paper 140
Paralysis 6, 114, 220, 233
– partial 224
Parasites, skin 213
Parrot lily 237
Parsley tree 141, 268
Pasnip tree 141
Pasta, wild basil and tomato sauce for 218
Patterson 83
Peach chutney, birdseed grass 27
Pear, wild 249
Pelargonium
– *acetosum* 178
– *antidystericam* 169
– *betulinum* 168
– *bowkeri* 180
– *capitatum* 178
– *cuculatum* 170
– *fragrans* 172
– *graveolens* 176
– *lobatum* 181
– *luridum* 197
– *peltatum* 171
– *quercifolium* 173
– *radens* 175
– *rapaceum* 182
– *tomentosum* 174
– *triste* 183
Pelargoniums, edible 179
Pellaea calomelanos 89
Pennywort 143
Pentanisia prunelloides 91, 263

Peperbos 26
Peppermint geranium 174
– cool drink 174
– jelly 174
Pepperweed 26
Pepperwort 26
Perdeboom 113
Perdebos 66
Perdespook 205
Perdestert 98
Perdevy 190
Perennials
– agapanthus 5
– bobbejaanstert 35
– *Ceratotheca triloba* 231
– clivia 68
– *Clivia miniata* 68
– *Dicoma anomala* 80
– *Dicoma zeyheri* 79
– *Dierama pendulum* 91
– doll's protea 79
– harebell 91
– *Pentanisia prunelloides* 263
– pincushion 147
– *Scabiosa columbaria* 147
– toy protea 80
– wild foxglove 231
– wild verbena 263
Perfume 25, 118
Perspiration
– copious 20
– increased 43
Pests, plant 228
Phthisis 102
Pig lily 17
Pig's ear cotyledon 145
Piles 67, 130, 230
Piles bush 66
Pillows
– scented 167, 176
– stuffing 30, 129, 174, 220, 252
Pimples 48, 86, 193, 207, 227, 260
– adolescent 112
– in babies 112
Pincushion 268
Pink eye 24
Pink ragwort 221
Plakkie 145
Plant pests 228
Planter warts 145
Plants, ailing 99
Plectranthus 149, 268
– *esculentus* 149
– *floribundus* 149
– *fruticosus* 149, 150
– *hirtus* 149
– insect-repelling pot-pourri 150
– *laxiflorus* 149

– *natalensis* 149
– *thunbergii* 150
– *urticoides* 149
Pleurisy 118
Plum blossom tree 249
Plumbago 151, 268
Plumbago auriculata 151
Pneumonia 57, 96, 110, 185, 239
Poeierkwasblom 137
Poison, fish 112
Poison arrows 205
Pollichia campestris 22
Pompon tree 153, 268
– fixative 154
Pond weed, Cape 207
Porkwood 165
Portulacaria afra 193
Post-nasal drip 133
Posy, spring 222
Pot scourer, horsetail 99
Pot-pourri 8, 14, 24, 42, 43, 44, 58, 65, 73, 77, 87, 94, 103, 104, 114, 117, 118, 119, 127, 141, 150, 154, 160, 164, 173, 174, 175, 176, 187, 198, 200, 212, 222, 223, 227, 234, 238, 258
– blue 151
– heather 95
– insect-repelling 115, 142
– jasmine 108
– lavender tree 117
– mauve 234
– plectranthus insect-repelling 150
– rose-scented geranium 177
– springtime 46
– springtime tulbaghia 235
– wild chamomile and Hottentotskooigoed 64
– wild flower 268
– wild gladiolus 238
Poultice
– for abscesses 73, 89, 102, 126, 170, 188, 200, 263
– for aching feet 71
– for arthritic swellings 205, 263
– for backache 33, 43, 126, 164
– for bites, sore and infected 165, 170, 193
– for bladder pains 63
– for boils 89, 102, 126, 188, 200, 227, 251, 263
– for breast disease 162
– for broken bones 139
– for bruises 71, 170, 174, 205
– for burns 207
– for chest 20, 110
– for cramp 111, 112
– for earache 146
– for gout 17

– for haemorrhoids 263
– for infected wounds and scratches 145
– for mumps 227
– for muscular pain 245
– for neuralgia 227
– for osteomylitis 200
– for painful areas 23
– for painful joints 43
– for piles 230
– for pimples 227
– for rashes 210
– for rheumatic joints 17, 25, 205, 263
– for scratches, inflamed 210
– for sores 73, 188, 193, 207, 251, 263
– for sprains 25, 71, 164, 174, 200, 205, 227, 245, 263
– for strains 205, 227
– for swellings 102, 240, 245, 263
– for toothache 146
– for ulcers 126
– for varicose ulcers, skin diseases, rashes, eczema, psoriasis ulcers 206
– for wounds 30, 137, 263
Porkbush 193
Poultry stuffings, to flavour 252
Pregnancy 6, 49, 191
Premenstrual tension 61
Pressed flower pictures 121, 175, 198, 231
Pretty morel 132
Prostate gland 43, 195
– infections 43, 98
Protea, doll's 79
Protection, plants for 24, 29, 34, 86, 114, 124, 131, 136, 146, 152, 157, 185, 206, 251
Purple ornamental basil 216
Psoriasis 206
Pteridium aquilinum 37
Puddings, to decorate 62, 121
Pulmonary tuberculosis 184, 214
Purgative 11, 13, 28, 66, 67, 80, 128, 143, 156, 166, 185, 187, 195, 233, 239
Purslane 132

Raasblaar 155
– seeds 268
Rabas 81
Rain ritual 133
Raisin tree 157
Raisinbush 157
Ranunculaceae family 40
Rashes 7, 24, 25, 38, 47, 48, 59, 118, 158, 164, 178, 193, 205, 206, 210, 224, 226, 251, 259, 260, 264
– in babies 112

– infected 91, 132, 210
Raspleaved geranium 175
Red aloe 10
Red dagga 223
Red spider 236
Relaxant honey 108
Relaxant tea 57
Religious ceremonies 220
Relish 264
Renosterbos 268
Renostertoppe 159
Resin leaf 212
Respiratory ailments 87, 120, 161, 214
Respiratory problems in animals 224
Restorative 97
Resurrection plant 161
Rhenoster bush 159
Rheumatic fever 59, 73
Rheumatic joints 17, 25, 103, 165, 205, 263
Rheumatism 17, 22, 23, 25, 33, 40, 42, 55, 59, 63, 73, 94, 126, 173, 186, 201, 213, 219, 220, 233, 248
Rheumatoid arthritis 263
Rhodesian foxglove 231
Rice flower 147
Ricinus communis 50, 52
Ringworm 7, 28, 80, 133, 142, 190
River crinum 72
River willow 59
Rivierlelie 72
Rivierwilg 59
Rooiboslelie 85
Rooilelie 237
Rope 19, 74, 131, 153, 155, 184, 195, 199
Rosemary, wild 252
Rose-scented geranium 176
– pot-pourri 177
– scones 177
Rosyntjiebos 157
Round-leaf buchu 42
Roundworm 213, 239
Rumex 188
Rutaceae family 42

Sacred basil 216
Sad geranium 183
Saddle sores 28
Sage
– aromatic 256
– blue 257
– brown 258
   dune 258
   sand 258
– creeping 259
– Free State 262
– narrow leaf 261
– small 259

– Transvaal 260
– wild 255
– wild giant 260
Sage wood 219
Sagewood 45
Salads, to decorate 62
Saliebossie 259
Saliehout 45
Salix
– *capensis* 59
– *gariepina* 59
– *mucronata* 59
Salmon geranium 197
Salt substitute 10, 188
Salve 265
Salvia
– *africana coerulea* 257
– *chamelaeagnea* 256
– *disermas* 260
– *panticulata* 256
– *repens* 259
– *rugosa* 260
– *stenophylla* 261
– *verbenaca* 262
Sambreelboom 111
Sand olive 163, 268
Sand sage 258
Sand salie 258
Sandbessies 157
Sandworm 28
Sanseviera
– *hyacinthoides* 130
– *thyrsiflora* 130
Sausage tree 165, 268
Scab 11, 28
Scabies 118, 147, 149
Scabiosa
– *africana* 148
– *incisa* 148
– *transvaalensis* 148
Scabious 147
Scabs, on heads of children 80
Scalds 100
Scalp
– conditions 142
– infections 118
– itches 59
– sores 59
Scarlet fever 33
Scented geraniums 61, 167, 268
Schotia
– *afra* 105
– *brachypetala* 104
– var. *angustifolia* 106
Sciatica 114
Scones, rose-scented geranium 177
Scorpion stings 85, 184, 223
Scrapes, *see* Grazes

Scratches 7, 20, 24, 29, 30, 40, 46, 117, 118, 132, 145, 158, 171, 178, 205, 210, 224, 251, 259, 260, 261, 264
Scratching, to prevent dogs 191
Screw worm, on cattle 85
Scrofula 73, 143
Scurvy 188
Sedative 132, 213
Sedge 139
Seeroogblom 72, 137, 205
Sekelbos 184
Senecio
   – *coronatus* 221
   – *elegans* 221
   – *purpureus* 221
   – *tamoides* 221
Sewejaartjies 76
Shiny leaf 32
Shock 100
Shortness of breath 142
Shrubs 132
   – Afrikaanse salie 256
   – agt-dae-geneesbossie 7
   – *Artemisia afra* 226
   – *Asclepias fruticosa* 128
   – barley sugar plant 22
   – bird's brandy 24
   – bitter apple 28
   – bitterwortel 29
   – blue sage 257
   – brown sage 258
   – buchu 42
   – Bushman's tea 51
   – bush-tick berry 53
   – cancer bush 55
   – Cape honeysuckle 57
   – china flower 44
   – *Chironia baccifera* 66
   – Christmas berry 66
   – *Clifforia odorata* 229
   – confetti bush 43
   – cross-berry 74
   – curry bush 76
   – *Cyclopia genistoides* 96
   – *Dichrostachys cinerea* subsp. *africana* 184
   – *Dodonaea viscosa* 163
   – *Elytropappus rhinocerotis* 159
   – Ericaceae family 93
   – *Eriocephalus* species 252
   – ginger bush 87
   – *Grewia flava* 157
   – *Grewia occidentalis* 74
   – heather 93
   – *Helichrysum* species 76
   – honeybush tea 96
   – *Iboza riparia* 87
   – klipdagga 225
   – lemon bush 118
   – *Leonotis leonitis* 225
   – *Leonotis leonurus* 223
   – *Lippia javanica* 118
   – melianthus 126
   – *melianthus major* 126
   – *melianthus minor* 126
   – melkbos 128
   – nastergal 132
   – plectranthus 149
   – plumbago 151
   – *Plumbago auriculata* 151
   – raisinbush 157
   – renosterbos 159
   – *Salvia africana coerulea* 257
   – *Salvia africana lutea* 258
   – *Salvia aurea* 258
   – *Salvia chamelaeagnea* 256
   – *Salvia panticulata* 256
   – sand olive 163
   – sickle bush 184
   – *Solanum nigrum* 132
   – *Sutherlandia frutescens* 55
   – *Tecomaria capensis* 57
   – *Tetradenia riparia* 87
   – wild dagga 223
   – wild rosemary 252
   – wilde als 226
   – wilde wingerd 229
Sickle bush 184
Sickle tree 184
Silver everlasting 79
Silverleafed vernonia 186, 268
Sinus 168, 204, 245
   – blocked 227, 230
   – congestion 142
   – headache 227, 230
   – stuffy 121
Skilpadknol 83
Skin
   – ailments 7, 54, 59, 66, 206, 227
   – to cleanse 227
   – dry 116, 124, 178
   – eruptions 30, 51, 85, 86, 114, 124, 184, 223
      in cattle 197
   – infections 24, 28, 264
   – irritations 147
   – itchy 223, 225
   – lesions, tuberculose 143
   – lotion 59
   – oily 90, 112, 171
   – rashes, *see* Rashes
   – to soften 89, 90, 149
   – softener 25, 178
   – wash 210
   – *see also* Cracked skin
Sleep, to aid 43, 63, 76, 78, 149, 176
Sleeplessness 57, 110, 205, 219, 220
Small knobwood tree 88, 113
Small sage 259
Smallpox 124
Snakebite 9, 68, 69, 81, 85, 110, 114, 184, 204, 213, 223, 225
Snakes, to keep away 15, 223, 233
Sneezing, to induce 204
Snow bush 252
Snuff 13, 128, 143
Soap making 38, 54
Societies, botanical 269–70
Soetdoring 199
Soil, to bind 14, 53
Sokoa oil 124
Solanceae family 28
Solanum
   – *nigrum* 132
   – *sodomem* 28
Sonchus oleraceus 133
Sooibrandbossie 263
Sore eye lily 137
Sore eyes 147, 148, 156, 185, 226
Sore feet 210, 220
Sore throat 15, 26, 51, 80, 102, 109, 110, 126, 133, 164, 171, 185, 190, 191, 193, 220, 226, 227, 229, 230, 248, 255, 257, 261
   – gargle, nastergal 134
Sores 7, 9, 24, 30, 40, 46, 48, 71, 80, 98, 110, 117, 126, 143, 145, 148, 165, 184, 193, 197, 207, 223, 224, 226, 251, 259, 260, 261, 263, 265
   – on animals 29, 80, 147, 149, 224
   – on genitalia 165
   – on heads of children 100
   – infected 77, 89, 205, 229
   – on legs 100
   – leprous 143
   – slow-healing 34, 81, 165
   – suppurating 73, 188
   – veld 38, 66, 80, 147, 149
   – venereal 66, 143, 147
Sorrel 188
Sorrel leaf 179
Sorrel, sweet and sour mayonnaise 189
Soup, waterblommetjie 209
Sour fig 91, 190
   – douche 191
   – jam 192
Sour thorn 199
Sow's thistle 132, 133
Spasms 63
Spearmint 245
   – cool drink 246
Spekboom 193
   – tomato bredie with 194
Spider bites, to relieve itch 191

Splint, for broken limbs 153, 249
Spots 193
Sprained ankle 22, 23, 50
Sprains 25, 71, 73, 85, 119, 137, 139, 153,
    156, 157, 164, 174, 200, 205, 227, 240,
    245, 263
Spring posy 222
Springtime pot-pourri 46
Springtime tulbaghia pot-pourri 235
Spurflower 149
St John's lily 68
Stalked bulbine 47
Stamina 135
Star flower 195
Steamed garlic chives 235
Sterblom 195
Sterility 80, 265
Sterkgras 26
Stiffness, prevention 23
Stings 9, 17, 38, 91, 118, 170, 184, 193,
    223, 224, 226, 249, 260
Stitch 78
Stock food 193
Stomach
    – ache 43, 87, 93, 113, 186, 250
        in children 128
    – ailments 13, 24, 30, 42, 53, 55, 63,
        163, 184, 196, 197, 214, 219, 231, 246,
        249, 257, 259
    – cramps 178, 227, 245, 249
    – pains 57
    – ulcers 27, 98
    – upsets 33, 46, 80, 87, 98, 113, 124,
        141, 204, 237, 244
Stork's bill geranium 197
Strains 85, 139, 205, 227
Strand salie 258
Strandblommetjie 221
Strength 38, 53, 71, 86, 97, 112, 135, 148
    – in frail children and old people 195
    – to restore 79
Stress 63, 78
String 19, 74, 131, 153, 184
Stubbleberry 132
Stuffy nose, see Blocked nose
Styptic 248
Suikertee 22, 96
Sunbirds, to attract 57, 126, 258
Sunburn 9, 163, 164, 190, 193, 207, 210,
    225
Sunstroke, to prevent 203
Superb lily 85
Suppression of urine 170
Suring 188
Survival food, see Famine food
Sutherlandia
    – frutescens 55
    – microphyulla 55

– tomentosa 55
Suurdoring 199
Suurvy 190
Swart vrouehaar 120
Swartolienhout 247
Swartstorm 80
Sweet and sour yellow sorrel
    mayonnaise 189
Sweet basil 216
Sweet thorn 199, 268
Sweetener 200
Swellings 33, 42, 102, 103, 240, 245, 263
    – tuberculose 33
Swollen glands 33
Swollen joints 22
Sydissel 133
Syphilis 7, 67, 143, 165, 185, 204
Syrup, amatungula 16

Talc 117, 148
    – for skin treatments 42
Tandpynblaar 40
Tandpynbossie 66
Tandpynwortel 201
Tanning 38, 153, 199
Tapaalwyn 10
Tape worm 13, 213
Tarchonanthus camphoratus 219
Tea
    – calming 77
    – hard fern 90
    – health-giving 51, 76, 96, 103
    – honeybush 96
    – Hottentots' 102
    – jasmine 108
    – lemon bush 118
    – purgative 9
    – refreshing 204
    – sleep-inducing 63, 176
    – tonic 162
Tecoma 57
Tee 61
Teebossie 81
Teesuikerbossie 22
Temperature 197
    – to bring down 197
    – in children 159
Tension 27, 93
Tetradenia riparia 87, 213
Thatching 38, 49
Thirst-quencher 20, 32, 157, 171, 193
Thorn tree 199
Threadworm 213
    – in horses and dogs 142
Throat infections, see Sore Throat
Thrush 100, 200, 204
    – vaginal 191, 257
Thunberg, Carl 40, 55, 96, 133

Thunder 195
Tick bites
    – infected 85, 103, 117, 197, 229
    – to relieve itch 191
Tick dip for animals 213
Ticks 11, 124
Tiger nuts 135
Tiger's milk 135
Tight chest 28, 121, 221, 227
    – see also Chest aliments
Tinderwood tree 212
Tiredness 116
Tomato bredie with spekboom 194
Tonic 7, 14, 56, 80, 95, 96, 116, 133, 148,
    162, 195, 220, 225, 260
    – digestive 43
Tontelbos 128
Tontelhout 212
Toothache 28, 80, 98, 114, 117, 130, 145,
    146, 149, 201, 227, 240
Toothache berry 66
Toothache root 201
Toothbrush 114, 164
Topiary 247
Toy protea 80
Toy sugar bush 79
Trachyandra
    – *ciliatum* 100
    – *revoluta* 100
Tradescant, John 167
Transvaal aalwyn 9
Transvaal basil 216, 217
Transvaal sage 260
Transvaal salie 260
Traveller's joy 203, 268
Tree fuchsia 104
Trees
    – *Acacia karroo* 199
    – boabab 19
    – blinkblaar 32
    – buddleja 45
    – Cape willow 59
    – *Clerodendrum glabrum* 212
    – *Combretum zeyheri* 155
    – coral tree 70
    – *Cussonia spicata* 111
    – *Dais cotinifolia* 153
    – *Dichrostachys cinerea* subsp.
        africana 184
    – *Dombeya rotundifolia* 249
    – *Erythrina lysistemon* 70
    – *Fagara capensis* 113
    – *Heteromorpha arborescens* 141
    – *Heteropyxis natalensis* 116
    – huilboerboon 104
    – kiepersol 111
    – *Kigelia africana* 165
    – knobwood 113

– lavender tree 116
– marula 123
– *Olea europaea* subsp. *africana* 247
– parsley tree 141
– pompon tree 153
– *Portulacaria afra* 193
– raasblaar 155
– *Salix mucronata* 59
– sausage tree 165
– *Sclerocarya birrea* subsp. *caffra* 123
– *Schotia brachypetala* 104
– sickle bush 184
– spekboom 193
– sweet thorn 199
– *Tarchonanthus camphoratus* 219
– *Vangueria infausta* 239
– white cat's whiskers 212
– wild camphor tree 219
– wild medlar 239
– wild olive 247
– wild pear 249
– *Zanthoxylum capense* 113
Trompetters 57
Trumpet lily 17
Tuberculose skin lesions 143
Tuberculosis 96, 102, 128, 133, 143, 190, 214, 233, 264
– pulmonary 214
Tubers, *see* Bulbs, corns and tubers
Tulbaghia 268
– *alliacea* 233
– *fragrans* 234
– *violacea* 233
Tulsi 216
Tumbleweed 205
Turk's cap 85
Tussie-mussie 222
Typha
– *capensis* 49
– *latifolia* 49
Typhoid fever 81, 109

Ulcerative colitis 187
Ulcers 81, 124, 126, 132, 133, 137, 148, 165
– gastric 14
– on horses and cattle 80
– internal 249
– intestinal 98
– leg 143
– mouth 48, 143, 226
– skin 147
– stomach 27, 98
Urethra infections 50, 185
Urinary infections 43, 185
Urine
– scant 221
– suppression of 170

Uterus ailments 55

Vaalbos 45, 219, 249
Vagina
– infection 230
– inflammation 98
– itch 230
– thrush 291, 257
Van der Stel, Adriaan 171
Van der Stel, Simon 159
Van Riebeeck 214
Vangueria infausta 239
Varicose ulcers 206
Varicose veins 80, 81, 130, 156, 250
Varicosities 73, 132
Varkblom 17
Varklelie 17
Varkoor 145
Varkoorblaar 145
Varkoortjies 143
Varkore 17
Veerkapok 252
Veld sores 66, 80, 147, 149
Veldvaalbos 219
Vellozia retinervis 35
Velloziaceae 137
Velskoenblaar 137
Venereal diseases 15, 49, 80, 112, 185, 204, 264, 265
Venereal sores 9, 11, 66, 143, 147
Vermifuge 188
Verninia oligocephala 186
Vernonia, silverleafed 186
Verucas 145
Vingerhoedblom 231
Virility 112
Vitality 97, 136, 148
Vitamin C 14, 124, 133, 239
Vlamlelie 85
Vlei poisoning 201
Vleikos 207
Vlieëbos 149
Vlieëbossie 150, 159
Voice, singing 80
Voice loss 257
Vomiting 43, 105, 176, 178, 197, 249
– to induce 28, 214
Vrouebossie 61
Vrouetee 61
Vrystaat salie 262
Vyerank 190
Vyfaartjies 80
Vygie 190

Wag-'n-bietjie 214
Wag-'n-bietjiebos 32
Wag-'n-bietjiedoring 32
Wait a bit 32

Walking sticks 114, 247
Warts 30, 128, 129, 145, 152
Wash
– fragrant 149, 255
– refreshing 22, 116, 248
– for wounds 40
Wash ball, hard fern 89
Wasp stings 48, 223, 225
Water
– baobab as source of 19
– to keep fresh 15
Water lily 210
Water onion 207
Water parsnip 201
Water retention 224
Waterblommetjie 207
– bredie, traditional 209
– soup 209
Watergrass 135, 139
Waternavel 143
Watersalie 87
Wateruintjie 207
Watervygen 207
Weakness 195
Weaning, to hasten 13
Weaving 19, 49, 131, 155, 195
Wedding bouquets 233
Weeping boer-bean 104, 268
Weeping schotia 104
Weeping willow 60
White arum 17
White cat's whiskers 212
White fly 228, 236
White thorn 199
Whitlows 205
Whooping cough 225, 226, 256, 257
Wild amaryllis 72
Wild asparagus 214
– boiled fresh 215
– tart 215
Wild basil 216, 268
– bathroom deodoriser 218
– and tomato sauce for pasta 218
Wild camphor tree 219, 268
Wild cardamom 113
Wild chamomile 63
– and Hottentotskooigoed pot-pourri 64
– massage oil 64
Wild cineraria 221, 268
Wild clematis 203
Wild commelina 265
Wild cotton 30, 128, 219
Wild dagga 223, 268
Wild everlasting 102
Wild fern 89
Wild flower pot-pourri 268
Wild foxglove 231, 268

– aphid spray 232
Wild garlic 233
  – dish, Elizabeth's 235
  – spray for aphids and fruitfly 236
Wild gentian 66
Wild geranium 61
Wild giant sage 260
Wild ginger 87
Wild gladiolus 237, 268
  – cupboard freshener 238
Wild grape 229
Wild lemon verbena 116
Wild lilac 45
Wild medlar 239
  – jam 240
Wild mint 241, 268
Wild olive 247
Wild pear 71, 249
Wild pineapple 251
Wild plum 157
Wild raisin 157
Wild rose geranium 178
Wild rosemary 11, 252, 268
  – bath vinegar or hair rinse 254
  – bean bake 253
Wild sage 255, 257, 268
Wild sage wood 219
Wild scabious 147
Wild sugar bush 22
Wild tea 118
Wild tradescantia 265
Wild verbena 91, 263
Wild verbena tree 116
Wild violet 143
Wild water mint 242
Wild wormwood 226

Wilde als 61, 102, 113, 118, 127, 226, 268
  – brandy 228
  – insecticide spray 228
  – moth repellent 228
Wilde basilikum 216
Wilde blomkool 100
Wilde druiwe 22
Wilde jasmyn 107
Wilde kamille 63
Wilde knoffel 233
Wilde knoflok 233
Wilde malva 167
Wilde mispel 239
Wilde olienhout 247
Wilde pynappel 251
Wilde roosmaryn 252
Wilde salie 245, 256, 257, 260, 262
Wilde suring 188
Wilde vioolblaar 143
Wilde wilgerboom 59
Wilde wingerd 229
  – cleansing douche 230
Wildedagga 223
Wildepieterseliebos 141
Wilderosyntjie 157
Wildesalie 45
Wilde-vye-rank 229
Willow, Cape 59
Wind breaker 52, 106, 163
Winter fodder 248
Witchcraft, protection against 221
Witdoring 199
Wood
  – carving 33, 60, 75, 104, 123, 166, 199, 219, 247, 249
  – fine quality 106

  – termite-proof 249
Wood sorrel 188
Woody nightshade 132
Woolly chamomile 65
Woolly long-leaf mint 246
Worms 13, 38, 61, 80, 113, 141, 142
  – in animals 213
  – in children 227
  – in dogs 188
Wormwood, wild 226
Worsboom 165
Wounds 7, 9, 10, 28, 30, 42, 49, 51, 55, 63, 73, 74, 78, 80, 98, 102, 103, 106, 110, 126, 132, 133, 137, 145, 148, 153, 157, 161, 170, 171, 184, 195, 224, 226, 229, 263
  – in cattle and sheep 103
  – dressing 7, 10
  – slow healing 143
  – suppurating 71
  – wash 40
Wurmbos 80

Xerophyta equisetoides 35
Xerophyta retinervis 35
Xysmalobium undulatum 30

Yellow star 195
Yellow wandering Jew 265

Zantedeschia aethiopica 17
Zanthoxylum capense 113
Ziziphus mucronata 32
Zulu basil 217
Zuurberg harebell 91